Advanced Topics in Quantum Mechanics

Quantum mechanics is one of the most successful theories in science, and is relevant to nearly all modern topics of scientific research. This textbook moves beyond the introductory and intermediate principles of quantum mechanics frequently covered in undergraduate and graduate courses, presenting in-depth coverage of many more exciting and advanced topics. The author provides a clearly structured text for advanced students, graduates, and researchers looking to deepen their knowledge of theoretical quantum mechanics. The book opens with a brief introduction covering key concepts and mathematical tools, followed by a detailed description of the Wentzel–Kramers–Brillouin (WKB) method. Two alternative formulations of quantum mechanics are then presented: Wigner's phase space formulation and Feynman's path integral formulation. The text concludes with a chapter examining metastable states and resonances. Step-by-step derivations, worked examples, and physical applications are included throughout.

Marcos Mariño is Professor of Mathematical Physics at the University of Geneva. He has held postdoctoral positions at Yale, Rutgers, and Harvard, and was a junior staff researcher at the CERN Theory Division. His research focuses on mathematical aspects of quantum field theory and string theory. He has previously authored the book *Instantons and Large N: An Introduction to Non-Perturbative Methods in Quantum Field Theory* (Cambridge University Press, 2015).

Advanced Topics in Quantum Mechanics

MARCOS MARIÑO

University of Geneva

CAMBRIDGE
UNIVERSITY PRESS

CAMBRIDGE
UNIVERSITY PRESS

University Printing House, Cambridge CB2 8BS, United Kingdom

One Liberty Plaza, 20th Floor, New York, NY 10006, USA

477 Williamstown Road, Port Melbourne, VIC 3207, Australia

314–321, 3rd Floor, Plot 3, Splendor Forum, Jasola District Centre, New Delhi – 110025, India

103 Penang Road, #05–06/07, Visioncrest Commercial, Singapore 238467

Cambridge University Press is part of the University of Cambridge.

It furthers the University's mission by disseminating knowledge in the pursuit of education, learning, and research at the highest international levels of excellence.

www.cambridge.org
Information on this title: www.cambridge.org/9781108495875
DOI: 10.1017/9781108863384

First published 2022

A catalogue record for this publication is available from the British Library.

Library of Congress Cataloging-in-Publication Data
Names: Mariño, Marcos, author.
Title: Advanced topics in quantum mechanics / Marcos Mariño, Université de Genève.
Description: Cambridge, United Kingdom ; New York, NY :
Cambridge University Press, 2022. | Includes bibliographical references and index.
Identifiers: LCCN 2021024889 (print) | LCCN 2021024890 (ebook) |
ISBN 9781108495875 (hardback) | ISBN 9781108863384 (epub)
Subjects: LCSH: Quantum theory.
Classification: LCC QC174.12 .M3595 2022 (print) | LCC QC174.12 (ebook) |
DDC 530.12–dc23
LC record available at https://lccn.loc.gov/2021024889
LC ebook record available at https://lccn.loc.gov/2021024890

ISBN 978-1-108-49587-5 Hardback

Contents

Preface

There is little doubt that quantum mechanics is one of the pillars of modern science. It provides the most fundamental description of the physical world that we have. Quantum mechanics is our modern ontology. For this reason, there is no shortage of books on the topic: from popular expositions to encyclopedic treatises, our science libraries are full of texts on the subject.

In view of this, why write yet another book on quantum mechanics?

This book grew out of two different sources. The first source was my teaching. Some years ago, I thought that it would be a good idea to teach an advanced course on quantum mechanics for graduate students in the University of Geneva. The goal of the course was to introduce Feynman's path integral formulation of quantum mechanics, as well as Wigner's formulation of quantum mechanics in phase space. These two formulations are very different from the conventional approach based on operators and Hilbert spaces. However, they are of fundamental interest to the serious practitioner, not only conceptually, but also due to their many applications (as is well known, the path integral formulation is widely used in quantum field theory and condensed matter physics). Given the paramount importance of quantum mechanics in our cognitive map of the world, I found it worthwhile to spend one semester studying these alternative formulations. The first drafts of this book started their life as typeset notes for the students in that course.

The other source of the book is my own research. In the last few years, I have felt that in order to make progress in fields as diverse as string theory, supersymmetric field theory, and topological quantum field theory, I needed tools from quantum mechanics, but at a level of rigor and sophistication that is not found in most of the standard textbooks. One particular subject that I needed for my research was the so-called exact WKB method. In this area, the gulf between textbooks and current developments seemed particularly deep. Most books on quantum mechanics spend some time explaining the WKB method; however, the traditional approach to this subject is confusing and imprecise, and the modern, correct treatment is only available in the specialized literature. Many practitioners of quantum mechanics are not even aware of these more precise formulations of the WKB method. In writing this book, I also wanted to provide an introduction to this subject for the nonexpert.

The book then contains, after a first introductory chapter presenting some useful topics, a very detailed chapter on the WKB method, and in particular on its exact version, which incorporates systematically nonperturbative effects. Among other things, the chapter explains the correct version of the WKB connection formulae, which, as Harris Silverstone pointed out in his seminal 1985 paper on the subject, are stated wrongly in traditional textbooks. It also includes a systematic discussion of the spectrum of energy levels in the double-well potential. Most theoretical physicists are familiar with this example, since it provides an elementary example of instantons and it is discussed in many references. However, as far as I know, no textbook discusses how to combine, in detail, perturbative and nonperturbative information to obtain the precise spectrum of the system. I also explain in a simple example the importance of complex turning points in the WKB analysis of spectra. The exact WKB method remains an active area in mathematical physics. It has led to many new recent developments in topics as diverse as supersymmetric gauge theory and quantum topology. I hope that the chapter of the book devoted to this topic will provide the required background to understand these developments (and maybe to participate in them).

The third chapter of the book presents the phase space approach to quantum mechanics. This is also a topic that is not covered in textbooks [with the notable exception of the recent short treatise by Curtright, Fairlie and Zachos [2014]]. After explaining the basics of this approach in detail, my treatment pays particular attention to its semiclassical aspects. I include for example a section on Berry's semiclassical approximation to the Wigner distribution (Berry, 1977), and I discuss different techniques to perform systematic semiclassical expansions.

The fourth chapter deals with the path integral approach to quantum mechanics. This approach is very well known since, as I mentioned above, it has become the natural language of quantum field theory and quantum many-body physics. In view of its importance, I think it is extremely useful to spend some time studying its original formulation in the context of nonrelativistic quantum mechanics. The chapter contains, in particular, a very careful discussion of the Gaussian integration over fluctuations, and its reformulation in terms of functional determinants. It introduces diagrammatic techniques that will be useful for the student interested in more advanced applications of path integrals. Section 4.11 of this chapter is on instantons, and I show explicitly that it leads to the same results obtained in Chapter 2 with the WKB method.

The final chapter is on metastable states and resonances. This is a particularly subtle subject, and I have tried to provide a practical introduction with many examples. The chapter introduces the techniques of complex dilatation, and it formulates the problem of resonances in the context of the exact WKB method and of the path integral approach discussed in previous chapters.

Chapter 2, on the WKB method, relies heavily on Borel resummation techniques, which I explain in some detail in Appendix A. The two other appendices collect useful formulae related to special functions and to Gaussian integration, respectively.

The style of the book is pedagogical, with many derivations spelled out in detail. It focuses on the more formal and mathematical aspects of the subject. This is due in part to the choice of topics, since I have focused on presenting tools for the advanced practitioner, and in part to my bias as a mathematical physicist. I should warn the reader that, in spite of my mathematical leanings, I do not provide a rigorous approach to the formal developments covered in the book. I have tried to find a happy middle course between the Scylla of cold and ineffectual mathematical rigor, and the Charybdis of handwaving and imprecise physics.

The book can provide the basis for an advanced course on quantum mechanics. I have used it regularly in a one-semester course at the master or doctoral level. In that course, I usually start by introducing the quantum-mechanical propagator appearing in the first chapter, and then I jump directly to the chapter on Feynman's path integral. After covering this material, students will be ready to quickly learn the path integral formalism of quantum field theory. Then, I cover the chapter on the phase space formulation of quantum mechanics. In the last part of the course, if time permits, I explain some selected topics on the WKB method. Of course, one could cover the whole book in a two-semester course. Although the book does not have an exercise section at the end of each chapter, it contains many examples throughout that should be regarded as solved exercises. Many of these examples have been used as exercises in my own course.

In preparing this book, I have been helped by many colleagues. I would like to mention in particular my former students Santiago Codesido, Szabolcs Zakany, and Yoan Emery. They were teaching assistants for my course and they worked out in detail some of the examples presented in the book. Ricardo Schiappa and Tomáš Procházka read preliminary versions of the book and gave me very valuable feedback. In particular, Tomáš pointed out many errors and inaccuracies, which I have tried to expurgate from the final version.

Books, even on quantum mechanics, are always written in a historical context. Although world events did not change the equations of the book, some of them had a deep impact on me. On October 1, 2017, while this book was being written, millions of people in Catalonia mobilized peacefully to organize an independence referendum from the Kingdom of Spain, and were brutally repressed by the police. As a consequence of these events, civil activists and elected representatives were imprisoned after show trials, or forced into exile.

This book is dedicated to the Catalan political prisoners and exiles.

1 Propagator and Resolvent

1.1 Introduction

The first chapter of this book is somewhat eclectic. It introduces concepts and tools that will be important later, and it also develops subjects that are not emphasized in many standard or more elementary courses of quantum mechanics.

We start by presenting the unitary time evolution operator and its integral kernel in the space representation, also known as the quantum-mechanical propagator. This important gadget reappears as the Moyal operator in the phase space formulation of Chapter 3, and it will be the central object in the path integral formulation of Chapter 4. The resolvent operator and Green's functions are also closely related to the quantum-mechanical propagator, and they play an important role in formal developments and in scattering theory. Although we do not present the general theory of scattering in this book, we dedicate a section to scattering in one-dimension, which illustrates the theory of the resolvent with explicit constructions. Some of the examples discussed in the section on one-dimensional scattering also provide our first examples of resonances, which we will study in much more detail in Chapter 5.

1.2 The Quantum-Mechanical Propagator

In quantum mechanics, time evolution between an initial time t_0 and a final time t_f is implemented by the unitary operator

$$\mathsf{U}(t_f, t_0), \tag{1.2.1}$$

which connects the quantum state of the system at time $t = t_0$ with the state at $t = t_f$:

$$\mathsf{U}(t_f, t_0)|\psi(t_0)\rangle = |\psi(t_f)\rangle. \tag{1.2.2}$$

It obviously satisfies the convolution law,

$$\mathsf{U}(t_f, t_0) = \mathsf{U}(t_f, t_1)\mathsf{U}(t_1, t_0). \tag{1.2.3}$$

Let H be the Hamiltonian operator of the quantum system. By using Schrödinger's equation,

$$i\hbar \frac{\mathrm{d}}{\mathrm{d}t}|\psi(t)\rangle = \mathsf{H}|\psi(t)\rangle, \tag{1.2.4}$$

we deduce the evolution equation for the operator $U(t, t_0)$:

$$i\hbar \frac{\partial U(t, t_0)}{\partial t} = H U(t, t_0),$$
(1.2.5)

with initial condition

$$U(t_0, t_0) = \mathbf{1}.$$
(1.2.6)

When the Hamiltonian H is time-independent (as we will assume most of the time in this book), we can solve the evolution equation to give

$$U(t_f, t_0) = e^{-\frac{i}{\hbar} H(t_f - t_0)}.$$
(1.2.7)

In this case, the evolution operator is invariant under time translation and only depends on the difference

$$T = t_f - t_0.$$
(1.2.8)

It is convenient to work in the position representation for the position operator. The corresponding eigenstates will be denoted by $|\boldsymbol{q}\rangle$, where $\boldsymbol{q} \in \mathbb{R}^d$. The integral kernel of the evolution operator in the position representation is called the *quantum-mechanical (QM) propagator*:

$$K(\boldsymbol{q}_f, \boldsymbol{q}_0; t_f, t_0) = \langle \boldsymbol{q}_f | U(t_f, t_0) | \boldsymbol{q}_0 \rangle.$$
(1.2.9)

The QM propagator can be regarded as a wavefunction at time $t = t_f$,

$$\psi(\boldsymbol{q}, t_f) = K(\boldsymbol{q}, \boldsymbol{q}_0; t_f, t_0),$$
(1.2.10)

which is obtained by evolving in time the state $|\boldsymbol{q}_0\rangle$ at $t = t_0$. This initial state is described by the wavefunction

$$\psi(\boldsymbol{q}, t_0) = \delta(\boldsymbol{q} - \boldsymbol{q}_0),$$
(1.2.11)

and it is an eigenfunction of the position operator with eigenvalue $\boldsymbol{q}_0$. The QM propagator has a direct physical interpretation: it gives the probability amplitude that a particle located is at the point $\boldsymbol{q}_f$ at time t_f, given that it was at the point $\boldsymbol{q}_0$ at time t_0.

The evolution operator and the QM propagator contain detailed information about the spectrum and eigenfunctions of the Hamiltonian. Indeed, let us assume that H has a discrete and nondegenerate spectrum E_n, $n \geq 0$, with orthonormal eigenfunctions $|\phi_n\rangle$. Then, we have the spectral decompositions,

$$U(t_f, t_0) = \sum_{n \geq 0} |\phi_n\rangle e^{-i E_n (t_f - t_0)/\hbar} \langle \phi_n|,$$
(1.2.12)

or, equivalently,

$$K(\boldsymbol{q}_f, \boldsymbol{q}_0; t_f, t_0) = \sum_{n \geq 0} \phi_n(\boldsymbol{q}_f) e^{-i E_n (t_f - t_0)/\hbar} \phi_n^*(\boldsymbol{q}_0).$$
(1.2.13)

Finally, let us note that the QM propagator is a solution of the time-dependent Schrödinger equation. If we denote the quantum counterparts of the canonical coordinates and momenta by the vectors of Heisenberg operators

$$\mathsf{q} = (\mathsf{q}_1, \ldots, \mathsf{q}_n), \qquad \mathsf{p} = (\mathsf{p}_1, \ldots, \mathsf{p}_n),$$
(1.2.14)

then the QM propagator satisfies

$$i\hbar\frac{\partial}{\partial t_f}K(\boldsymbol{q}_f,\boldsymbol{q}_0;t_f,t_0) = \mathsf{H}\left(\boldsymbol{q}_f,-i\hbar\frac{\partial}{\partial \boldsymbol{q}_f}\right)K(\boldsymbol{q}_f,\boldsymbol{q}_0;t_f,t_0), \tag{1.2.15}$$

as well as

$$i\hbar\frac{\partial}{\partial t_f}K(\boldsymbol{q}_f,\boldsymbol{q}_0;t_f,t_0) = \mathsf{H}\left(\boldsymbol{q}_0,i\hbar\frac{\partial}{\partial \boldsymbol{q}_0}\right)K(\boldsymbol{q}_f,\boldsymbol{q}_0;t_f,t_0). \tag{1.2.16}$$

The evolution operator in quantum mechanics is closely related to other useful quantities. The first one is the (unnormalized) density operator for the canonical ensemble. This is defined by

$$\rho(\beta) = \mathrm{e}^{-\beta\mathsf{H}}, \tag{1.2.17}$$

and we recall that $k_B\beta$ is the inverse temperature of the system. The density operator can be obtained from the evolution operator by the so-called *Euclidean continuation* or *Wick rotation*:

$$T = -iu, \qquad u = \beta\hbar. \tag{1.2.18}$$

Therefore, if we know the evolution operator, we know the density operator, and if we know the QM propagator, we know the integral kernel of the density operator, sometimes called the *density matrix*,

$$\langle \boldsymbol{q}|\rho(\beta)|\boldsymbol{q}'\rangle = \rho(\boldsymbol{q},\boldsymbol{q}';\beta). \tag{1.2.19}$$

More precisely, for a theory with a time-independent Hamiltonian, we have

$$\rho(\boldsymbol{q},\boldsymbol{q}';\beta) = K(\boldsymbol{q},\boldsymbol{q}';-i\hbar\beta,0). \tag{1.2.20}$$

Note that the unnormalized density matrix satisfies the differential equation

$$-\frac{\partial}{\partial\beta}\rho(\boldsymbol{q},\boldsymbol{q}';\beta) = \mathsf{H}\left(\boldsymbol{q},-i\hbar\frac{\partial}{\partial \boldsymbol{q}}\right)\rho(\boldsymbol{q},\boldsymbol{q}';\beta), \tag{1.2.21}$$

which is sometimes called the *Bloch equation*. This is the counterpart of (1.2.15). The initial condition is

$$\rho(\beta = 0) = \mathbf{1}. \tag{1.2.22}$$

This is of course the analogue of (1.2.5). The canonical partition function is easily obtained from the unnormalized density operator,

$$Z(\beta) = \mathrm{Tr}\,\rho(\beta) = \int_{\mathbb{R}^d}\mathrm{d}\boldsymbol{q}\,\rho(\boldsymbol{q},\boldsymbol{q};\beta). \tag{1.2.23}$$

The density matrix is even more useful than the QM propagator when extracting the spectral information, since it has the spectral decomposition

$$\rho(\boldsymbol{q},\boldsymbol{q}';\beta) = \sum_{n\geq 0}\phi_n(\boldsymbol{q})\mathrm{e}^{-\beta E_n}\phi_n^*(\boldsymbol{q}'). \tag{1.2.24}$$

This means that we can extract the energies and eigenfunctions by performing a low temperature expansion (i.e. by considering the limit $\beta \to \infty$). The leading-order term gives the energy and wavefunction of the ground state, the next-to-leading term contains information about the first excited state, and so on.

Let us now present three important examples where the quantum-mechanical propagator can be computed in closed form: the free particle, the harmonic oscillator, and a particle in a linear potential.

Example 1.2.1 *Propagator for the free particle.* The QM propagator can easily be computed for a free particle in one dimension, with Hamiltonian

$$H = \frac{p^2}{2m}. \tag{1.2.25}$$

Indeed, we have

$$
\begin{aligned}
K(q_f, q_0; t_f, t_0) &= \langle q_f | \exp\left[-\frac{i(t_f - t_0)p^2}{2m\hbar} \right] | q_0 \rangle \\
&= \int_{\mathbb{R}} \langle q_f | p \rangle \langle p | \exp\left[-\frac{i(t_f - t_0)p^2}{2m\hbar} \right] | q_0 \rangle \mathrm{d}p \\
&= \frac{1}{2\pi\hbar} \int_{\mathbb{R}} e^{ip(q_f - q_0)/\hbar} e^{-\frac{i(t_f - t_0)p^2}{2m\hbar}} \, \mathrm{d}p \\
&= \left(\frac{m}{2\pi i\hbar(t_f - t_0)} \right)^{1/2} \exp\left[\frac{im}{2\hbar(t_f - t_0)}(q_f - q_0)^2 \right],
\end{aligned} \tag{1.2.26}
$$

where we have used the Gaussian integral formula (C.1) and the result for the plane waves

$$\langle q | p \rangle = \psi_p(q) = \frac{1}{\sqrt{2\pi\hbar}} e^{ipq/\hbar}. \tag{1.2.27}$$

From (1.2.26), we can also deduce the canonical density matrix for a free particle:

$$\rho(q, q'; \beta) = \left(\frac{m}{2\pi\beta\hbar^2} \right)^{1/2} \exp\left[-\frac{m}{2\beta\hbar^2}(q - q')^2 \right]. \tag{1.2.28}$$

The generalization to D dimensions is straightforward. We have, for example,

$$K(\boldsymbol{q}_f, \boldsymbol{q}_0; t_f, t_0) = \left(\frac{m}{2\pi i\hbar(t_f - t_0)} \right)^{D/2} \exp\left[\frac{im}{2\hbar(t_f - t_0)}(\boldsymbol{q}_f - \boldsymbol{q}_0)^2 \right]. \tag{1.2.29}$$

In this way, one finds the well-known expression for the thermal partition function of a free particle in D dimensions:

$$Z(\beta) = \left(\frac{m}{2\pi\beta\hbar^2} \right)^{D/2} V_D, \tag{1.2.30}$$

where V_D is the volume where the particle lives. □

Example 1.2.2 *Propagator for the harmonic oscillator.* Let us consider a quantum harmonic oscillator in one dimension, with Hamiltonian

$$H = \frac{p^2}{2m} + \frac{m\omega^2 q^2}{2}. \tag{1.2.31}$$

We will set $t_0 = 0$, which we can always do when H is time independent. Let us recall that the time-dependent Heisenberg operators associated to q, p are given by

$$q_H(t) = e^{iHt/\hbar} \, q \, e^{-iHt/\hbar}, \qquad p_H(t) = e^{iHt/\hbar} \, p \, e^{-iHt/\hbar}. \tag{1.2.32}$$

They satisfy the Heisenberg equation of motion (EOM), which reads in this case as

$$\dot{q}_H(t) = \frac{1}{m}p_H(t), \qquad \dot{p}_H(t) = -m\omega^2 q_H(t), \tag{1.2.33}$$

and can be integrated to give

$$q_H(t) = \cos(\omega t)q_H(0) + \frac{1}{m\omega}\sin(\omega t)p_H(0),$$
$$p_H(t) = -m\omega\sin(\omega t)q_H(0) + \cos(\omega t)p_H(0). \tag{1.2.34}$$

We also recall that $q_H(0) = q$, $p_H(0) = p$ are the operators in the Schrödinger representation. Let us also denote

$$|q_0(t)\rangle = e^{-iHt/\hbar}|q_0\rangle. \tag{1.2.35}$$

We want to calculate

$$K(q_f, q_0; t, 0) = \langle q_f | q_0(t)\rangle. \tag{1.2.36}$$

We first note that

$$\langle q_f|q_H(-t)|q_0(t)\rangle = \langle q_f|e^{-iHt/\hbar}\, q\, e^{iHt/\hbar}e^{-iHt/\hbar}|q_0\rangle = q_0 K(q_f, q_0; t, 0). \tag{1.2.37}$$

On the other hand, by using the explicit solution of $q_H(-t)$, we find

$$\langle q_f|q_H(-t)|q_0(t)\rangle = \cos(\omega t)\langle q_f|q_H(0)|q_0(t)\rangle - \frac{1}{m\omega}\sin(\omega t)\langle q_f|p_H(0)|q_0(t)\rangle$$
$$= \cos(\omega t)q_f K(q_f, q_0; t, 0) + \frac{i\hbar}{m\omega}\sin(\omega t)\frac{\partial}{\partial q_f}K(q_f, q_0; t, 0). \tag{1.2.38}$$

Putting both results together, we obtain the following differential equation for the propagator:

$$\frac{\partial}{\partial q_f}K(q_f, q_0; t, 0) = \frac{m\omega}{i\hbar\sin(\omega t)}\left(q_0 - q_f\cos(\omega t)\right)K(q_f, q_0; t, 0), \tag{1.2.39}$$

whose solution is

$$K(q_f, q_0; t, 0) = \mathcal{N}(t)\exp\left[\frac{im\omega}{\hbar\sin(\omega t)}\left(\frac{1}{2}q_f^2\cos(\omega t) - q_f q_0\right)\right]. \tag{1.2.40}$$

Here, $\mathcal{N}(t)$ is an undetermined function of t. To find this function, we use that

$$i\hbar\frac{\partial}{\partial t}K(q_f, q_0; t, 0) = \langle q_f|H|q_0(t)\rangle = \left(-\frac{\hbar^2}{2m}\frac{\partial^2}{\partial q_f^2} + \frac{m\omega^2}{2}q_f^2\right)K(q_f, q_0; t, 0). \tag{1.2.41}$$

Plugging (1.2.40) into this equation, we obtain

$$\frac{\partial \mathcal{N}}{\partial t} = \left(-\frac{\omega}{2}\cot(\omega t) - \frac{im\omega^2}{2\hbar\sin^2(\omega t)}q_0^2\right)\mathcal{N}(t), \tag{1.2.42}$$

which is easily integrated to

$$\mathcal{N}(t) = \frac{C}{\sqrt{\sin(\omega t)}}\exp\left(\frac{im\omega}{2\hbar}\cot(\omega t)q_0^2\right). \tag{1.2.43}$$

The quantum propagator can then be expressed as,

$$K(q_f, q_0; t, 0) = \frac{C}{\sqrt{\sin(\omega t)}} \exp\left[\frac{im\omega}{2\hbar\sin(\omega t)}\left(\left(q_f^2 + q_0^2\right)\cos(\omega t) - 2q_f q_0\right)\right].$$

(1.2.44)

The constant C can be determined by considering the free particle limit $\omega \to 0$. In this limit, we should recover the result (1.2.26). This fixes

$$C = \sqrt{\frac{m\omega}{2\pi i\hbar}},$$

(1.2.45)

and we finally obtain

$$K(q_f, q_0; t_f, t_0) = \sqrt{\frac{m\omega}{2\pi i\hbar\sin(\omega T)}} \exp\left[\frac{im\omega}{2\hbar\sin(\omega T)}\left(\left(q_f^2 + q_0^2\right)\cos(\omega T) - 2q_f q_0\right)\right],$$

(1.2.46)

where we used time translation invariance, and T is given in (1.2.8).   □

Example 1.2.3 *Propagator for the linear potential.* Let us consider now the quantum Hamiltonian

$$H = \frac{p^2}{2m} - Fq,$$

(1.2.47)

which corresponds to a linear potential in one dimension. We can compute the QM propagator by using a method similar to that in the previous example. The Heisenberg EOM are

$$\dot{q}(t) = \frac{p(t)}{m}, \qquad \dot{p}(t) = F,$$

(1.2.48)

which can be integrated immediately to

$$q(t) = \frac{Ft^2}{2m} + \frac{t}{m}p + q,$$

(1.2.49)

$$p(t) = Ft + p,$$

where $q = q(0)$, $p = p(0)$. Using this explicit solution we obtain

$$\langle q_f | q(-t) | q_0(t) \rangle = \left(\frac{Ft^2}{2m} + \frac{i\hbar t}{m}\frac{\partial}{\partial q_f} + q_f\right) K(q_f, q_0; t, 0),$$

(1.2.50)

and we find the equation

$$\frac{\partial}{\partial q_f} \log K(q_f, q_0; t, 0) = \frac{im}{t\hbar}\left(q_f - q_0 + \frac{Ft^2}{2m}\right).$$

(1.2.51)

This can be integrated as

$$K(q_f, q_0; t, 0) = \mathcal{N}(t) \exp\left[\frac{im}{t\hbar}\left(\frac{q_f^2}{2} - q_f q_0\right) + \frac{iFq_f t}{2\hbar}\right].$$

(1.2.52)

To determine $\mathcal{N}(t)$, we again use the analogue of (1.2.41), which reads in this case as

$$i\hbar \frac{\partial}{\partial t} K(q_f, q_0; t, 0) = \langle q_f | H | q_0(t) \rangle = \left(-\frac{\hbar^2}{2m} \frac{\partial^2}{\partial q_f^2} - F q_f \right) K(q_f, q_0; t, 0). \quad (1.2.53)$$

Plugging (1.2.52) into this equation, we find

$$\frac{\partial}{\partial t} \log \mathcal{N}(t) = -\frac{1}{2t} + \frac{iF q_0}{2\hbar} - \frac{imq_0^2}{2\hbar t^2} - \frac{iF^2 t^2}{8m\hbar}, \quad (1.2.54)$$

so that

$$\mathcal{N}(t) = \frac{C}{\sqrt{t}} \exp\left(\frac{imq_0^2}{2\hbar t} + \frac{iF q_0 t}{2\hbar} - \frac{iF^2 t^3}{24m\hbar} \right). \quad (1.2.55)$$

Again, the integration constant can be obtained by comparing the full result to the free particle limit, when $F \to 0$. Finally, one finds that,

$$K(q_f, q_0; t, 0) = \sqrt{\frac{m}{2\pi i\hbar t}} \exp\left[\frac{im}{2t\hbar} \left(q_f - q_0 \right)^2 + \frac{iFt}{2\hbar} (q_f + q_0) - \frac{iF^2 t^3}{24m\hbar} \right]. \quad (1.2.56)$$

$\square$

An interesting aspect of the examples we have just considered is that both the exponent and the prefactor of the QM propagator are related to quantities in classical mechanics. Let us consider a classical path of trajectory, $q(t)$, satisfying the boundary conditions

$$q(t_0) = q_0, \qquad q(t_f) = q_f. \quad (1.2.57)$$

The classical action is a functional of the trajectory obtained by integrating the Lagrangian,

$$S(q(t)) = \int_{t_0}^{t_f} L(q(t), \dot{q}(t)) dt. \quad (1.2.58)$$

As is well known from classical mechanics, this functional has an extremum when the trajectory $q(t)$ solves the classical EOM. Indeed, if we perform a variation $\delta q(t)$ preserving the boundary conditions (1.2.57), one has

$$\frac{\delta S}{\delta q(t)} = -\frac{d}{dt} \left(\frac{\partial L}{\partial \dot{q}} \right) + \frac{\partial L}{\partial q}. \quad (1.2.59)$$

Let us denote by $q_c(t)$ the solution to the Lagrange EOM with the boundary conditions (1.2.57) (which we assume to exist and to be unique, for simplicity). Then, we have

$$\left. \frac{\delta S}{\delta q(t)} \right|_{q(t) = q_c(t)} = 0. \quad (1.2.60)$$

In the following, we will denote by S_c the value of the classical action on the classical trajectory:

$$S_c = S(q_c(t)). \quad (1.2.61)$$

It is a function of the boundary data q_f, q_0, t_f, and t_0.

In the case of the free particle in one dimension, the action reads

$$S(q(t)) = \frac{m}{2} \int_{t_0}^{t_f} (\dot{q}(t))^2 \, dt. \tag{1.2.62}$$

The classical trajectory has constant velocity, equal to

$$\dot{q}_c(t) = \frac{q_f - q_0}{t_f - t_0}. \tag{1.2.63}$$

The classical action is in this case,

$$S_c = \frac{m}{2} \frac{(q_f - q_0)^2}{t_f - t_0}. \tag{1.2.64}$$

We can then write (1.2.26) as

$$K(q_f, q_0; t_f, t_0) = \frac{1}{\sqrt{2\pi i \hbar}} \left(-\frac{\partial^2 S_c}{\partial q_f \partial q_0} \right)^{1/2} e^{iS_c/\hbar}. \tag{1.2.65}$$

This result also holds for the harmonic oscillator and for the particle in a linear potential. Let us verify it for the harmonic oscillator. The solution to the classical EOM that satisfies the boundary conditions is

$$q_c(t) = q_0 \cos(\omega(t - t_0)) + \frac{q_f - q_0 \cos(\omega T)}{\sin(\omega T)} \sin(\omega(t - t_0)). \tag{1.2.66}$$

The classical action evaluated at this path is:

$$S_c = \int_{t_0}^{t_f} \left(\frac{m\dot{q}_c^2}{2} - \frac{m\omega^2 q_c^2}{2} \right) dt = \frac{m}{2} \dot{q}_c(t) q_c(t) \Big|_{t_0}^{t_f} - \frac{m}{2} \int_{t_0}^{t_f} q_c \left(\ddot{q}_c + \omega^2 q_c \right) dt$$

$$= \frac{m}{2} \left(\dot{q}_c(t_f) q_c(t_f) - \dot{q}_c(t_0) q_c(t_0) \right) = \frac{m\omega}{2\sin(\omega T)} \left(\left(q_f^2 + q_0^2 \right) \cos(\omega T) - 2q_f q_0 \right). \tag{1.2.67}$$

It is now easy to verify that the quantum propagator (1.2.46) also has the structure (1.2.65).

In the conventional formulation of quantum mechanics, the result (1.2.65) is somewhat surprising and far from obvious. Why does the calculation of a quantum-mechanical propagator involve the classical action of Lagrangian mechanics? We will find an *a priori* explanation of this structure in Chapter 4, in the context of the path integral formulation of quantum mechanics.

1.3 Resolvent and Green's Functions

The evolution operator $U(t, t')$ is closely related to the Green's functions of the time-dependent Schrödinger operator

$$i\hbar \frac{\partial}{\partial t} - H. \tag{1.3.68}$$

Let us define

$$G_{\pm}(t - t') = \pm \theta \left(\pm (t - t') \right) U(t, t'). \tag{1.3.69}$$

The functions $G_\pm(t)$ are called *retarded* (respectively, advanced) Green's functions. Obviously,

$$U(t, t') = G_+(t - t') - G_-(t - t'). \quad (1.3.70)$$

From (1.2.5) it follows that $G_\pm$ satisfy the Schrödinger equation with a delta function,

$$\left(i\hbar\frac{\partial}{\partial t} - H\right)G_\pm(t - t') = i\hbar\delta(t - t'), \quad (1.3.71)$$

which is the defining equation for a Green's function.

After a Fourier transform w.r.t. t, this equation will become algebraic. We then introduce the Fourier transforms of the Green's functions:

$$G_\pm(E) = \frac{1}{i\hbar}\int_{\mathbb{R}} e^{iEt/\hbar}G_\pm(t)dt, \quad (1.3.72)$$

with inverses,

$$G_\pm(t) = \frac{i}{2\pi}\int_{\mathbb{R}} e^{-iEt/\hbar}G_\pm(E)dE. \quad (1.3.73)$$

Let us first consider $G_+(E)$. Since $G_+(t) = 0$ for $t < 0$, we find

$$G_+(E) = \frac{1}{i\hbar}\int_0^\infty e^{iEt/\hbar}G_+(t)dt. \quad (1.3.74)$$

In order to evaluate the integral over t, one introduces as a regularization the damping factor $e^{-\epsilon t/\hbar}$, where $\epsilon > 0$ and small. This is equivalent to shifting the energy:

$$E \rightarrow E + i\epsilon. \quad (1.3.75)$$

In this way, one finds

$$G_+(E) = \lim_{\epsilon\to 0^+}\frac{1}{i\hbar}\int_0^\infty e^{i(E+i\epsilon)t/\hbar}e^{-iHt/\hbar}dt = \lim_{\epsilon\to 0^+}\frac{1}{E + i\epsilon - H}. \quad (1.3.76)$$

Note that, with the above regularization, the poles of $G_+(E)$ are in the lower half-plane $\text{Im}(E) < 0$. Therefore, when $t < 0$, the integral (1.3.73) can be computed by closing the contour in the upper half plane, and no contributions will appear, guaranteeing that $G_+(t)$ vanishes. A similar calculation shows that

$$G_-(E) = \lim_{\epsilon\to 0^+}\frac{1}{E - i\epsilon - H}. \quad (1.3.77)$$

These results suggest we introduce of the *resolvent operator*, defined by

$$G(E) = \frac{1}{E - H}. \quad (1.3.78)$$

Here, E is in general complex, so the resolvent is an operator-valued function on the complex plane. It follows from the above discussion that the operators $G_\pm(E)$ are the limits of $G(E)$ when E approaches the real axis from above or from below in the complex plane, respectively:

$$G_\pm(E) = \lim_{\epsilon\to 0^+}G(E \pm i\epsilon). \quad (1.3.79)$$

This indicates that, as a function on the complex E plane, the resolvent has a branch cut on the real axis. We will see in a moment that this is the case already for the free particle in one dimension, and in Section 1.4 we will determine the analytic structure of the resolvent.

The resolvent is an operator, and in order to explore it in detail it is more useful to consider functions associated to it, such as its integral kernel or its trace. Its integral kernel is

$$G(x, y; E) = \langle x | (E - H)^{-1} | y \rangle, \tag{1.3.80}$$

and its trace will simply be denoted by

$$G(E) = \text{Tr } G(E). \tag{1.3.81}$$

It follows from (1.3.78) that the resolvent satisfies

$$(E - H)G(E) = 1, \tag{1.3.82}$$

which is a time-independent Schrödinger equation with a delta source. If we consider a standard one-dimensional Hamiltonian of the form

$$H = \frac{p^2}{2m} + V(q), \tag{1.3.83}$$

we find that the integral kernel of the resolvent satisfies

$$\left(\frac{\hbar^2}{2m} \frac{d^2}{dx^2} - V(x) + E \right) G(x, y; E) = \delta(x - y). \tag{1.3.84}$$

Example 1.3.1 *Resolvent for the free particle in one dimension.* For a free particle,

$$H = \frac{p^2}{2m}, \tag{1.3.85}$$

and the resolvent is given by

$$G(x, y; E) = \left\langle x \left| \left(E - \frac{p^2}{2m} \right)^{-1} \right| y \right\rangle. \tag{1.3.86}$$

By introducing the resolution of identity,

$$\int_{\mathbb{R}} dp |p\rangle\langle p| = \mathbf{1}, \tag{1.3.87}$$

we find

$$G(x, y; E) = \int_{\mathbb{R}} \langle x | p \rangle \frac{1}{E - p^2/(2m)} \langle p | y \rangle dp = \frac{1}{2\pi\hbar} \int_{\mathbb{R}} \frac{e^{ip(x-y)/\hbar}}{E - \frac{p^2}{2m}} dp. \tag{1.3.88}$$

Since the answer only depends on the difference $x - y$ (which is expected, due to translation invariance), we will denote

$$G(x - y; E) \equiv G(x, y; E). \tag{1.3.89}$$

The integral in (1.3.88) depends on the value of E. If E is *not* on the positive real axis, the integral is well defined and can be computed by using the residue theorem. Let us suppose, for example, that $E < 0$. In this case, there are two poles on the imaginary axis at

$$\pm i p_0 = \pm i \sqrt{2m|E|} \tag{1.3.90}$$

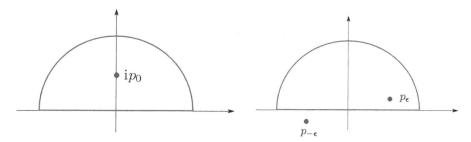

(Left) When $E < 0$, the poles are on the imaginary axis, and their position does not depend on the ϵ prescription. (Right) When $E > 0$, the poles for $G_+(x;E)$ are as shown in the figure.

(see Figure 1.1 [left]). Depending on whether $x > 0$ or $x < 0$, we can close the contour on the upper (respectively, lower) half plane and use Jordan's lemma. We obtain,

$$G(x; E) = -\frac{1}{\hbar}\sqrt{\frac{m}{-2E}}e^{-\sqrt{-2mE}\,|x|/\hbar}, \qquad E < 0. \tag{1.3.91}$$

Let us now suppose that $E > 0$. In this case, there are singularities in the integration contour and it is crucial to shift $E \to E \pm i\epsilon$, as expected from the previous discussion. We then consider,

$$G_\pm(x, y; E) = \lim_{\epsilon \to 0^+}\frac{1}{2\pi\hbar}\int_{\mathbb{R}}\frac{e^{ip(x-y)/\hbar}}{E \pm i\epsilon - \frac{p^2}{2m}}. \tag{1.3.92}$$

Let us first consider $G_+(x, y; E)$. The poles are no longer on the real axis, but at

$$p_{\pm\epsilon} = \pm\sqrt{2mE} \pm i\epsilon \tag{1.3.93}$$

(see Figure 1.1 [right]). When $x > 0$, we close the contour integral in the upper half plane, and we pick up the residue of the pole with + sign. When $x < 0$, we close the contour on the lower half plane. We find,

$$G_+(x; E) = -\frac{i}{\hbar}\sqrt{\frac{m}{2E}}e^{i\sqrt{2mE}\,|x|/\hbar}, \qquad E > 0. \tag{1.3.94}$$

If we compute $G_-(x; E)$, the poles will be at

$$\pm\sqrt{2mE} \mp i\epsilon, \tag{1.3.95}$$

and

$$G_-(x; E) = \frac{i}{\hbar}\sqrt{\frac{m}{2E}}e^{-i\sqrt{2mE}\,|x|/\hbar}, \qquad E > 0. \tag{1.3.96}$$

We conclude that the integral kernel of the resolvent is the following function on the complex plane E:

$$G(x; E) = -\frac{i}{\hbar}\sqrt{\frac{m}{2E}}e^{i\sqrt{2mE}\,|x|/\hbar}, \tag{1.3.97}$$

where $\sqrt{E}$ has a branch cut along $[0, \infty)$. In particular, we have

$$\lim_{\epsilon \to 0^+}\sqrt{E \pm i\epsilon} = \pm\sqrt{E}, \qquad E > 0, \tag{1.3.98}$$

so that we recover the functions $G_\pm(x; E)$ as the limits of $G(x; E)$ as we approach the positive real axis. In addition,

$$\sqrt{E} = i\sqrt{-E}, \qquad E < 0, \tag{1.3.99}$$

and we recover our previous result (1.3.91). A useful way to write down the resolvent is to introduce a complex variable k such that

$$E = \frac{\hbar^2 k^2}{2m} = \frac{\hbar^2 |k|^2 e^{2i\alpha}}{2m}. \tag{1.3.100}$$

This parametrization is such that the complex plane of E is covered as $\alpha \in [0, \pi]$, i.e. as $\mathrm{Im}(k) > 0$. This is called the *physical sheet*. When the complex energy is of the form $E + i\epsilon$, with $E > 0$, we take k real and positive:

$$k = \frac{\sqrt{2mE}}{\hbar}. \tag{1.3.101}$$

As we rotate in the E plane towards the negative axis, we have that $\alpha = \pi/2$, so that $E = -|E|$ is negative. In this case, k is chosen to be

$$k = |k|e^{i\pi/2} = i|k| = i\frac{\sqrt{2m|E|}}{\hbar}. \tag{1.3.102}$$

Finally, as $\alpha = \pi$, the complex energy is of the form $E - i\epsilon$, with $E > 0$, and

$$k = -\frac{\sqrt{2mE}}{\hbar}. \tag{1.3.103}$$

With these conventions for k, the resolvent can be written as

$$G(x; E) = -\frac{im}{\hbar^2 k} e^{ik|x|}. \tag{1.3.104}$$

$\square$

1.4 Analytic Properties of the Resolvent

In Section 1.3 we saw that the resolvent has a nontrivial analytic structure, as a function on the complex E plane. We will now obtain general information about this analytic structure, in terms of spectral information of the Hamiltonian H.

In general, the Hamiltonian H will have both a discrete spectrum, with eigenvalues E_n and eigenvectors $|\phi_n\rangle$, and a continuous spectrum with generalized eigenvectors $|\lambda\rangle$ and eigenvalues $E(\lambda)$. Here, λ is a label for these continuous states (it could be, for example, the momentum). The resolution of the identity for these states gives

$$\mathbf{1} = \sum_n |\phi_n\rangle\langle\phi_n| + \int \frac{d\lambda}{N(\lambda)} |\lambda\rangle\langle\lambda|, \tag{1.4.105}$$

where

$$\langle\lambda|\lambda'\rangle = N(\lambda)\delta(\lambda - \lambda'). \tag{1.4.106}$$

Note that the factor of $1/N(\lambda)$ guarantees that

$$\left(\int \frac{d\lambda}{N(\lambda)} |\lambda\rangle\langle\lambda|\right) |\lambda'\rangle = |\lambda'\rangle. \tag{1.4.107}$$

Let us now multiply the above identity by $(z - H)^{-1}$. We obtain,

$$G(z) = \sum_n \frac{|\phi_n\rangle\langle\phi_n|}{z - E_n} + \int \frac{d\lambda}{N(\lambda)} \frac{|\lambda\rangle\langle\lambda|}{z - E(\lambda)}. \tag{1.4.108}$$

Let us introduce the following projector of the continuous states onto those with energy E,

$$P_E = \int \frac{d\lambda}{N(\lambda)} |\lambda\rangle \delta(E - E(\lambda))\langle\lambda|. \tag{1.4.109}$$

In terms of this projector, we can write

$$G(z) = \sum_n \frac{|\phi_n\rangle\langle\phi_n|}{z - E_n} + \int dE \frac{P_E}{z - E}. \tag{1.4.110}$$

In order to write an equation for complex functions of z (instead of operator-valued functions of z), we consider averages of the resolvent over an arbitrary unit norm state $|u\rangle$:

$$G_u(z) = \langle u|G(z)|u\rangle. \tag{1.4.111}$$

We then obtain the spectral decomposition,

$$G_u(z) = \sum_n \frac{|\langle u|\phi_n\rangle|^2}{z - E_n} + \int dE \frac{\langle u|P_E|u\rangle}{z - E}. \tag{1.4.112}$$

Note that

$$\langle u|P_E|u\rangle = \int \frac{d\lambda}{N(\lambda)} \delta(E - E(\lambda)) |\langle u|\lambda\rangle|^2. \tag{1.4.113}$$

Let z be a point in the complex plane at a nonzero distance δ from the real axis. It is clear that

$$\left|\frac{1}{z - E}\right| \le \frac{1}{\delta}, \tag{1.4.114}$$

for any real E, and we find the upper bound

$$|G_u(z)| \le \frac{1}{\delta} \left\{ \sum_n |\langle u|\phi_n\rangle|^2 + \int dE \langle u|P_E|u\rangle \right\} \le \frac{1}{\delta}, \tag{1.4.115}$$

since the term inside the brackets is just the norm of u, which is one by assumption. We conclude that $G_u(z)$ is bounded away from the real axis. Similarly, one finds that $G'_u(z)$ is bounded. Since $G_u(z)$ is given by a convergent infinite sum of analytic functions, it is an analytic function of z away from the real axis. A similar argument shows that $G_u(z)$ is analytic for any real value of z that is not an eigenvalue and does not belong to the continuous spectrum.

It is also clear that, for a generic u such that $\langle u|\phi_n\rangle \ne 0$, the point $z = E_n$ corresponding to the discrete spectrum is a *simple pole* of the resolvent. Let us now study what happens when z approaches a value of E in the continuous spectrum. We set $z = E \pm i\epsilon$ and we calculate

$$\lim_{\epsilon \to 0} G_u(E \pm i\epsilon) = \sum_n \frac{\left|\langle u|\phi_n\rangle\right|^2}{E \pm i\epsilon - E_n} + \int dE' \frac{\langle u|P_{E'}|u\rangle}{E \pm i\epsilon - E'}. \tag{1.4.116}$$

We now use the following equality of distributions,

$$\lim_{\epsilon \to 0} \frac{1}{x \pm i\epsilon} = P\frac{1}{x} \mp \pi i \delta(x), \tag{1.4.117}$$

where P denotes the principal part. We find,

$$\lim_{\epsilon \to 0} G_u(E \pm i\epsilon) = \sum_n \frac{\left|\langle u|\phi_n\rangle\right|^2}{E - E_n} + P\int dE' \frac{\langle u|P_{E'}|u\rangle}{E - E'} \mp \pi i\langle u|P_E|u\rangle. \tag{1.4.118}$$

There is a discontinuity in the function when we approach a continuous eigenvalue from above or from below the real axis. We conclude that $G_u(E)$ has a *branch cut* along the continuous spectrum, and the discontinuity is given by

$$G_u(E + i\epsilon) - G_u(E - i\epsilon) = -2\pi i\langle u|P_E|u\rangle, \tag{1.4.119}$$

or, equivalently,

$$\text{Im}\,(G_u(E + i\epsilon)) = -\pi\langle u|P_E|u\rangle. \tag{1.4.120}$$

Plugging this result back into (1.4.112), we obtain

$$G_u(z) = \sum_n \frac{\left|\langle u|\phi_n\rangle\right|^2}{z - E_n} - \frac{1}{\pi}\int_C \frac{\text{Im}\,(G_u(E + i\epsilon))}{z - E}\,dE, \tag{1.4.121}$$

where the integral is along the branch cut C of the function.

This equality is simply a manifestation of Cauchy's theorem. Indeed, let us consider a counter-clockwise contour C in the complex plane, as shown in Figure 1.2,

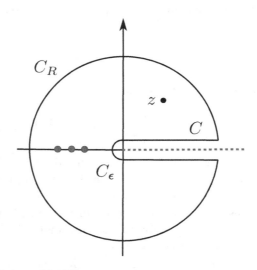

The contour C used in the integral (1.4.122).

and enclosing a point $z \in \mathbb{C} \backslash \mathbb{R}$. For simplicity, we will assume that the branch cut of $G_u(E)$ starts at $E = 0$. Let us consider the integral

$$\oint_C f(z')\mathrm{d}z', \qquad (1.4.122)$$

where the function

$$f(z') = \frac{G_u(z')}{z' - z}, \qquad (1.4.123)$$

is analytic along the contour. The integral can be written as,

$$\oint_C f(z')\mathrm{d}z' = \int_{C_R} f(z')\mathrm{d}z' + \int_{C_\epsilon} f(z')\mathrm{d}z' + \int_\epsilon^\infty (f(x + \mathrm{i}\epsilon) - f(x - \mathrm{i}\epsilon))\mathrm{d}x, \qquad (1.4.124)$$

where C_ϵ is small contour of radius ϵ around $z' = 0$, and C_R is a large contour of radius R. On the other hand, we can evaluate the integral by using Cauchy's residue theorem. There is a pole at $z' = z$, with residue $G_u(z)$. In addition, $G_u(z)$ has poles at $z = E_n$, $n \geq 0$, due to bound states, with residues R_n, and we have

$$\oint_C f(z')\mathrm{d}z' = 2\pi\mathrm{i}\left(G_u(z) + \sum_{n=0}^N \frac{R_n}{E_n - z}\right), \qquad (1.4.125)$$

where $E_0, \ldots, E_N$ are the poles contained in the contour. Let us assume that the integrals around C_R and C_ϵ vanish as $R \to \infty$ and $\epsilon \to 0$ (this can be proved rigorously in some cases, such as in the one-dimensional situation considered below). We conclude that

$$G_u(z) = \sum_{n=0}^\infty \frac{R_n}{z - E_n} + \frac{1}{\pi} \int_0^\infty \frac{\mathrm{Im}(G_u(x + \mathrm{i}\epsilon))}{x - z}\mathrm{d}x. \qquad (1.4.126)$$

This is precisely (1.4.121).

Although we have focused on the quantity $G_u(z)$, similar conclusions hold for other quantities, such as the integral kernel of the resolvent $G(\boldsymbol{x}, \boldsymbol{y}; E)$ or its trace $G(E)$. In the case of the integral kernel, we have the spectral decomposition

$$G(\boldsymbol{x}, \boldsymbol{y}; z) = \sum_n \frac{\phi_n(\boldsymbol{x})\phi_n^*(\boldsymbol{y})}{z - E_n} + \int \mathrm{d}E \frac{\langle \boldsymbol{x}|P_E|\boldsymbol{y}\rangle}{z - E}. \qquad (1.4.127)$$

Note that the simple poles of the integral kernel give information about the bound state spectrum, while their residues give information about the corresponding eigenfunctions. For the continuous spectrum, we have that

$$\langle \boldsymbol{x}|P_E|\boldsymbol{y}\rangle = \int \frac{\mathrm{d}\lambda}{N(\lambda)} \langle \boldsymbol{x}|\lambda\rangle\langle\lambda|\boldsymbol{y}\rangle\delta(E - E(\lambda)). \qquad (1.4.128)$$

Let us suppose that, for each value of E, there is a finite set of values of λ, λ_E, such that $E(\lambda_E) = E$. Then, this integral can be evaluated by using that

$$\delta(E - E(\lambda)) = \sum_{\lambda_E} \frac{1}{E'(\lambda_E)}\delta(\lambda - \lambda_E). \qquad (1.4.129)$$

Let us denote by

$$\langle \boldsymbol{x}|\lambda\rangle = \psi_\lambda(\boldsymbol{x}), \qquad (1.4.130)$$

the wavefunction corresponding to the state $|\lambda\rangle$. Then, we obtain

$$\langle x|\mathsf{P}_E|y\rangle = \sum_{\lambda_E} \frac{1}{N(\lambda_E)|E'(\lambda_E)|} \psi_{\lambda_E}(x)\psi^*_{\lambda_E}(y). \tag{1.4.131}$$

As before, this function gives the discontinuity of the integral kernel of the resolvent across the cut:

$$G(x, y; E + i\epsilon) - G(x, y; E - i\epsilon) = -2\pi i\langle x|\mathsf{P}_E|y\rangle. \tag{1.4.132}$$

In the following two examples, we verify this general result in two cases where the resolvent can be computed explicitly.

Example 1.4.1 *The one-dimensional free particle.* In Example 1.3.1 we computed the resolvents $G_\pm(x; E)$ explicitly in the case of a one-dimensional free particle. The discontinuity is given by

$$\begin{aligned}
G_+(x; E) - G_-(x; E) &= -\frac{i}{\hbar}\sqrt{\frac{m}{2E}}e^{i\sqrt{2mE}|x|/\hbar} - \frac{i}{\hbar}\sqrt{\frac{m}{2E}}e^{-i\sqrt{2mE}|x|/\hbar} \\
&= -\frac{i}{\hbar}\sqrt{\frac{2m}{E}}\cos\left(\frac{\sqrt{2mE}}{\hbar}x\right).
\end{aligned} \tag{1.4.133}$$

Let us now evaluate the r.h.s. of (1.4.132), by using (1.4.131). We label the continuum spectrum by the momentum $\lambda = p$. Given an energy E, there are two momenta corresponding to it,

$$p_E = \pm\sqrt{2mE}. \tag{1.4.134}$$

The associated wavefunctions are given in (1.2.27), and they are normalized in such a way that $N(p) = 1$. Finally, we note that

$$E(p) = \frac{p^2}{2m}, \qquad E'(p) = \frac{p}{m}. \tag{1.4.135}$$

We then obtain from (1.4.131) (we set $y = 0$ by exploiting translation invariance)

$$-2\pi i\langle x|\mathsf{P}_E|0\rangle = -2\pi i\frac{m}{|p_E|}\frac{1}{2\pi\hbar}\left(e^{i|p_E|x/\hbar} + e^{-i|p_E|x/\hbar}\right), \tag{1.4.136}$$

which is precisely (1.4.133).   □

Example 1.4.2 Let us consider the following one-dimensional Hamiltonian,

$$\mathsf{H} = 2\cosh\left(\frac{a\mathsf{p}}{\hbar}\right), \qquad a \in \mathbb{R}_{>0}, \tag{1.4.137}$$

which acts on wavefunctions as a difference operator

$$(\mathsf{H}\psi)(x) = \psi(x + ia) + \psi(x - ia). \tag{1.4.138}$$

The parameter a has dimensions of length. The Hamiltonian (1.4.137) is a deformation of the standard free particle Hamiltonian, in the sense that for $ap \ll \hbar$ one finds

$$\mathsf{H} \approx 2 + \frac{a^2\mathsf{p}^2}{2\hbar^2}. \tag{1.4.139}$$

This Hamiltonian appears in various contexts in mathematical physics. It is clear that the plane waves $\psi_p(x)$ in (1.2.27) are eigenfunctions of H, with eigenvalue

$$E = 2\cosh\left(\frac{ap}{\hbar}\right). \tag{1.4.140}$$

We will now calculate explicitly the resolvent of H. Clearly, H is diagonal in the momentum basis,

$$\langle p|(E-H)^{-1}|p'\rangle = \frac{1}{E - 2\cosh(ap/\hbar)}\delta(p - p'), \tag{1.4.141}$$

so we can write

$$G(x - y; E) = \langle x|(E-H)^{-1}|y\rangle = \int_{\mathbb{R}} \frac{dp}{2\pi\hbar}\frac{e^{ip(x-y)/\hbar}}{E - 2\cosh(ap/\hbar)}. \tag{1.4.142}$$

This integral can be computed using Cauchy's residue theorem. First, we change variables to $\xi = ap/\hbar$, so that

$$G(x; E) = \int_{\mathbb{R}} \frac{d\xi}{2\pi a}\frac{e^{i\xi x/a}}{E - 2\cosh\xi}. \tag{1.4.143}$$

In addition, we parametrize $E \in \mathbb{C}\backslash[2, \infty)$ as

$$E = 2\cosh k, \qquad 0 < \mathrm{Im}(k) \le \pi. \tag{1.4.144}$$

In this parametrization, the cut $[2, \infty)$ is covered twice as the variable k runs along the real line. We consider the contour C_R in the ξ plane shown in Figure 1.3. There are two poles inside the contour: $\xi = k$ and $\xi = 2\pi i - k$. Then, we have

$$\lim_{R\to\infty}\oint_{C_R} \frac{dz}{2\pi a}\frac{e^{izx/a}}{E - 2\cosh z} = \left(1 - e^{-2\pi x/a}\right)G(x; E)$$

$$= \frac{i}{a}\left(\mathrm{Res}_{z=k}\frac{e^{izx/a}}{E - 2\cosh z} + \mathrm{Res}_{z=2\pi i-k}\frac{e^{izx/a}}{E - 2\cosh z}\right)$$

$$= -\frac{i}{2a\sinh k}\left(e^{ikx/a} - e^{-ikx/a-2\pi x/a}\right), \tag{1.4.145}$$

and we conclude that

$$G(x; E) = -\frac{i}{2a\sinh k}\left(\frac{e^{ikx/a}}{1 - e^{-2\pi x/a}} + \frac{e^{-ikx/a}}{1 - e^{2\pi x/a}}\right). \tag{1.4.146}$$

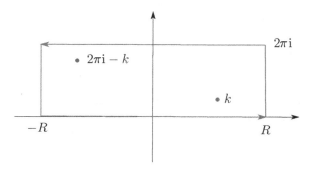

Figure 1.3   The contour C_R used in the integral (1.4.145).

The discontinuity is easily calculated: since

$$E - 2 \approx k^2, \tag{1.4.147}$$

we see that $E \pm i\epsilon$, $E > 2$, correspond to k positive and negative, as in Example 1.3.1 for the free particle. We then find

$$G(x; E + i\epsilon) - G(x; E - i\epsilon) = -\frac{i}{a \sinh k} \cos(kx/a). \tag{1.4.148}$$

This is again in precise agreement with the general formulae (1.4.131), (1.4.132), since in this case

$$\langle x|\mathsf{P}_E|0 \rangle = \frac{1}{2\pi a \sinh k} \cos(kx/a). \tag{1.4.149}$$

□

1.5 Scattering and Resolvent in One Dimension

In the case of standard Hamiltonians in one dimension, we can often construct the resolvent explicitly by considering special solutions to the Schrödinger equation. We will now present some fundamental results on this topic.

Let us consider a one-dimensional potential, $V(x)$, that has compact support or decreases rapidly at infinity, i.e.

$$V(x) \to 0, \qquad |x| \to \infty. \tag{1.5.150}$$

Then, we have a continuous spectrum of positive energies and we can consider scattering states. We will write the energy of a state in the continuum as

$$E = \frac{\hbar^2 k^2}{2m}, \tag{1.5.151}$$

where k defines the wave vector of the state. In this section we will set $m = 1/2$, $\hbar = 1$ to simplify our notation. Let us now introduce the two *Jost functions*, denoted by $f_1(x, k)$, $f_2(x, k)$. They are solutions to the Schrödinger equation

$$-\psi''(x) + V(x)\psi(x) = k^2 \psi(x), \tag{1.5.152}$$

for real k, and they are uniquely characterized by their asymptotic behavior at infinity,

$$\begin{aligned} f_1(x, k) &= e^{ikx} + \cdots, & x \to \infty, \\ f_2(x, k) &= e^{-ikx} + \cdots, & x \to -\infty. \end{aligned} \tag{1.5.153}$$

They can, however, be extended analytically to complex k, provided that

$$\text{Im}(k) > 0, \tag{1.5.154}$$

i.e. as long as k belongs to the physical sheet (we recall that the concept of physical sheet was introduced in Example 1.3.1).

Let us now assume that k is real and nonzero. Uniqueness of the solutions implies that

$$\overline{f_1(x, k)} = f_1(x, -k), \qquad \overline{f_2(x, k)} = f_2(x, -k), \tag{1.5.155}$$

since both solve the Schrödinger equation with asymptotics $\exp(\mp ikx)$ as $x \to \pm\infty$. The pairs $f_1(x, k)$ and $f_1(x, -k)$, as well as $f_2(x, k)$ and $f_2(x, -k)$, are bases of solutions to the Schrödinger equation. One way to check this is to consider their Wronskians. We recall that the Wronskian of two functions is given by

$$W(f, g) = f(x)g'(x) - g(x)f'(x). \tag{1.5.156}$$

The Wronskian of two solutions to the Schrödinger equation is a constant, and it vanishes if and only if the two solutions are linearly dependent. A simple computation indeed shows that

$$W(f_{1,2}(x, k), f_{1,2}(x, -k)) = \mp 2ik, \tag{1.5.157}$$

therefore justifying our previous statement. Since $f_1(x, k)$ and $f_1(x, -k)$ are a basis, we can express $f_2(x, k)$ as a linear combination of them:

$$f_2(x, k) = a(k)f_1(x, -k) + b(k)f_1(x, k). \tag{1.5.158}$$

The coefficients $a(k)$ and $b(k)$ are sometimes called *transition coefficients*. They can be written in terms of Wronskians as follows:

$$a(k) = \frac{i}{2k}W(f_1(x, k), f_2(x, k)), \qquad b(k) = \frac{i}{2k}W(f_2(x, k), f_1(x, -k)). \tag{1.5.159}$$

In addition, when k is real they satisfy

$$\overline{a(k)} = a(-k), \qquad \overline{b(k)} = b(-k). \tag{1.5.160}$$

We can also write,

$$f_1(x, k) = a(k)f_2(x, -k) - b(-k)f_2(x, k), \tag{1.5.161}$$

which follows from the Wronskian relations. Plugging (1.5.158) in here, we find the normalization condition

$$|a(k)|^2 - |b(k)|^2 = 1. \tag{1.5.162}$$

From (1.5.158) and (1.5.161) we also deduce the asymptotic behavior of the Jost functions. For $f_1(x, k)$, we have

$$f_1(x, k) \approx \begin{cases} e^{ikx}, & x \to \infty, \\ a(k)e^{ikx} - b(-k)e^{-ikx}, & x \to -\infty, \end{cases} \tag{1.5.163}$$

while for $f_2(x, k)$ we find,

$$f_2(x, k) \approx \begin{cases} a(k)e^{-ikx} + b(k)e^{ikx}, & x \to \infty, \\ e^{-ikx}, & x \to -\infty. \end{cases} \tag{1.5.164}$$

We can also extend the coefficients $a(k)$, $b(k)$ to the complex k-plane. In order to do that, we consider Jost functions for complex values of k. In this more general setting, they satisfy

$$\overline{f_1(x, k)} = f_1(x, -\overline{k}), \qquad \overline{f_2(x, k)} = f_2(x, -\overline{k}). \tag{1.5.165}$$

Although $a(k)$ cannot vanish along the real line, due to the normalization condition (1.5.162), it can vanish if $\text{Im}(k) \neq 0$. Let k_0 be a zero of $a(k)$ with a positive imaginary part:

$$a(k_0) = 0, \qquad \text{Im}(k_0) > 0. \tag{1.5.166}$$

In this case, the Wronskian in (1.5.159) vanishes, therefore $f_1(x, k_0)$, $f_2(x, k_0)$ are linearly dependent:

$$f_1(x, k_0) = c_0 f_2(x, k_0). \tag{1.5.167}$$

On the other hand, if $\text{Im}(k_0) > 0$, $f_1(x, k)$ decays exponentially as $x \to \infty$, and $f_2(x, k)$ decays exponentially as $x \to -\infty$. Since they are proportional to each other, $f_1(x, k)$ is a solution to the Schrödinger equation that decays exponentially as $|x| \to \infty$, and in particular it is square integrable. We conclude that, if k_0 is a zero of $a(k)$ with $\text{Im}(k_0) > 0$, we have a bound state with energy $E = k_0^2$. However, since energies are real, we must have

$$k_0 = i\kappa_0, \qquad \kappa_0 > 0. \tag{1.5.168}$$

It follows from (1.5.165) that the Jost functions are real.

The integral kernel of the resolvent can be written in terms of Jost functions, through the formula:

$$G(x, y; k) = \frac{1}{2ika(k)} \left(f_1(x, k) f_2(y, k) \theta(x - y) + f_1(y, k) f_2(x, k) \theta(y - x) \right). \tag{1.5.169}$$

In this equation, $\theta(x)$ is the Heaviside function. To prove this, we first note that

$$\partial_x^2 G(x, y; k) = \frac{1}{2ika(k)} \left(f_1''(x, k) f_2(y, k) \theta(x - y) + f_1(y, k) f_2''(x, k) \theta(y - x) \right)$$
$$+ \frac{1}{2ika(k)} \left(f_1'(x, k) f_2(y, k) - f_1(y, k) f_2'(x, k) \right) \delta(x - y), \tag{1.5.170}$$

where we have used that

$$f(x)\delta'(x) = -f'(0)\delta(x) + f(0)\delta'(x). \tag{1.5.171}$$

The last term in (1.5.170) can be written as

$$-\frac{1}{2ika(k)} W(f_1(x, k), f_2(x, k))\delta(x - y) = \delta(x - y), \tag{1.5.172}$$

where we have used the first equation in (1.5.159). Since $f_1(x, k)$, $f_2(x, k)$ solve the Schrödinger equation with energy k^2, we conclude that

$$\left(k^2 + \frac{\partial^2}{\partial x^2} - V(x) \right) G(x, y; k) = \delta(x - y). \tag{1.5.173}$$

It is instructive to verify in the explicit expression (1.5.169) the general analytic properties of the resolvent discussed in Section 1.4. First of all, the discrete spectrum should lead to poles in the resolvent, and the residue at this pole should be given by

$$\text{Res}_{\lambda = k_0^2} G(x, y; k) = \psi_{k_0}^*(x)\psi_{k_0}(y), \tag{1.5.174}$$

where $\psi_{k_0}(x)$ is the normalized wavefunction with energy $E = k_0^2$. As required, the expression (1.5.169) has poles at the zeros of $a(k)$, which as we have just seen, correspond to bound states. We also have (we assume for simplicity that $x \geq y$)

$$\text{Res}_{\lambda = k_0^2} G(x, y; k) = \lim_{k \to k_0} (k^2 - k_0^2) G(x, y; k) = \frac{1}{ia'(k_0)} f_1(x, k_0) f_2(y, k_0),$$

$$(1.5.175)$$

where we used (1.5.167) and we expanded $a(k)$ near $k = k_0$:

$$a(k) = a'(k_0)(k - k_0) + \cdots .$$

$$(1.5.176)$$

This assumes that k_0 is a simple zero of $a(k)$, as we will justify shortly. The normalized wavefunction $\psi_{k_0}(x)$ can be written as

$$\psi_{k_0}(x) = \frac{f_2(x, k_0)}{\|f_2(\cdot, k_0)\|},$$

$$(1.5.177)$$

where $\| \cdot \|$ is the standard L^2 norm. Consistency between (1.5.174) and (1.5.175) requires that

$$a'(k_0) = -ic_0 \|f_2(\cdot, k_0)\|^2,$$

$$(1.5.178)$$

where c_0 is the constant in (1.5.167). We will now show that the property (1.5.178) follows from a detailed analysis of the Jost solutions. Let y be a solution to the Schrödinger equation,

$$-y'' + V(x)y = k^2 y.$$

$$(1.5.179)$$

Taking a derivative w.r.t. k, which we denote by $\dot{}$, we find

$$-\dot{y}'' + V(x)\dot{y} = 2ky + k^2 \dot{y}.$$

$$(1.5.180)$$

Jost functions satisfy these two equations, so we can write

$$(-f_1'' + V(x)f_1 - k^2 f_1)\dot{f}_2 = 0,$$
$$(-\dot{f}_2'' + V(x)\dot{f}_2 - 2kf_2 - k^2 \dot{f}_2)f_1 = 0.$$

$$(1.5.181)$$

Subtracting both equations, we find

$$f_1 \dot{f}_2'' - f_1'' \dot{f}_2 = -2k f_1 f_2,$$

$$(1.5.182)$$

or,

$$\frac{d}{dx} W(f_1, \dot{f}_2) = -2k f_1 f_2.$$

$$(1.5.183)$$

Exchanging the indices, we find

$$\frac{d}{dx} W(\dot{f}_1, f_2) = 2k f_1 f_2.$$

$$(1.5.184)$$

These relations can be integrated to obtain,

$$W(f_1, \dot{f}_2)\Big|_{-A}^{x} = -2k \int_{-A}^{x} f_1(y, k) f_2(y, k) dy,$$

$$W(\dot{f}_1, f_2)\Big|_{x}^{A} = 2k \int_{x}^{A} f_1(y, k) f_2(y, k) dy,$$

$$(1.5.185)$$

and after subtracting the second equation from the first, we obtain

$$W(\dot{f}_1, f_2)(x) + W(f_1, \dot{f}_2)(x) = -2k \int_{-A}^{A} f_1(y, k) f_2(y, k) \mathrm{d}k \tag{1.5.186}$$
$$+ W(\dot{f}_1, f_2)(A) + W(f_1, \dot{f}_2)(-A).$$

The l.h.s. of this equation can be evaluated by taking a derivative w.r.t. k in the first equality of (1.5.159):

$$W(\dot{f}_1, f_2)(x) + W(f_1, \dot{f}_2)(x) = -2i a(k) - 2i k a'(k). \tag{1.5.187}$$

The r.h.s. of (1.5.186) must be independent of A, therefore we can evaluate it in the limit $A \to \infty$. To compute $W(\dot{f}_1, f_2)(A)$ in this limit, we note that, when $x \to \infty$,

$$\dot{f}_1(x, k) \approx i x \, \mathrm{e}^{ikx}, \tag{1.5.188}$$

while from (1.5.158) we have

$$f_2(x, k) \approx a(k) \mathrm{e}^{-ikx} + b(k) \mathrm{e}^{ikx}. \tag{1.5.189}$$

Finally we obtain

$$W(\dot{f}_1, f_2)(x) = \dot{f}_1 f_2' - \dot{f}_1' f_2 \approx 2kxa(k) - ia(k) - ib(k) \mathrm{e}^{2ikx}, \qquad x \to \infty. \tag{1.5.190}$$

A similar calculation gives

$$W(f_1, \dot{f}_2)(x) = f_1 \dot{f}_2' - f_1' \dot{f}_2 \approx -2kxa(k) - ia(k) + ib(-k) \mathrm{e}^{-2ikx}, \qquad x \to -\infty, \tag{1.5.191}$$

which can be used to evaluate $W(f_1, \dot{f}_2)(-A)$ for $A \to \infty$. If we now set $k = k_0$, $a(k_0) = 0$, and $k = i\kappa_0$ with $\kappa_0 > 0$, so the exponentials appearing in (1.5.190), (1.5.191) vanish asymptotically. Therefore, the r.h.s. of (1.5.186), for $k = k_0$ and in the limit $A \to \infty$, simply gives

$$-2k_0 \int_{\mathbb{R}} f_1(y, k_0) f_2(y, k_0) \mathrm{d}y. \tag{1.5.192}$$

On the other hand, by evaluating (1.5.187) at $k = k_0$, we obtain

$$W(\dot{f}_1(x, k_0), f_2(x, k_0)) + W(f_1(x, k_0), \dot{f}_2(x, k_0)) = -2i k_0 a'(k_0). \tag{1.5.193}$$

We conclude that

$$a'(k_0) = -i \int_{\mathbb{R}} f_1(x, k_0) f_2(x, k_0) \mathrm{d}x, \tag{1.5.194}$$

which can be written as (1.5.178). As a consequence of this formula we verify that the zeros of $a(k)$ are simple, since the r.h.s. of (1.5.178) is nonzero.

We can also compute the discontinuity of the resolvent explicitly in the continuous part of the spectrum and verify the general result (1.4.132). The branch cut corresponds to the two determinations of $\sqrt{k^2} = \pm k$, as in the free particle case studied in Example 1.3.1. We find,

$$G(x, y; E + i\epsilon) - G(x, y; E - i\epsilon) = \frac{f_1(x, k) f_2(y, k)}{2i k a(k)} + \frac{f_1(x, -k) f_2(y, -k)}{2i k a(-k)}, \tag{1.5.195}$$

where $k > 0$ and we have assumed for simplicity that $x \geq y$. We now use

$$f_1(x, -k) = \frac{1}{a(k)} f_2(x, k) - \frac{b(k)}{a(k)} f_1(x, k), \qquad (1.5.196)$$

which follows from (1.5.158), as well as

$$f_2(y, k) = \frac{1}{a(-k)} f_1(y, -k) + \frac{b(k)}{a(-k)} f_2(y, -k), \qquad (1.5.197)$$

which follows from (1.5.161). By applying this, the r.h.s. of (1.5.195) reads,

$$\frac{f_1(x, k)}{2ika(k)} \left(\frac{1}{a(-k)} f_1(y, -k) + \frac{b(k)}{a(-k)} f_2(y, -k) \right)$$
$$+ \frac{f_2(y, -k)}{2ika(-k)} \left(\frac{1}{a(k)} f_2(x, k) - \frac{b(k)}{a(k)} f_1(x, k) \right) \qquad (1.5.198)$$
$$= \frac{1}{2ik|a(k)|^2} \left(f_1(x, k)\overline{f_1(y, k)} + f_2(x, k)\overline{f_2(y, k)} \right).$$

We now introduce the functions

$$\psi_i(x, k) = \frac{1}{a(k)} f_i(x, k), \qquad i = 1, 2, \qquad k > 0, \qquad (1.5.199)$$

which correspond in this example to the wavefunctions appearing in (1.4.131). The labels $i = 1, 2$ correspond to the two choices $\pm k$ associated to $E = k^2$. We can write (1.5.195) as

$$G(x, y; E + i\epsilon) - G(x, y; E - i\epsilon) = \frac{1}{2ik} \sum_{i=1}^{2} \psi_i(x, k)\overline{\psi_i(y, k)}. \qquad (1.5.200)$$

The functions $\psi_i(x, k)$ satisfy the orthogonality property

$$\int_{\mathbb{R}} \psi_i(x, k)\overline{\psi_j(x, p)} dx = N(k)\delta_{ij}\delta(k - p), \qquad i = 1, 2, \qquad (1.5.201)$$

with

$$N(k) = 2\pi. \qquad (1.5.202)$$

The property (1.5.201) can be proved by using that,

$$\frac{d}{dx} W\left(\psi_i(x, k), \overline{\psi_j(x, p)}\right) = (k^2 - p^2)\psi_i(x, k)\overline{\psi_j(x, p)}, \qquad (1.5.203)$$

which follows from the fact that $\psi_i(x, k)$, $\overline{\psi_j(x, p)}$ satisfy the Schrödinger equation with energies k^2, p^2, respectively. Let us check (1.5.201) for $i = j = 1$. We find,

$$\int_{-A}^{A} \psi_i(x, k)\overline{\psi_j(x, p)} dx$$
$$= \frac{1}{(k^2 - p^2)a(k)\overline{a(p)}} \left\{ W\left(f_1(A, k), \overline{f_1(A, p)}\right) - W\left(f_1(-A, k), \overline{f_1(-A, p)}\right) \right\}. \qquad (1.5.204)$$

To evaluate the Wronskian at $\pm A$, with A very large, we can use the asymptotic form of the Jost function in (1.5.163). One finds, after using (1.5.160),

$$\frac{1}{(k^2 - p^2)} \left\{ W\left(f_1(A, k), \overline{f_1(A, p)}\right) - W\left(f_1(-A, k), \overline{f_1(-A, p)}\right) \right\}$$

$$= \frac{i}{k - p} \left\{ -e^{i(k-p)A} + a(k)\overline{a(p)}e^{-i(k-p)A} - b(p)\overline{b(k)}e^{i(k-p)A} \right\} \qquad (1.5.205)$$

$$+ \frac{i}{k + p} \left\{ -a(k)b(p)e^{-i(k+p)A} + \overline{a(p)b(k)}e^{i(k+p)A} \right\} .$$

As $A \to \infty$, this becomes a distribution. To determine it, we use the Riemann–Lebesgue theorem, which says that

$$\lim_{k \to \infty} \int_{\mathbb{R}} f(x)e^{ikx}dx = 0 \qquad (1.5.206)$$

for any square-integrable function $f(x)$. Since $k, p > 0$, the last line of (1.5.205) leads to a zero distribution as $A \to \infty$, while the second line leads to a distribution supported on the locus where $k = p$. We can therefore replace $a(p), b(p)$ by $a(k)$, $b(k)$ and use (1.5.162) to conclude that

$$\lim_{A \to \infty} \frac{1}{(k^2 - p^2)a(k)\overline{a(p)}} \left\{ W\left(f_1(A, k), \overline{f_1(A, p)}\right) - W\left(f_1(-A, k), \overline{f_1(-A, p)}\right) \right\}$$

$$= \lim_{A \to \infty} 2\frac{\sin((k - p)A)}{k - p} = 2\pi\delta(k - p).$$

$$(1.5.207)$$

We conclude that (1.4.132) is satisfied, since in this case

$$-2\pi i \langle x | P_E | y \rangle = -\frac{i}{2k} \sum_{i=1}^{2} \psi_i(x, k)\overline{\psi_i(y, k)}. \qquad (1.5.208)$$

The functions $\psi_i(x, k)$ play an important role in scattering theory in one dimension, and they are sometimes called *scattering solutions*. Note that they are Jost functions but with a different normalization. It follows from (1.5.163) and (1.5.164) that these solutions have the following asymptotic behavior:

$$\psi_1(x, k) \approx \begin{cases} s_{11}e^{ikx}, & x \to \infty, \\ e^{ikx} + s_{21}e^{-ikx}, & x \to -\infty, \end{cases} \qquad (1.5.209)$$

where

$$s_{11} = \frac{1}{a(k)}, \qquad s_{21} = -\frac{b(-k)}{a(k)}, \qquad (1.5.210)$$

while

$$\psi_2(x, k) \approx \begin{cases} e^{-ikx} + s_{12}e^{ikx}, & x \to \infty, \\ s_{22}e^{-ikx}, & x \to -\infty, \end{cases} \qquad (1.5.211)$$

where

$$s_{12} = \frac{b(k)}{a(k)}, \qquad s_{22} = \frac{1}{a(k)}. \qquad (1.5.212)$$

The coefficients s_{ij}, $i, j = 1, 2$ have a very clear scattering interpretation. The solution $\psi_1(x, k)$ describes a particle coming from $-\infty$ (where it is free), moving to the right, and interacting with the potential. s_{11} and s_{21} are the amplitudes s of the transmitted and reflected waves, respectively. Similarly, $\psi_2(x, k)$ describes a particle coming from ∞ (where is free), moving to the left, and interacting with the potential. s_{12} and s_{22} are the amplitudes of the reflected and transmitted waves, respectively. The transmission and reflection coefficients are given by

$$T(k) = |s_{11}|^2 = |s_{22}|^2, \qquad R(k) = |s_{21}|^2 = |s_{12}|^2, \qquad (1.5.213)$$

where we used the second relation in (1.5.160). Note that the relation (1.5.162) implies that

$$T(k) + R(k) = 1, \qquad (1.5.214)$$

which expresses the conservation of probability for the Schrödinger equation. The coefficients s_{ij}, $i, j = 1, 2$ can be put inside a two by two matrix, called the *S-matrix*,

$$S = \begin{pmatrix} s_{11} & s_{12} \\ s_{21} & s_{22} \end{pmatrix}. \qquad (1.5.215)$$

This matrix satisfies

$$SS^\dagger = \frac{1}{|a(k)|^2} \begin{pmatrix} 1 & b(k) \\ -b(k) & 1 \end{pmatrix} \begin{pmatrix} 1 & -b(k) \\ b(k) & 1 \end{pmatrix} = \mathbf{1}, \qquad (1.5.216)$$

where we used (1.5.160) and (1.5.162), i.e. it is *unitary*. The unitarity property of the S-matrix expresses the conservation of probability (1.5.214). Note that zeros of $a(k)$ become poles of S.

Example 1.5.1 *The single delta potential.* Perhaps the simplest solvable scattering problem in one dimension is the delta function potential,

$$V(x) = g\delta(x). \qquad (1.5.217)$$

The wavefunctions are continuous but their derivatives jump at the origin. The discontinuity in the derivative can be easily obtained by integrating the Schrödinger equation. Take $0 < \epsilon \ll 1$. Then,

$$\psi'(\epsilon) - \psi'(-\epsilon) = \int_{-\epsilon}^{\epsilon} \psi''(x)\mathrm{d}x = \int_{-\epsilon}^{\epsilon} \left(-k^2 + g\delta(x) \right) \psi(x)\mathrm{d}x = g\psi(0) + \mathcal{O}(\epsilon). \qquad (1.5.218)$$

We will distinguish two regions in this problem, region I with $x < 0$, and region II with $x > 0$. Let us obtain the Jost function $f_1(x, k)$. The asymptotic formula (1.5.163) now describes the Jost function exactly, and we have

$$f_1(x, k) = \begin{cases} e^{ikx}, & x > 0, \\ a(k)e^{ikx} - b(-k)e^{-ikx}, & x < 0. \end{cases} \qquad (1.5.219)$$

Imposing continuity of $f_1(x, k)$ one finds,

$$a(k) - b(-k) = 1, \qquad (1.5.220)$$

while the discontinuity of the first derivative, (1.5.218), gives

$$a(k) + b(-k) = 1 - \frac{g}{ik}. \tag{1.5.221}$$

These two equations determine the transition coefficients,

$$a(k) = 1 - \frac{g}{2ik}, \qquad b(k) = \frac{g}{2ik}. \tag{1.5.222}$$

As a check, note that they verify the conjugation conditions (1.5.160) and the normalization condition (1.5.162). Similarly, the second Jost function agrees with its asymptotic expression (1.5.164):

$$f_2(x, k) = \begin{cases} a(k)e^{-ikx} + b(k)e^{ikx}, & x > 0, \\ e^{-ikx}, & x < 0. \end{cases} \tag{1.5.223}$$

As an application of this result, let us suppose that $g = -\lambda$, $\lambda > 0$, so the potential supports bound states. The coefficient $a(k)$ vanishes precisely for

$$k_0 = \frac{i\lambda}{2}, \tag{1.5.224}$$

which has $\text{Im}(k_0) > 0$. This corresponds to a bound state with energy

$$E_0 = -\frac{\lambda^2}{4}, \tag{1.5.225}$$

and with a normalizable wavefunction

$$\psi_{k_0}(x) \propto e^{-\lambda|x|/2}. \tag{1.5.226}$$

□

Example 1.5.2 *The double delta potential.* Our second example is the double delta potential,

$$V(x) = B\left(\delta(x - a) + \delta(x + a)\right). \tag{1.5.227}$$

As in the previous example, the wavefunction is continuous, but its first derivative has discontinuities at $x = \pm a$. The Jost function $f_1(x, k)$ has the form,

$$f_1(x, k) = \begin{cases} e^{ikx}, & x \geq a, \\ A_1 e^{ikx} + A_2 e^{-ikx}, & -a \leq x \leq a, \\ a(k)e^{ikx} - b(-k)e^{-ikx}, & x \leq a. \end{cases} \tag{1.5.228}$$

Imposing the continuity of the function and the discontinuity of its first derivative at $x = \pm a$, we obtain the values of A_1, A_2, $a(k)$, and $b(k)$. To write down the result for $a(k)$, it is useful to introduce the following functions:

$$\phi_0(\mu) = \frac{1}{\mu}\left(\mu + i\beta e^{i\mu}\cos\mu\right), \qquad \phi_1(\mu) = \frac{1}{\mu}\left(\mu + \beta e^{i\mu}\sin\mu\right), \tag{1.5.229}$$

where

$$\mu = ka, \qquad \beta = Ba. \tag{1.5.230}$$

In terms of these functions, one finds

$$a(k) = 2\frac{\phi_0(\mu)\phi_1(\mu)}{\phi_0(\mu)\phi_1(-\mu) + \phi_0(-\mu)\phi_1(\mu)}. \tag{1.5.231}$$

We can use this result to determine, for example, the bound states of this potential. The zeros of $a(k)$ are given by zeros of $\phi_0(\mu)$ or of $\phi_1(\mu)$ with $k = i\kappa$, $\kappa > 0$. Zeros of $\phi_0(\mu)$ with $\mu = i\gamma_0$ satisfy the equation

$$1 + e^{-2\gamma_0} = -\frac{2\gamma_0}{\beta}. \tag{1.5.232}$$

A simple graphical analysis of this equation shows the result is that, when $\beta < 0$, there is one single root with $\gamma_0 > 0$. This is a first bound state in this potential. Let us now look for zeros of $\phi_1(\mu)$ with $\mu = i\gamma_1$. They satisfy

$$1 - e^{-2\gamma_1} = -\frac{2\gamma_1}{\beta}. \tag{1.5.233}$$

This again has one solution with $\gamma_1 > 0$ provided $\beta < -1$. We conclude that in the double delta potential there is a bound state for $-1 \leq \beta < 0$, and two bound states when $\beta < -1$. $\qquad\square$

Example 1.5.3 *The Pöschl–Teller potential.* The Pöschl–Teller potential is given by

$$V(x) = -\frac{\hbar^2}{2m}\frac{\alpha^2\lambda(\lambda - 1)}{\cosh^2(\alpha x)}. \tag{1.5.234}$$

We will consider three different situations:

1. When $\lambda > 1$ we have a potential well.
2. When $1/2 \leq \lambda < 1$, we have a low barrier.
3. When

$$\lambda = \frac{1}{2} + i\ell, \qquad \ell > 0, \tag{1.5.235}$$

we have a high barrier.

As we will see, the low and high barriers have different physical properties. The Schrödinger equation in the presence of the potential (1.5.234) reads

$$\psi''(x) + \left(k^2 + \frac{\alpha^2\lambda(\lambda - 1)}{\cosh^2(\alpha x)}\right)\psi(x) = 0, \tag{1.5.236}$$

where we have set $\hbar = 2m = 1$. Let us perform the change of variables

$$y = \tanh(\alpha x), \tag{1.5.237}$$

and let us write the wavefunction as

$$\psi(x) = (1 + y)^{i\bar{k}/2}(1 - y)^{-i\bar{k}/2}\phi(y), \qquad \bar{k} = \frac{k}{\alpha}. \tag{1.5.238}$$

Then, the Schrödinger equation reads

$$(1 - y^2)\phi''(y) + 2(i\bar{k} - y)\phi'(y) + \lambda(\lambda - 1)\phi(y) = 0. \tag{1.5.239}$$

Performing the further change

$$z = \frac{1 + y}{2}, \tag{1.5.240}$$

we find a hypergeometric equation

$$z(1 - z)\phi''(z) + \left(i\bar{k} - 2z + 1\right)\phi'(z) + \lambda(\lambda - 1)\phi(z) = 0. \tag{1.5.241}$$

By using the general theory of these equations, we obtain two independent solutions. The first one

$$z^{-i\bar{k}}{}_2F_1\left(\lambda - i\bar{k}, 1 - \lambda - i\bar{k}; 1 - i\bar{k}; z\right), \tag{1.5.242}$$

is appropriate near $z = 0$, which corresponds to $x \to -\infty$. The second solution

$$z^{-i\bar{k}}{}_2F_1\left(\lambda - i\bar{k}, 1 - \lambda - i\bar{k}; 1 - i\bar{k}; 1 - z\right), \tag{1.5.243}$$

is appropriate near $z = 1$, which corresponds to $x \to \infty$. Since

$$1 - \tanh^2(\alpha x) \approx 4e^{-2\alpha|x|}, \qquad |x| \to \infty, \tag{1.5.244}$$

we conclude that the Jost functions of the problem are

$$f_1(x, k) = 2^{i\bar{k}}(1 - \tanh^2(\alpha x))^{-i\bar{k}/2}{}_2F_1\left(\lambda - i\bar{k}, 1 - \lambda - i\bar{k}; 1 - i\bar{k}; \frac{1 - \tanh(ax)}{2}\right),$$

$$f_2(x, k) = 2^{i\bar{k}}(1 - \tanh^2(\alpha x))^{-i\bar{k}/2}{}_2F_1\left(\lambda - i\bar{k}, 1 - \lambda - i\bar{k}; 1 - i\bar{k}; \frac{1 + \tanh(ax)}{2}\right). \tag{1.5.245}$$

Let us now calculate the coefficients $a(k)$ and $b(k)$. To do this, we can for example determine the asymptotics of $f_1(x, k)$ as $x \to -\infty$. This can be done by using the transformation properties of the hypergeometric functions, which imply in particular that (see eq. (9.5.7) of Lebedev (1972)):

$$_2F_1\left(\lambda - i\bar{k}, 1 - \lambda - i\bar{k}; 1 - i\bar{k}; 1 - z\right)$$
$$= \frac{\Gamma(1 - i\bar{k})\Gamma(i\bar{k})}{\Gamma(1 - \lambda)\Gamma(\lambda)}{}_2F_1\left(\lambda - i\bar{k}, 1 - \lambda - i\bar{k}; 1 - i\bar{k}; z\right)$$
$$+ z^{i\bar{k}}\frac{\Gamma(1 - i\bar{k})\Gamma(-i\bar{k})}{\Gamma(\lambda - i\bar{k})\Gamma(1 - \lambda - i\bar{k})}{}_2F_1\left(1 - \lambda, \lambda; 1 + i\bar{k}; z\right). \tag{1.5.246}$$

We then have that, when $x \to -\infty$,

$$f_1(x, k) \approx \frac{\Gamma(1 - i\bar{k})\Gamma(i\bar{k})}{\Gamma(1 - \lambda)\Gamma(\lambda)}e^{-ikx} + \frac{\Gamma(1 - i\bar{k})\Gamma(-i\bar{k})}{\Gamma(\lambda - i\bar{k})\Gamma(1 - \lambda - i\bar{k})}e^{ikx}, \tag{1.5.247}$$

and we conclude that

$$a(k) = \frac{\Gamma(1 - i\bar{k})\Gamma(-i\bar{k})}{\Gamma(\lambda - i\bar{k})\Gamma(1 - \lambda - i\bar{k})}, \qquad b(k) = -\frac{\Gamma(1 + i\bar{k})\Gamma(-i\bar{k})}{\Gamma(1 - \lambda)\Gamma(\lambda)}. \tag{1.5.248}$$

From here one calculates the transmission coefficient $T(k)$ as

$$T(k) = \frac{\sinh^2(\pi k/\alpha)}{\sinh^2(\pi k/\alpha) + \sin^2(\pi\lambda)}. \tag{1.5.249}$$

The bound states for the potential well can also be found from the above explicit formula for $a(k)$. The zeros occur at the poles of the denominator. There are two types of poles, given by

$$
\begin{aligned}
k_1(n) &= -i\alpha(n + \lambda), & n &\in \mathbb{Z}_{\geq 0}, \\
k_2(n) &= i\alpha(\lambda - 1 - n), & n &\in \mathbb{Z}_{\geq 0}.
\end{aligned}
\tag{1.5.250}
$$

Bound states require $\mathrm{Im}(k) > 0$. It is easy to see that this only happens when we have a well (i.e. $\lambda > 1$) and is due to the poles at $k = k_2(n)$. In addition, there is a maximal possible value of n:

$$
n_{\max} = \begin{cases} [\lambda - 1], & \text{if } \lambda \notin \mathbb{Z}, \\ \lambda - 2, & \text{if } \lambda \in \mathbb{Z}, \end{cases}
\tag{1.5.251}
$$

where $[\cdot]$ denotes the integer part. Therefore, there are only a finite number of bound states. For example, if $\lambda = 2$, we have exactly one bound state with $n = 0$. Note that in both the well and the low barrier case we have poles in which k is purely imaginary and $\mathrm{Im}(k) < 0$. The corresponding states are called *antibound states*. They are eigenfunctions of the Schrödinger equation with a real eigenvalue, but they do not belong to $L^2(\mathbb{R})$ (on the contrary, they increase exponentially at infinity). Finally, in the case of a high barrier, the poles lead to eigenfunctions with complex eigenvalues. These states are called *resonances* and we will discuss them in detail in Chapter 5.

□

Another interesting result that follows from the analysis of Jost functions is an explicit expression for the trace of the resolvent in terms of the function $a(k)$, when $E \in \mathbb{C}\backslash[0, \infty]$ in the physical sheet (i.e. $\mathrm{Im}(k) > 0$). The trace of the resolvent is itself divergent, but it can be regularized by subtracting the result for the free particle, i.e. the case in which the potential vanishes. Quantities for the free particle will be denoted by a 0 subindex. Since for the free particle one has

$$
f_1(x, k) = \mathrm{e}^{ikx}, \qquad f_2(x, k) = \mathrm{e}^{-ikx} = f_1(x, -k),
\tag{1.5.252}
$$

we have that

$$
a_0(k) = 1, \qquad b_0(k) = 0,
\tag{1.5.253}
$$

and

$$
\mathrm{Tr}\,(\mathrm{G}(k) - \mathrm{G}_0(k)) = \frac{1}{2ika(k)} \int_{\mathbb{R}} (f_1(x, k)f_2(x, k) - a(k))\,\mathrm{d}x.
\tag{1.5.254}
$$

The integral can be evaluated using the results (1.5.186), (1.5.187), (1.5.190), and (1.5.191). We obtain, for A large,

$$
\int_{-A}^{A} f_1(x, k)f_2(x, k)\,\mathrm{d}x \approx 2Aa(k) + i\dot{a}(k).
\tag{1.5.255}
$$

In this calculation, we have used the fact that, for $\mathrm{Im}(k) > 0$, the exponentials appearing in (1.5.190), (1.5.191) vanish as $A \to \infty$. We then obtain,

$$
\mathrm{Tr}\,(\mathrm{G}(k) - \mathrm{G}_0(k)) = \frac{1}{2k}\frac{\mathrm{d}}{\mathrm{d}k}\log a(k).
\tag{1.5.256}
$$

1.6 Bibliographical Notes

In this book we assume some previous knowledge of quantum mechanics. There are many excellent textbooks that provide the appropriate background. My favourite, recent introduction is the book by Konishi and Paffuti (2009), which contains a fantastic collection of solved exercises and touches on many different (and modern) topics. In my view, the best references for an advanced treatment of the subject are the two-volume set by Galindo and Pascual (1990), and the book by Takhtajan (2008). They provide careful treatments of all the subjects they cover, and they are rigorous without being pedantic.

The quantum-mechanical propagator is introduced in most advanced textbooks of quantum mechanics, such as, for example, Galindo and Pascual (1990). The examples discussed in Section 1.2 of this chapter are based on Galitski et al. (2013). A useful introduction to resolvent operators is presented in Cohen-Tannoudji et al. (2012) and Konishi and Paffuti (2009). Example 1.4.2 is based on Faddeev and Takhtajan (2014). The presentation of scattering theory in one dimension closely follows Galindo and Pascual (1990) and, in particular, Takhtajan (2008). The double delta potential is analyzed in detail in Galindo and Pascual (1990), and useful references for the Pöschl–Teller potential are Galindo and Pascual (1990) and Çevik et al. (2016).

2 The WKB Method and Nonperturbative Effects

2.1 Introduction

The Schrödinger equation can only be solved in closed form in a very small number of cases. When a closed form solution is not available, one needs approximation schemes to calculate eigenvalues and eigenfunctions. Among these approximations, semiclassical techniques play an important role. In these techniques, $\hbar$ is regarded as a small parameter, and the relevant quantities are then computed as formal power series in $\hbar$. In 1926, the one-dimensional Schrödinger equation was analyzed from that point of view in papers by Wentzel, Kramers, and Brillouin. The resulting set of approximation techniques is called the WKB method, and it plays a fundamental role in quantum mechanics.

The WKB method produces solutions to the Schrödinger equation that are formal power series in $\hbar$, which lead in turn to power series for the eigenvalues. The leading-order approximation for the eigenvalues reproduces the Bohr–Sommerfeld quantization condition, which played a crucial role in the early quantum theory. The WKB analysis shows that this quantization condition provides a general asymptotic formula for the energies of bound states in the limit of large quantum numbers.

However, the power series obtained with the WKB method are not convergent, but asymptotic, and their correct mathematical handling is very subtle. This led to much confusion, in particular in the context of the connection formulae relating WKB solutions on the two sides of a turning point in the potential. These were only clarified when the WKB method was analyzed by using more sophisticated asymptotic techniques, involving Borel resummation and including exponentially small corrections in $\hbar$. Such corrections turn out to be crucial in quantum mechanics, because they control important phenomena, such as tunneling through a potential barrier. The correct connection formulae were obtained in the early 1980s, in independent work by André Voros and Harris Silverstone, following rigorous mathematical analysis.

An additional outcome of the careful treatment of the WKB method is that one can transform it into a systematic device to obtain exact results. This is achieved by the appropriate inclusion of exponentially small, "tunneling" effects that are nonperturbative. The resulting formalism is often called the *exact WKB method*, in contrast to the more traditional WKB approximation. The exact WKB method requires an extension of the Schrödinger equation to the complex plane and, for this reason, it is also called the complex WKB method. Since the pioneering work of Voros and Silverstone, the exact WKB method has been developed intensively and has led in particular to exact quantization conditions for one-dimensional potentials in quantum mechanics. It played an important role in the development of the theory of resurgence due to Jean Écalle and it has found applications in many other contexts in mathematical physics.

In this chapter we will present a pedagogical introduction to the exact WKB method in one dimension by using traditional physics tools, and in the spirit of Silverstone's paper on the connection formulae. Most textbooks in quantum mechanics have inherited the confusion surrounding the traditional WKB method, so our analysis will be very detailed and with the highest possible physical rigor. We will show that the exact WKB method makes it possible to include nonperturbative effects in $\hbar$ in a systematic way, and we will illustrate it using various applications. In particular, we will derive exact quantization conditions for the double-well potential, which were first conjectured by J. Zinn-Justin by using instanton techniques. We will also discuss the WKB approximation in higher dimensions, as applicable to integrable systems, and we will illustrate it in detail in the advanced example of the Toda lattice. In this chapter we will make extensive use of the theory of Borel resummation, which we have summarized in Appendix A. The reader not familiar with this theory is encouraged to study this appendix before venturing into this chapter.

2.2 The WKB Approximation in One Dimension

The WKB method is a systematic approximation scheme to solve the time-independent Schrödinger equation

$$-\frac{\hbar^2}{2m}\psi''(x) + V(x)\psi(x) = E\psi(x). \tag{2.2.1}$$

The result of this approximation is an expression for the wavefunction as a formal power series in $\hbar$. In particular, as we will see, in the analysis of bound state problems, the WKB method gives estimates of the energy spectrum when $\hbar$ is small as compared to the classical action of the problem. This happens when the quantum number labelling the bound state is very large. The resulting approximation often provides an accurate description of the energy spectrum, and it has the advantage that it can be formulated in terms of classical quantities.

In the WKB method, we treat $\hbar$ as a small parameter and we perform a systematic expansion of all quantities in a power series in $\hbar$, around $\hbar = 0$. However, if we write (2.2.1) as

$$\hbar^2\psi''(x) + p^2(x)\psi(x) = 0, \qquad p(x) = \sqrt{2m(E - V(x))}, \tag{2.2.2}$$

it is clear that we cannot send $\hbar \to 0$ directly, since in this limit the equation becomes algebraic. Let us however write the wavefunction as

$$\psi(x) = \exp\left[\frac{i}{\hbar}\int^x Y(x')dx'\right]. \tag{2.2.3}$$

When we do this, we transform the Schrödinger equation into a Riccati equation for $Y(x)$,

$$Y^2(x) - i\hbar \frac{dY(x)}{dx} = p^2(x), \qquad (2.2.4)$$

which can be solved in power series in $\hbar$:

$$Y(x) = \sum_{k=0}^{\infty} Y_k(x)\hbar^k. \qquad (2.2.5)$$

We will regard this as a *formal* power series, i.e. we will not address issues of convergence for the time being. The functions $Y_k(x)$ can be computed recursively as

$$Y_0(x) = \pm p(x),$$

$$Y_1(x) = \frac{i}{2} \frac{Y_0'(x)}{Y_0(x)},$$

$$Y_{n+1}(x) = \frac{1}{2Y_0(x)} \left(i \frac{dY_n(x)}{dx} - \sum_{k=1}^{n} Y_k(x) Y_{n+1-k}(x) \right), \qquad n \geq 1. \qquad (2.2.6)$$

The two choices of sign for $Y_0(x)$ will give the two independent solutions of the Schrödinger equation. Let us split the formal series $Y(x)$ into two series of even and odd powers of $\hbar$,

$$P(x) = \sum_{k=0}^{\infty} Y_{2k}(x)\hbar^{2k}, \qquad Y_{\text{odd}}(x) = \sum_{k=0}^{\infty} Y_{2k+1}(x)\hbar^{2k+1}, \qquad (2.2.7)$$

so that

$$Y(x) = P(x) + Y_{\text{odd}}(x). \qquad (2.2.8)$$

The Riccati equation (2.2.4) splits into two different equations:

$$\begin{aligned} \text{even:} \quad & Y_{\text{odd}}^2(x) + P^2(x) - i\hbar Y_{\text{odd}}'(x) = p^2(x), \\ \text{odd:} \quad & 2Y_{\text{odd}}(x)P(x) - i\hbar P'(x) = 0. \end{aligned} \qquad (2.2.9)$$

The second equation can be solved in closed form:

$$Y_{\text{odd}}(x) = \frac{i\hbar}{2} \frac{P'(x)}{P(x)} = \frac{i\hbar}{2} \frac{d}{dx} \log P(x). \qquad (2.2.10)$$

Therefore,

$$\frac{i}{\hbar} \int^x Y(x')dx' = -\frac{1}{2} \log P(x) + \frac{i}{\hbar} \int^x P(x')dx', \qquad (2.2.11)$$

and the corresponding wavefunction is

$$\psi(x) = \frac{1}{\sqrt{P(x)}} \exp\left(\frac{i}{\hbar} \int^x P(x')dx' \right). \qquad (2.2.12)$$

The very first solutions to the recursive equation (2.2.6) are

$$Y_0(x) = \pm p(x),$$

$$Y_1(x) = \frac{\mathrm{i}}{2}\frac{Y_0'(x)}{Y_0(x)},$$

$$Y_2(x) = \frac{3}{8}\frac{(Y_0'(x))^2}{Y_0^3(x)} - \frac{1}{4}\frac{Y_0''(x)}{Y_0^2(x)},$$

$$Y_3(x) = -\frac{3\mathrm{i}}{4}\frac{(Y_0'(x))^3}{Y_0^5(x)} + \frac{3\mathrm{i}}{4}\frac{Y_0'(x)Y_0''(x)}{Y_0^4(x)} - \frac{\mathrm{i}}{8}\frac{Y_0'''(x)}{Y_0^3(x)}$$

$$= \frac{\mathrm{d}}{\mathrm{d}x}\left[\frac{3\mathrm{i}}{16}\frac{(Y_0'(x))^2}{Y_0^4(x)} - \frac{\mathrm{i}}{8}\frac{Y_0''(x)}{Y_0^3(x)}\right],$$ (2.2.13)

$$Y_4(x) = -\frac{297}{128}\frac{(Y_0'(x))^4}{Y_0^7(x)} + \frac{99}{32}\frac{(Y_0'(x))^2 Y_0''(x)}{Y_0^6(x)} - \frac{13}{32}\frac{(Y_0''(x))^2}{Y_0^5(x)}$$

$$- \frac{5}{8}\frac{Y_0'(x)Y_0'''(x)}{Y_0^5(x)} + \frac{1}{16}\frac{Y_0^{(4)}(x)}{Y_0^4(x)}.$$

Note that, as a consequence of (2.2.10), the functions $Y_{2k+1}(x)$ with $k \geq 0$ are total derivatives. It is useful to introduce the function

$$S(x) = \int^x P(x')\mathrm{d}x',$$ (2.2.14)

which is defined up to an integration constant. The wavefunction (2.2.12) can be written as

$$\psi(x) = \frac{1}{\sqrt{S'(x)}}\exp\left(\frac{\mathrm{i}}{\hbar}S(x)\right).$$ (2.2.15)

It is easy to see from the first equation in (2.2.9) that $S(x)$ satisfies the non-linear differential equation

$$(S'(x))^2 + \frac{\hbar^2}{2}\{S(x), x\} = 2m(E - V(x)),$$ (2.2.16)

where

$$\{S(x), x\} = \frac{S'''(x)}{S'(x)} - \frac{3}{2}\left(\frac{S''(x)}{S'(x)}\right)^2 = -2\sqrt{S'(x)}\frac{\mathrm{d}^2}{\mathrm{d}x^2}\left(\frac{1}{\sqrt{S'(x)}}\right),$$ (2.2.17)

is the Schwarzian derivative of $S(x)$. The function $S(x)$ has the asymptotic expansion

$$S(x) = \sum_{n\geq 0} S_n(x)\hbar^{2n},$$ (2.2.18)

where

$$S_0(x) = \int^x p(x')\mathrm{d}x'.$$ (2.2.19)

There are in fact two different solutions of the form (2.2.15), depending on the choice of sign for the momentum $p(x)$ in $Y_0(x)$, which leads to two different choices of sign for $S(x)$. Any solution to the Schrödinger equation (normalizable or not) has an asymptotic expansion that can be written as a linear combination of the form

$$\frac{A_+}{\sqrt{S'(x)}} \exp\left(\frac{i}{\hbar} S(x)\right) + \frac{A_-}{\sqrt{S'(x)}} \exp\left(-\frac{i}{\hbar} S(x)\right), \tag{2.2.20}$$

where $S(x)$ is defined by the formal power series (2.2.18).

In the traditional WKB method, the solutions to the Schrödinger equation are approximated by taking the leading-order term of the series for $S(x)$, i.e. one approximates

$$S(x) \approx S_0(x). \tag{2.2.21}$$

This is equivalent to neglecting the contribution of the Schwarzian derivative in (2.2.16). In this case, approximate wavefunctions, sometimes called *basic WKB solutions*, are obtained. They have the form

$$\psi_\pm(x) = \frac{1}{\sqrt{p(x)}} \exp\left\{\pm \frac{i}{\hbar} \int^x p(x')dx'\right\}. \tag{2.2.22}$$

We can then approximate a general solution to the Schrödinger equation using a linear combination of the basic WKB solutions. It is useful to distinguish the solution in the classically allowed zone from that in the classically forbidden zone. In the classically forbidden zone, the momentum $p(x)$ is imaginary and the basic WKB solutions involve real exponentials. If we denote

$$p_1(x) = \sqrt{2m(V(x) - E)}, \tag{2.2.23}$$

we have

$$\psi(x) = \frac{A}{\sqrt{p_1(x)}} e^{\frac{1}{\hbar} \int_{x_0}^x p_1(x')dx'} + \frac{B}{\sqrt{p_1(x)}} e^{-\frac{1}{\hbar} \int_{x_0}^x p_1(x')dx'}. \tag{2.2.24}$$

In the classically allowed region, the basic WKB function solutions are oscillatory and can be written as

$$\psi(x) = \frac{A}{\sqrt{p(x)}} e^{\frac{i}{\hbar} \int_{x_0}^x p(x')dx'} + \frac{B}{\sqrt{p(x)}} e^{-\frac{i}{\hbar} \int_{x_0}^x p(x')dx'}. \tag{2.2.25}$$

Example 2.2.1 *WKB expansion in the linear potential.* An example that will play an important role is the linear potential

$$V(x) = -kx, \qquad k > 0. \tag{2.2.26}$$

Let us first solve the Schrödinger equation exactly. We have

$$-\frac{\hbar^2}{2m} \psi''(x) - (kx + E)\psi(x) = 0. \tag{2.2.27}$$

Let us introduce the variable

$$\phi = c(x - x_0), \tag{2.2.28}$$

where

$$x_0 = -\frac{E}{k}, \tag{2.2.29}$$

is the turning point, and

$$c = (2mk)^{1/3}. \tag{2.2.30}$$

Then, by comparing (2.2.27) to equation (B.1), which defines the Airy function, we find the general solution to (2.2.27) as

$$\psi(x) = a\mathrm{Ai}\left(-\hbar^{-2/3}\phi\right) + b\mathrm{Bi}\left(-\hbar^{-2/3}\phi\right), \tag{2.2.31}$$

where a, b are constants. Let us now apply the WKB method. It is useful to choose units in such a way that the parameter c in (2.2.30) is set to one. We first note that

$$p(x) = \phi^{1/2}, \tag{2.2.32}$$

and it is easy to see that the recursion (2.2.6) is solved by

$$Y_n(x) = \xi_n i^n \phi^{1/2-3n/2}, \tag{2.2.33}$$

where the coefficients ξ_n satisfy

$$\xi_{n+1} = \frac{1}{4}(1-3n)\xi_n - \frac{1}{2}\sum_{k=1}^{n} \xi_k \xi_{n+1-k}, \qquad n \geq 1, \tag{2.2.34}$$

with the initial conditions

$$\xi_0 = 1, \qquad \xi_1 = \frac{1}{4}. \tag{2.2.35}$$

The first coefficients are

$$\xi_2 = -\frac{5}{32}, \qquad \xi_3 = \frac{15}{64}, \qquad \xi_4 = -\frac{1105}{2048}. \tag{2.2.36}$$

The two WKB solutions in the region $E > V(x)$ are independent asymptotic expansions in $\hbar$ that solve the Airy equation (after an appropriate rescaling of the variables). By comparing the asymptotic expansion of the Airy functions in (B.32) and (B.35) to the WKB wavefunction, we should have, up to an x-independent constant,

$$\exp\left(\frac{i}{\hbar}\int^x Y(x')\mathrm{d}x'\right) \propto \frac{1}{\phi^{1/4}}\beta\,(i\zeta), \tag{2.2.37}$$

where

$$\zeta = \frac{2}{3\hbar}\phi^{3/2}, \tag{2.2.38}$$

and $\beta(\zeta)$ is defined in (B.31). Indeed, let us define the coefficients $\widetilde{c}_k$, $k \geq 1$, as

$$\log\left(\sum_{k\geq 0} c_k t^k\right) = \sum_{k\geq 1} \widetilde{c}_k t^k, \tag{2.2.39}$$

where the coefficients c_k are defined in (B.31). One finds, for the first few coefficients,

$$\widetilde{c}_1 = \frac{5}{72}, \qquad \widetilde{c}_2 = \frac{5}{144}, \qquad \widetilde{c}_3 = \frac{1105}{31104}. \tag{2.2.40}$$

Then, by comparing the exponent in (2.2.3) with the definition (B.31), we conclude that

$$\frac{i}{\hbar} \sum_{n \geq 0} Y_n \hbar^n = \frac{d}{dx} \left(i\zeta - \frac{1}{4} \log(x - x_0) + \sum_{k \geq 1} \widetilde{c}_k (i\zeta)^{-k} \right), \qquad (2.2.41)$$

which implies the relationship

$$\xi_n = (-1)^{n+1}(n-1) \left(\frac{3}{2} \right)^n \widetilde{c}_{n-1}, \qquad n \geq 2. \qquad (2.2.42)$$

This is easily checked for the very first values of n by using the explicit results quoted above.

Note that the function $S(x)$ in (2.2.14) is only defined up to an integration constant. However, the comparison with the asymptotic expansions of the Airy functions suggests a natural choice in the case of the linear potential, namely

$$S(x) = -\frac{i\hbar}{2} \log \left(\frac{\beta(i\zeta)}{\beta(-i\zeta)} \right). \qquad (2.2.43)$$

This has the formal power series expansion

$$\frac{1}{\hbar} S(x) = \zeta + \sum_{n \geq 1} (-1)^n \widetilde{c}_{2n-1} \zeta^{-2n+1}. \qquad (2.2.44)$$

These expressions for $S(x)$ will be important in the derivation of the connection formulae. $\qquad \square$

2.3 The Uniform WKB Method

One important drawback of the standard WKB approximation is that it *breaks down* whenever $p(x) = 0$. These are the *turning points* of the classical motion. Therefore, the basic WKB solutions for the wavefunctions only make sense far from these points. In order to analyze the wavefunctions near the turning points, we need a more careful treatment of the wavefunction. Such a treatment is provided by the *uniform WKB approximation*. The basic idea of this approximation is the following. In the standard WKB method, we use the ansatz (2.2.15) in order to implement the $\hbar$ expansion. However, there are other ansaztes that offer the same possibility, and that might better fit the problem at hand. Indeed, let us consider the following ansatz for a wavefunction satisfying Schrödinger's equation:

$$\psi(x) = \frac{1}{\sqrt{\phi'(x)}} f \left(\phi(x) \right), \qquad (2.3.45)$$

where $f(\phi)$ satisfies the second-order differential equation

$$f''(\phi) + \frac{1}{\hbar^2} \Pi^2(\phi) f(\phi) = 0. \qquad (2.3.46)$$

Then, $\phi(x)$ has to satisfy the nonlinear differential equation

$$\Pi^2(\phi)(\phi'(x))^2 + \frac{\hbar^2}{2} \{\phi(x), x\} = 2m(E - V(x)), \qquad (2.3.47)$$

where $\{\phi(x), x\}$ is the Schwarzian derivative introduced in (2.2.17). To derive (2.3.47), we just calculate

$$\psi'(x) = -\frac{1}{2}\frac{\phi''(x)}{\phi'(x)}\psi(x) + \sqrt{\phi'(x)}f'(\phi), \qquad (2.3.48)$$

and

$$\psi''(x) = -\frac{1}{2}\{\phi(x), x\}\psi(x) - \hbar^{-2}(\phi'(x))^2\Pi^2(\phi)\psi(x). \qquad (2.3.49)$$

Plugging the resulting expression in the Schrödinger equation, we immediately find (2.3.47). The ansatz (2.2.15) is recovered when $f(\phi)$ has the form of a plane wave,

$$f(\phi) = e^{i\phi/\hbar}, \qquad (2.3.50)$$

so it solves (2.3.46) with the choice $\Pi^2(\phi) = 1$. In general, however, other choices are possible. In all cases, we implement the small $\hbar$ approximation by solving (2.3.47) for $\phi(x)$ in a power series in $\hbar$:

$$\phi(x) = \sum_{n\geq 0}\hbar^{2n}\phi_n(x). \qquad (2.3.51)$$

The function $\phi(x)$ appearing in the uniform WKB method has the meaning of a generalized coordinate, which might be more convenient to study the problem at hand. For example, near a point where the potential behaves linearly (or quadratically), it might be useful to consider a uniform WKB approximation in which $\Pi^2(\phi)$ is a linear (respectively, a quadratic) function of ϕ. In this way, the auxiliary wavefunction $f(\phi)$ can be solved exactly, and then the deviations from an exactly linear or quadratic potential are encoded in the nontrivial dependence of $\phi(x)$ on the original coordinate x.

The linear case, which is particularly useful to study a potential near a turning point, will be studied in detail in Section 2.4. In the quadratic case, one chooses

$$\Pi^2(\phi) = \hbar^2(2v - \phi^2). \qquad (2.3.52)$$

The function $f(\phi)$ solves the equation,

$$f''(\phi) + (2v - \phi^2)f(\phi) = 0, \qquad (2.3.53)$$

whose general solution can be written as a linear combination of parabolic cylinder functions,

$$f(\phi) = AD_{v-1/2}\left(\sqrt{2}\phi\right) + BD_{-v-1/2}\left(-i\sqrt{2}\phi\right). \qquad (2.3.54)$$

As in the case of the harmonic oscillator, it is often required that $f(\phi)$ decays when $\phi \to \pm\infty$. Since

$$D_{-v-1/2}(-iz) \sim z^{-v-1/2}e^{+z^2/4}, \qquad z \to +\infty, \qquad (2.3.55)$$

we must set $B = 0$. To analyze the limit $\phi \to -\infty$, we use the asymptotic behavior

$$D_{v-1/2}(z) \sim z^{v-1/2}e^{-z^2/4}\left(1 + \mathcal{O}(z^{-2})\right) + \frac{i\sqrt{2\pi}}{\Gamma\left(-v+\frac{1}{2}\right)}e^{\pi iv}z^{-v-1/2}e^{z^2/4}\left(1 + \mathcal{O}(z^{-2})\right),$$

$$(2.3.56)$$

for $z \to -\infty$. Therefore, $f(\phi)$ grows at infinity unless

$$\frac{1}{\Gamma\left(\frac{1}{2} - v\right)} = 0. \tag{2.3.57}$$

This implies

$$v = n + \frac{1}{2}, \qquad n = 0, 1, 2, \ldots, \tag{2.3.58}$$

which is the standard quantization condition for the harmonic oscillator.

Example 2.3.1 Let us consider the potential

$$V(x) = \frac{1}{2}x^2 + gx^p, \tag{2.3.59}$$

where $p \geq 4$ is an *even* integer. Near $x = 0$ this is approximately a harmonic oscillator, so in a neighborhood of this point it might be useful to use the quadratic form of the uniform WKB approximation. By rescaling

$$x = \lambda^{-1/2}z, \tag{2.3.60}$$

where λ is chosen such that

$$g = \frac{1}{2}\lambda^{p/2-1}, \tag{2.3.61}$$

we can put the Schrödinger equation in the form

$$-\xi^2\psi''(z) + \left(z^2 + z^p - 2\xi\mathcal{E}\right)\psi(z) = 0, \tag{2.3.62}$$

where

$$\xi = \frac{\lambda\hbar}{\sqrt{m}}, \qquad \mathcal{E} = \frac{\sqrt{m}}{\hbar}E. \tag{2.3.63}$$

In this example, ξ plays the role of $\hbar$ in the model. In addition, it might be used for a perturbative expansion in the coupling constant g, as we will see.

Let us now study this potential with the uniform WKB method and a quadratic ansatz for $\Pi^2(\phi)$, similar to what we had in (2.3.52):

$$\Pi^2(\phi) = \xi^2\left(2v - \phi^2\right). \tag{2.3.64}$$

The wavefunction $\psi(x)$ is expressed in terms of $f(\phi)$ as in (2.3.45). Let us now write

$$\phi(z) = \frac{1}{\sqrt{\xi}}u(z). \tag{2.3.65}$$

The nonlinear equation (2.3.47) for $u(z)$ reads in this case,

$$u'(z)^2u^2(z) = z^2 + z^p - 2\xi\left(\mathcal{E} - vu'(z)^2\right) + \frac{\xi^2}{2}\{u, z\}. \tag{2.3.66}$$

The standard WKB approach consists in expanding ϕ in powers of $\hbar$. We will now follow a slightly different approach, which combines WKB with perturbation theory in ξ. It is convenient to rewrite (2.3.66) as an equation for

$$U(z) = \frac{1}{2}u^2(z). \tag{2.3.67}$$

In terms of this new variable, the differential equation (2.3.66) reads

$$(U'(z))^2 = z^2 + z^p - 2\xi \left(\mathcal{E} - v \frac{(U'(z))^2}{2U(z)} \right) + \frac{\xi^2}{2} \{\sqrt{U}, z\}. \tag{2.3.68}$$

We will solve (2.3.68) in a power series expansion in ξ by using the ansatz

$$u(z) = \sum_{n \geq 0} u_n(z) \xi^n, \qquad \mathcal{E} = \sum_{n \geq 0} \mathcal{E}_n(v) \left(\frac{1}{2} \xi^{p/2-1} \right)^n. \tag{2.3.69}$$

Note that

$$U(z) = \sum_{n \geq 0} U_n(z) \xi^n, \tag{2.3.70}$$

with $U_0(z) = u_0^2(z)/2$, $U_1(z) = u_0(z)u_1(z)$, and so on. At leading order we find,

$$U_0'(z) = \left(z^2 + z^p \right)^{1/2}, \tag{2.3.71}$$

which we can solve as

$$U_0(z) = \int_0^z (s^2 + s^p)^{1/2} \mathrm{d}s = \frac{z^2}{2} \, {}_2F_1 \left(-\frac{1}{2}, q; q + 1; -z^{2/q} \right), \tag{2.3.72}$$

where

$$q = \frac{2}{p - 2}. \tag{2.3.73}$$

When solving for $u(z)$, we take the sign in the square root in such a way that, for small z, we have

$$u_0(z) = z + \mathcal{O}(z^2). \tag{2.3.74}$$

The behavior of $u_0(z)$ near $z = 0$ is required by (2.3.66) and fixes the integration constant in (2.3.72). We can now solve for the higher order corrections. For example, for $U_1(z)$, we find the equation

$$U_1'(z) = -\frac{\mathcal{E}_0(v)}{u_0(z)u_0'(z)} + v \frac{u_0'(z)}{u_0(z)}. \tag{2.3.75}$$

We will require the functions $u_n(z)$ to be regular at the origin, i.e. to have a Taylor expansion around $z = 0$. This fixes the values of the coefficients $\mathcal{E}_n(v)$. For example, it is easy to see from (2.3.75) that regularity of $u_1(z)$ at $z = 0$ forces

$$\mathcal{E}_0(v) = v. \tag{2.3.76}$$

In this way, one can solve for the $u_n(z)$, $\mathcal{E}_n(v)$ recursively. For $p = 4$ one finds, after setting $m = \hbar = 1$,

$$\mathcal{E} = v + \frac{3}{8}(1 + 4v^2)g - \frac{1}{16}(67v + 68v^3)g^2 + \cdots \tag{2.3.77}$$

To determine v, we use square integrability of the original wavefunction $\psi(x)$. From (2.3.71) we find,

$$u_0(z) \sim \frac{2}{\sqrt{p + 2}} |z|^{\frac{p+2}{4}}, \qquad |z| \to \infty. \tag{2.3.78}$$

It is easy to see from (2.3.66) that higher order corrections in ξ give lower order corrections to this behavior as $|z| \to \infty$, so $u(z)$ grows at infinity as the right-hand side (r.h.s.) of (2.3.78). Therefore, if we want $\psi(x)$ to decay at infinity, we must require that $f(\phi)$, given by (2.3.54), also decays at infinity. As we have seen, this requires $B = 0$ in (2.3.54), as well as the quantization of the parameter ν as in (2.3.58). When this quantization condition is plugged in (2.3.77), we obtain the perturbative series for the energy levels of the quartic oscillator, as a function of the energy level n.   □

2.4 The Voros–Silverstone Connection Formula

The standard WKB approximation breaks down at the turning points of the classical motion. Therefore, the main problem in the WKB method is to find *connection formulae* relating the WKB solutions in the two regions adjacent to a turning point, i.e. relating the solution in a classically allowed region to the solution in an adjacent, classically allowed region. This is a delicate problem since one has to pay attention to subleading exponentials in the asymptotic expansion, and incomplete analysis has, elsewhere, led to many confusing issues. It requires the all-orders WKB method and an explicit use of resummation techniques, which are summarized in Appendix A.

Near a standard turning point at $x = x_0$, we have

$$V(x_0) = E, \tag{2.4.79}$$

therefore near $x = x_0$ we can approximate

$$V(x) - E \approx -k(x - x_0). \tag{2.4.80}$$

We will assume that we have a first order turning point, so that $k \neq 0$. We will also assume that the turning point x_0 is such that $E - V(x)$ is positive when x is slightly above x_0, and negative when x is slightly below x_0, as in Figure 2.1. Locally, we are solving the Schrödinger equation in the presence of a linear potential. As we know from Example 2.2.1, this problem is solved using Airy functions, and this suggests approximating our wavefunction by Airy functions near the turning point. This is indeed the standard strategy to find the connection formulae in many textbooks. Here, we are interested in controlling in detail the resulting asymptotic expansion in $\hbar$, so we will follow a slightly different route: we will use an ansatz of the form (2.3.45), in which f is a linear combination of Airy functions, i.e. we will set

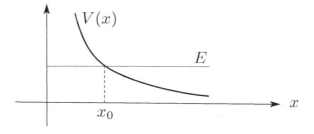

Figure 2.1   A turning point.

$$f(\phi) = a \operatorname{Ai}\left(-\hbar^{-2/3}\phi\right) + b \operatorname{Bi}\left(-\hbar^{-2/3}\phi\right), \qquad (2.4.81)$$

so that

$$\psi(x) = \frac{1}{\sqrt{\phi'(x)}}\left\{a \operatorname{Ai}\left(-\hbar^{-2/3}\phi(x)\right) + b \operatorname{Bi}\left(-\hbar^{-2/3}\phi(x)\right)\right\}. \qquad (2.4.82)$$

This is a particular case of the uniform WKB method, in which we make the choice

$$\Pi^2(\phi) = \phi. \qquad (2.4.83)$$

Therefore, the nonlinear equation (2.3.47) for $\phi(x)$ reads,

$$(\phi'(x))^2\phi(x) + \frac{\hbar^2}{2}\{\phi, x\} = 2m(E - V(x)). \qquad (2.4.84)$$

We solve $\phi(x)$ in a power series in $\hbar^2$:

$$\phi(x) = \sum_{n \geq 0} \hbar^{2n}\phi_n(x). \qquad (2.4.85)$$

The leading-order term satisfies

$$\left(\phi_0'(x)\right)^2 \phi_0(x) = 2m(E - V(x)). \qquad (2.4.86)$$

Note that $\phi_0(x)$ is positive when $E > V(x)$, i.e. to the right of the turning point, and negative to the left of the turning point. One can solve

$$\phi_0(x) = \left[\frac{3}{2}\int_{x_0}^{x} p(x')\mathrm{d}x'\right]^{2/3}. \qquad (2.4.87)$$

The equation for the next order, $\phi_1(x)$, is given by

$$(\phi_0'(x))^2\phi_1(x) + 2\phi_0(x)\phi_0'(x)\phi_1'(x) + \frac{1}{2}\{\phi_0(x), x\} = 0. \qquad (2.4.88)$$

This is a linear, first order ordinary differential equation (ODE) that can be integrated immediately. It is easy to see that the equations governing the higher order $\phi_n(x)$ are of the same form, namely

$$(\phi_0'(x))^2\phi_n(x) + 2\phi_0(x)\phi_0'(x)\phi_n'(x) + \cdots = 0, \qquad (2.4.89)$$

where the $\cdots$ stand for a function involving $\phi_{n'}(x)$ with $n' < n$. It also follows from these equations that the $\phi_n(x_0)$ are analytic at $x = x_0$, provided $V(x)$ is analytic there, as we will assume. We will choose the integration constant for $\phi_n(x_0)$ in such a way that they vanish at the turning point, i.e.

$$\phi_n(x_0) = 0. \qquad (2.4.90)$$

We can think about $\phi(x)$ as the modification of $x - x_0$ that is needed in an arbitrary potential in order that the Airy functions, when expressed in terms of $\phi(x)$, satisfy exactly the Schrödinger equation. Note that, if

$$V(x) = E - k(x - x_0), \qquad (2.4.91)$$

(2.4.84) is solved exactly by

$$\phi(x) = c(x - x_0), \qquad c^3 = 2mk, \tag{2.4.92}$$

as it follows from the analysis in Example 2.2.1.

We will now use the asymptotic expansion of the Airy functions (including exponentially small contributions) to obtain precise matching conditions. These expansions are encoded in the formal power series $\beta(\pm i\zeta)$, which already appeared in (2.2.37), in Example 2.2.1. Let us define the formal power series

$$\beta_\pm = \beta\left(\pm\frac{2i}{3\hbar}\phi^{3/2}(x)\right), \tag{2.4.93}$$

where $\beta(\zeta)$ is defined in (B.31). Our goal is to reconstruct the formal WKB solutions (2.2.15) by using these ingredients. The result (2.2.43) for the linear potential suggests we define the formal power series

$$S(x, \hbar) = -\frac{i\hbar}{2}\log\left(\frac{\beta_+}{\beta_-}\right). \tag{2.4.94}$$

It is easy to see that $S(x, \hbar)$ can be written as a formal power series in $\hbar^2$, of the form (2.2.18). One finds, for example, that

$$S_0(x) = \frac{2}{3}\left(\phi_0(x)\right)^{3/2}. \tag{2.4.95}$$

We will now show that, as a formal power series,

$$S'(x, \hbar) = P(x), \tag{2.4.96}$$

where $P(x)$ is the WKB series introduced in (2.2.7). In other words, we will show that $S(x, \hbar)$ is indeed given by (2.2.14) (with a specific choice of integration constant). To see this, we first calculate

$$S'(x, \hbar) = \frac{\phi'(x)\sqrt{\phi(x)}}{2}\frac{\beta'_+\beta_- + \beta'_-\beta_+}{\beta_+\beta_-}. \tag{2.4.97}$$

The numerator in the r.h.s. can be evaluated in closed form by using the Wronskian formula (B.36). Indeed, let us evaluate the left-hand side (l.h.s.) of this Wronskian identity by setting $z \to -z$ and using the asymptotic formulae for $|\arg(z)| < \pi/3$. One finds,

$$\beta'_+\beta_- + \beta'_-\beta_+ = 2, \tag{2.4.98}$$

and we conclude that

$$S'(x, \hbar) = \phi'(x)\sqrt{\phi(x)}\frac{1}{\beta_+\beta_-}. \tag{2.4.99}$$

We also note that, from the definition (2.4.94),

$$\exp\left(\pm\frac{i}{\hbar}S(x, \hbar)\right) = \left(\frac{\beta_+}{\beta_-}\right)^{\pm 1/2}. \tag{2.4.100}$$

Therefore,

$$\frac{1}{\sqrt{S'(x, \hbar)}}\exp\left(\pm\frac{i}{\hbar}S(x, \hbar)\right) = \frac{1}{\sqrt{\phi'(x)}\phi^{1/4}(x)}\beta_\pm. \tag{2.4.101}$$

The r.h.s. is a linear combination of the asymptotic expansions of the functions $\text{Ai}(-z)$, $\text{Bi}(-z)$, where z is given by $\hbar^{-2/3}\phi(x)$ [as in (2.2.38)]. In other words, it is the asymptotic expansion of a function of the form (2.4.82), and as a formal power series in $\hbar$ it satisfies the Schrödinger equation. We conclude that $S(x,\hbar)$ is indeed given by the function (2.2.14).

Note, however, that the definition (2.4.94) also specifies the integration constants that are not fixed by (2.2.14). An equivalent way to fix the integration constants is to note that $S_n(x)$, as defined by (2.4.94), can be understood as a regularization of the integral

$$\int_{x_0}^{x} Y_{2n}(x')\mathrm{d}x'. \tag{2.4.102}$$

This regularization is implemented as follows. We first note that $Y_{2n}(x)$, near $x = x_0$, behaves as

$$Y_{2n}(x) \sim (x - x_0)^{1/2-3n}, \qquad n \geq 0. \tag{2.4.103}$$

This follows from the result for these functions, (2.2.33), in the case of the linear potential. Therefore, $Y_{2n}(x)$ has a branch cut near $x = x_0$. Let us define the sign of $Y_{2n}(x)$ so that $\pm Y_{2n} > 0$ when $x > x_0$ in the first sheet (respectively, in the second sheet). We now consider the integration path C from x_0 to x shown in Figure 2.2. The continuous line indicates that the path takes place in the first sheet, while the path with a dashed line takes place in the second sheet. We then define,

$$\int_{x_0}^{x} Y_{2n}(x')\mathrm{d}x' = \frac{1}{2}\int_{C} Y_{2n}(x')\mathrm{d}x'. \tag{2.4.104}$$

An equivalent definition is

$$\int_{x_0}^{x} Y_{2n}(x')\mathrm{d}x' = \lim_{\epsilon \to 0}\left(\int_{x_0+\epsilon}^{x} Y_{2n}(x')\mathrm{d}x' - \text{Re}\int_{C_\epsilon} Y_{2n}(z)\mathrm{d}z\right), \tag{2.4.105}$$

where the path C_ϵ is given by a semicircle of radius ϵ, centered at $x = x_0$, and oriented counterclockwise. To see that the regularized integral fixes the integration constant as in (2.4.94), we start from the expression

$$S(x,\hbar) = \int_{x_0}^{x} P(x)\mathrm{d}x, \tag{2.4.106}$$

where the integral is regularized as we just have explained. By changing variables from x to $\phi(x)$, noting that $\phi(x_0) = 0$, and using the expression (2.4.99), we find,

$$S(x,\hbar) = \int_{0}^{\phi(x)} \frac{\sqrt{u}}{\beta\,(\mathrm{i}\zeta_u)\,\beta\,(-\mathrm{i}\zeta_u)}\mathrm{d}u, \qquad \zeta_u = \frac{2}{3\hbar}u^{3/2}. \tag{2.4.107}$$

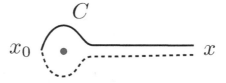

Regularization of the integral (2.4.102).

It is easy to show, by combining (2.2.41) with (2.2.10), that

$$\frac{1}{\beta\,(\mathrm{i}\zeta)\,\beta\,(-\mathrm{i}\zeta)} = 1 + \sum_{n\geq 1}(-1)^{n-1}(2n-1)\widetilde{c}_{2n-1}\zeta^{-2n}. \qquad (2.4.108)$$

Therefore, the calculation of the regularized integrals in (2.4.107) reduces to the calculation in the linear case, namely

$$\int_0^\phi u^{1/2-3n}\,\mathrm{d}u = -\frac{2}{3}\frac{\phi^{3/2-3n}}{2n-1}, \qquad (2.4.109)$$

and we recover the expansion of (2.4.94) (see e.g. (2.2.44)).

Let us summarize what we have done so far. The uniform WKB ansatz for the wavefunction (2.4.82) expresses the wavefunction $\psi(x)$ in terms of a new function $\phi(x)$. The function $\phi(x)$ admits an $\hbar$ expansion that leads to a semiclassical approximation for the wavefunction. This semiclassical expansion is related to the expansion in (2.2.18) by the equation (2.4.94).

We can now obtain the all-orders connection formulae. In the classically allowed region $E > V(x)$, $\phi(x)$ is positive, and the argument of the Airy functions in (2.4.82) is negative, so that the asymptotic expansion of $\psi(x)$ reads, up to an overall constant,

$$\psi(x) \sim \frac{1}{\sqrt{\phi'(x)}\phi^{1/4}(x)}\left[(b-\mathrm{i}a)\mathrm{e}^{\mathrm{i}\pi/4}\beta\left(\frac{2\mathrm{i}}{3\hbar}\phi^{3/2}\right) + (b+\mathrm{i}a)\mathrm{e}^{-\mathrm{i}\pi/4}\beta\left(-\frac{2\mathrm{i}}{3\hbar}\phi^{3/2}\right)\right].$$

$$(2.4.110)$$

In the classically forbidden region, $\phi(x)$ is negative, so that $\hbar^{-2/3}(-\phi) > 0$ and the argument of the Airy functions in (2.4.82) is positive. The variable ζ in (B.33) is given by

$$\zeta = \frac{2}{3\hbar}(-\phi)^{3/2}, \qquad (2.4.111)$$

and the asymptotic expansion reads

$$\psi(x) \sim \frac{1}{\sqrt{\phi'(x)}(-\phi(x))^{1/4}}\left[2b\beta\left(\frac{2}{3\hbar}(-\phi)^{3/2}\right) + (a\pm\mathrm{i}b)\beta\left(-\frac{2}{3\hbar}(-\phi)^{3/2}\right)\right].$$

$$(2.4.112)$$

The signs $\pm$ correspond to the choice of Borel lateral resummation of the asymptotic series, as explained in Appendix B.1 (specifically after (B.35)). Therefore, the asymptotic expansions above have to be understood in a generalized sense, in which subleading, exponentially small terms are included. After the asymptotic series have been resummed by using appropriate lateral Boral resummations, one obtains exact expressions for the wavefunctions.

Now, we want to express the above asymptotic expansions in terms of the function $S(x,\hbar)$, in order to make contact with the conventional WKB expansion. To do this, we simply use the relation (2.4.101). This gives, in the classically allowed region,

$$\psi(x) \sim (S'(x,\hbar))^{-1/2}\left[(b-\mathrm{i}a)\mathrm{e}^{\mathrm{i}S/\hbar+\mathrm{i}\pi/4} + (b+\mathrm{i}a)\mathrm{e}^{-\mathrm{i}S/\hbar-\mathrm{i}\pi/4}\right]. \qquad (2.4.113)$$

Let us note that, near $x = x_0$,

$$S_0(x) \sim \frac{2}{3}c^{3/2}(x-x_0)^{3/2}, \qquad (2.4.114)$$

as one easily deduces from (2.4.95). There is a branch cut in the classically forbidden region $x - x_0 < 0$, and in order to write down the wavefunctions it is useful to make a choice of sign for the above multivalued function. Our choice is implemented by setting $x - x_0 = e^{i\pi}(x_0 - x)$. Remember that, in the elementary WKB method, it was useful to introduce the real function $p_1(x)$ in (2.2.23) in order to write wavefunctions in the forbidden region. In the same way, in the forbidden region it is useful to introduce the real function

$$Q(x, \hbar) = iS(e^{\pi i}(x_0 - x) + x_0). \tag{2.4.115}$$

It behaves near $x = x_0$ as

$$Q(x, \hbar) \approx \frac{2}{3}c^{3/2}ie^{3\pi i/2}(x_0 - x)^{3/2} = \frac{2}{3}c^{3/2}(x_0 - x)^{3/2}. \tag{2.4.116}$$

With the above choice of sign, we have

$$\pm\frac{2i}{3\hbar}\phi^{3/2} = \pm\frac{2i}{3\hbar}e^{3\pi i/2}(-\phi)^{3/2} = \pm\frac{2}{3\hbar}(-\phi)^{3/2}. \tag{2.4.117}$$

We now write,

$$(S'(x, \hbar))^{-1/2} \exp(\pm iS(x, \hbar)/\hbar) = (-iQ'(x, \hbar))^{-1/2}e^{\pm Q(x,\hbar)/\hbar}. \tag{2.4.118}$$

On the other hand, by using (2.4.101) and implementing carefully our choice of sign, we find

$$(S'(x, \hbar))^{-1/2} \exp(\pm iS(x, \hbar)/\hbar) = e^{-\pi i/4}(-\phi)^{-1/4}(\phi')^{-1/2}\beta\left(\pm\frac{2}{3\hbar}(-\phi)^{3/2}\right). \tag{2.4.119}$$

We conclude that

$$(-\phi)^{-1/4}(\phi')^{-1/2}\beta\left(\pm\frac{2}{3\hbar}(-\phi)^{3/2}\right) = (-Q'(x, \hbar))^{-1/2}e^{\pm Q(x,\hbar)/\hbar}. \tag{2.4.120}$$

We can now write the asymptotic expansion of the wave function in the forbidden region, (2.4.112), in terms of the function $Q(x, \hbar)$, which is the appropriate one for the conventional WKB expansion,

$$\psi(x) \sim (-Q'(x, \hbar))^{-1/2}\left\{2b\, e^{Q(x,\hbar)/\hbar} + (a \pm ib)e^{-Q(x,\hbar)/\hbar}\right\}. \tag{2.4.121}$$

By putting together (2.4.113) and (2.4.121) we obtain the connection formula in the form

$$(S'(x, \hbar))^{-1/2}\left[(b - ia)e^{iS/\hbar + i\pi/4} + (b + ia)e^{-iS/\hbar - i\pi/4}\right]$$
$$\longleftrightarrow (-Q'(x, \hbar))^{-1/2}\left\{2b\, e^{Q(x,\hbar)/\hbar} + (a \pm ib)e^{-Q(x,\hbar)/\hbar}\right\}. \tag{2.4.122}$$

This is called the *Voros–Silverstone connection formula*. Another useful form is obtained if we express

$$a = \cos\eta, \qquad b = \sin\eta. \tag{2.4.123}$$

Then, (2.4.122) can be written as

$$(S'(x, \hbar))^{-1/2}\cos\left\{\frac{S(x, \hbar)}{\hbar} - \frac{\pi}{4} + \eta\right\}$$
$$\longleftrightarrow (-Q'(x, \hbar))^{-1/2}\left\{\sin\eta\, e^{Q(x,\hbar)/\hbar} + \frac{1}{2}e^{\pm i\eta}e^{-Q(x,\hbar)/\hbar}\right\}. \tag{2.4.124}$$

Another form of the connection formula can be obtained by writing $S(x, \hbar)$, $Q(x, \hbar)$ in terms of regularized integrals, involving the function $P(x)$ introduced in (2.2.7). Note that $P(x)$ is equal to $S'(x, \hbar)$. Similarly, we introduce the function

$$P_1(x) = -Q'(x, \hbar) = p_1(x) + \mathcal{O}(\hbar^2), \tag{2.4.125}$$

which is the all-orders counterpart of the function $p_1(x)$ in (2.2.23). Then, we can write (2.4.124) as

$$\frac{1}{\sqrt{P(x)}} \cos\left\{ \frac{1}{\hbar} \int_{x_0}^{x} P(x') \mathrm{d}x' - \frac{\pi}{4} + \eta \right\}$$
$$\longleftrightarrow \frac{1}{\sqrt{P_1(x)}} \left\{ \sin\eta\, e^{\frac{1}{\hbar} \int_{x}^{x_0} P_1(x')\mathrm{d}x'} + \frac{1}{2} e^{\pm i\eta} e^{-\frac{1}{\hbar} \int_{x}^{x_0} P_1(x')\mathrm{d}x'} \right\}. \tag{2.4.126}$$

This formula has been derived when the classically allowed region is to the right of the turning point, i.e. $x > x_0$ in the integration in the l.h.s., while in the integration of the r.h.s. we have $x < x_0$. When the classically allowed region is to the left of the turning point, it is easy to see that the connection formula has to be modified as

$$\frac{1}{\sqrt{P(x)}} \cos\left\{ \frac{1}{\hbar} \int_{x}^{x_0} P(x) \mathrm{d}x' - \frac{\pi}{4} + \eta \right\}$$
$$\longleftrightarrow \frac{1}{\sqrt{P_1(x)}} \left\{ \sin\eta\, e^{\frac{1}{\hbar} \int_{x_0}^{x} P_1(x')\mathrm{d}x'} + \frac{1}{2} e^{\pm i\eta} e^{-\frac{1}{\hbar} \int_{x_0}^{x} P_1(x')\mathrm{d}x'} \right\}, \tag{2.4.127}$$

where now $x < x_0$ in the integration in the l.h.s., while in the integration of the r.h.s. we have $x > x_0$.

Traditionally, an approximation to the above connection formulae is used, in which the resummed asymptotic series are truncated to the basic WKB solutions (2.2.22). When this is done, however, one loses control over the subdominant exponentials, and one has to introduce *ad hoc* prescriptions. In this book we will mostly use the exact connection formulae to avoid those ambiguities.

2.5 The Bohr–Sommerfeld Quantization Condition

Historically, one of the first successes of the WKB method was to provide a derivation of the Bohr–Sommerfeld (BS) quantization condition for bound states, which played an important role in early quantum theory. The WKB method also shows that the BS quantization condition is only a approximation, and in principle one can use the all-orders WKB expansions and their resummation to obtain systematic corrections to it. We will now derive an all-orders version of the BS quantization condition that provides an asymptotic expansion for the energy levels as power series in $\hbar$.

Let us assume that we have a potential well with two turning points, x_- and x_+, as shown in Figure 2.3. There are three regions to consider. In the classically forbidden region *III*, the square-integrable solution to the Schrödinger equation is given by

$$\psi_{III}(x) = \frac{1}{2} \frac{C}{\sqrt{P_1(x)}} \exp\left\{ -\frac{1}{\hbar} \int_{x_+}^{x} P_1(x') \mathrm{d}x' \right\}, \qquad x > x_+, \tag{2.5.128}$$

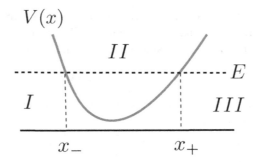

A particle of energy *E* in a potential *V(x)* with two turning points. The regions *I* and *III* are classically forbidden, while region *II* is classically allowed.

which decreases as $x \to \infty$. Here, C is an overall constant that might be fixed by normalization of the wavefunction. We now use the connection formula (2.4.127) with $\eta = 0$, to obtain

$$\psi_{II}(x) = \frac{C}{\sqrt{P(x)}} \cos\left(\frac{1}{\hbar}\int_x^{x_+} P(x')dx' - \frac{\pi}{4}\right), \qquad x_- < x < x_+. \quad (2.5.129)$$

In order to find the wavefunction in region *I*, we write (2.5.129) as

$$\psi_{II}(x) = \frac{C}{\sqrt{P(x)}}\left[\sin\left(\frac{1}{\hbar}\int_{x_-}^{x_+} P(x')dx'\right)\cos\left(\frac{1}{\hbar}\int_{x_-}^{x} P(x')dx' - \frac{\pi}{4}\right)\right.$$
$$\left. - \cos\left(\frac{1}{\hbar}\int_{x_-}^{x_+} P(x')dx'\right)\cos\left(\frac{1}{\hbar}\int_{x_-}^{x} P(x')dx' - \frac{\pi}{4} - \frac{\pi}{2}\right)\right]. \quad (2.5.130)$$

Let us use the connection formula again. From (2.4.126) with $\eta = 0$, we deduce that,

$$\frac{2}{\sqrt{P(x)}} \cos\left(\frac{1}{\hbar}\int_{x_-}^{x} P(x')dx' - \frac{\pi}{4}\right) \to \frac{1}{\sqrt{P_1(x)}} \exp\left\{-\frac{1}{\hbar}\int_x^{x_-} P_1(x')dx'\right\}, \quad (2.5.131)$$

while (2.4.126) with $\eta = -\pi/2$ gives

$$\frac{1}{\sqrt{P(x)}} \cos\left(\frac{1}{\hbar}\int_{x_-}^{x} P(x')dx' - \frac{\pi}{4} - \frac{\pi}{2}\right)$$
$$\to -\frac{1}{\sqrt{P_1(x)}}\left[\exp\left\{\frac{1}{\hbar}\int_x^{x_-} P_1(x')dx'\right\} \pm \frac{i}{2}\exp\left\{-\frac{1}{\hbar}\int_x^{x_-} P_1(x')dx'\right\}\right]. \quad (2.5.132)$$

We conclude that

$$\psi_I(x) = \frac{C}{\sqrt{P_1(x)}} \cos\left(\frac{1}{\hbar}\int_{x_-}^{x_+} P(x')dx'\right) \exp\left\{\frac{1}{\hbar}\int_x^{x_-} P_1(x')dx'\right\}$$
$$\pm \frac{1}{2}\frac{iC}{\sqrt{P_1(x)}} \exp\left\{\mp\frac{i}{\hbar}\int_{x_-}^{x_+} P(x')dx'\right\}\exp\left\{-\frac{1}{\hbar}\int_x^{x_-} P_1(x')dx'\right\},$$
$$x < x_-. \quad (2.5.133)$$

Since the wavefunction in region I should decrease as $x \to -\infty$, the coefficient of the exponentially increasing wavefunction in the first line of (2.5.133) should vanish, i.e.

$$\cos\left(\frac{1}{\hbar}\int_{x_-}^{x_+} P(x')\mathrm{d}x'\right) = 0. \tag{2.5.134}$$

We obtain in this way

$$\int_{x_-}^{x_+} P(x')\mathrm{d}x' = \pi\hbar\left(n + \frac{1}{2}\right), \qquad n \in \mathbb{Z}_{\geq 0}. \tag{2.5.135}$$

In the l.h.s. of this equation, we have a formal power series in $\hbar^2$. The correct interpretation of this series, and how to implement it in practice, will be discussed in some detail in this chapter. We can however approximate this series by its leading-order term. After taking into account (2.4.125), one obtains

$$\int_{x_-}^{x_+} \sqrt{2m(E - V(x))} \approx \pi\hbar\left(n + \frac{1}{2}\right), \qquad n \in \mathbb{Z}_{\geq 0}. \tag{2.5.136}$$

This is the BS quantization condition for the discrete energy levels in the potential $V(x)$. The corresponding approximate wavefunction in the classically allowed region is obtained by setting $P(x) \approx p(x)$ in (2.5.129), and we have

$$\psi_{II}(x) \approx \frac{C}{\sqrt{p(x)}}\cos\left(\frac{1}{\hbar}\int_x^{x_+} p(x')\mathrm{d}x' - \frac{\pi}{4}\right), \tag{2.5.137}$$

which is a linear combination of basic WKB solutions. As x varies between x_- and x_+, the phase of the cosine in the r.h.s. varies between $-\pi/4$ and $n\pi + \pi/4$, so the cosine has n zeros in the interval $[x_-, x_+]$. Therefore, the approximate WKB wavefunction corresponding to an energy level with quantum number n (the nth excited state) has n nodes inside the interval of classically allowed motion. Although we have derived this result in the leading WKB approximation, it is a property of the exact eigenfunctions, which is sometimes called the *oscillation theorem*.

The normalization constant C appearing in (2.5.137) can be computed as follows. Since the wavefunction decays exponentially in the classically forbidden regions, we can normalize it by restricting ourselves to the classically allowed region. We can also approximate the square cosine function by $1/2$. In this way, we find the condition

$$\frac{C^2}{2}\int_{x_-}^{x_+} \frac{\mathrm{d}x}{p(x)} \approx 1. \tag{2.5.138}$$

This determines the value of C in terms of the above integral, which is in turn related to the period T of the classical motion in the potential,

$$\int_{x_-}^{x_+} \frac{\mathrm{d}x}{p(x)} = \frac{T}{2m}. \tag{2.5.139}$$

We can then write,

$$C \approx 2\sqrt{\frac{m}{T}}. \tag{2.5.140}$$

The BS quantization condition (2.5.136) leads to a constraint relating the energy E to the quantum number n, whose solution gives the energy levels $E(n) \equiv E_n$. This is the leading approximation to the spectrum of energies when the classical action appearing in the l.h.s. of (2.5.136) is large as compared to $\hbar$. This happens when n is large, and the BS quantization condition gives excellent approximations to the energy eigenvalues for large quantum numbers. It also has a beautiful semiclassical interpretation, which goes as follows. The r.h.s. of (2.5.136) is the volume of the available region in phase space for a system of energy E. Indeed, if

$$\mathcal{R}(E) = \{(x, p) \in \mathbb{R}^2 : H(x, p) \leq E\}, \tag{2.5.141}$$

is the available region in phase space, its volume is

$$\mathrm{vol}_0(E) = \int_{\mathcal{R}(E)} \omega = 2 \int_{x_-}^{x_+} p(x)\mathrm{d}x, \tag{2.5.142}$$

where $\omega = \mathrm{d}p \wedge \mathrm{d}x$ is the canonical symplectic form on phase space. The BS quantization condition can now be written as

$$\mathrm{vol}_0(E) \approx 2\pi\hbar \left(n + \frac{1}{2}\right), \tag{2.5.143}$$

says that each cell of volume $2\pi\hbar$ in $\mathcal{R}(E)$ corresponds, roughly, to a bound state of the system. Note, in particular, that a quantum system with a discrete spectrum requires a region $\mathcal{R}(E)$ of finite volume in the classical theory.

Example 2.5.1 *BS approximation for the harmonic oscillator.* Let us set $m = 1$ and consider the harmonic oscillator potential $V(x) = x^2/2$ (we also set $\omega = 1$). Then, the integral in the l.h.s. of the BS quantization condition (2.5.136) is

$$\int_{-x_+}^{x_+} \sqrt{2(E - x^2/2)}\,\mathrm{d}x = 2E \int_{-1}^{1} \sqrt{1 - u^2}\,\mathrm{d}u = \pi E, \tag{2.5.144}$$

where $x_+ = \sqrt{2E}$. We obtain

$$\pi E \approx \hbar\pi \left(n + \frac{1}{2}\right), \tag{2.5.145}$$

which is already the exact answer. In fact, it can be shown that, in the case of the harmonic oscillator, higher order corrections in $\hbar^2$ to the BS quantization condition vanish, and the BS approximation is exact. This is of course quite exceptional.

Let us now consider the basic WKB wavefunctions for the bound states. In the classically forbidden region $x > x_+$, we have to compute

$$\int_{x_+}^{x} p_1(x')\mathrm{d}x' = \sqrt{2E} \int_{x_+}^{x} \mathrm{d}x' \sqrt{\left(\frac{x'}{x_+}\right)^2 - 1} = E\left\{\frac{x}{x_+}\sqrt{\frac{x^2}{x_+^2} - 1} - \cosh^{-1}\left(\frac{x}{x_+}\right)\right\}. \tag{2.5.146}$$

The function in this region reads,

$$\psi_{III}(x) \approx \frac{C}{2\sqrt{p_1(x)}} \exp\left[-\frac{E}{\hbar}\left\{\frac{x}{x_+}\sqrt{x^2 - x_+^2} - \cosh^{-1}\left(\frac{x}{x_+}\right)\right\}\right], \tag{2.5.147}$$

where C is a normalization constant that we will determine in a moment. Let us now consider the classically allowed region. We have,

$$\int_{x}^{x_+} p(x')\mathrm{d}x' = \frac{\pi\hbar}{2}\left(n + \frac{1}{2}\right) - E\left\{\frac{x}{x_+}\sqrt{1 - \frac{x^2}{x_+^2}} + \sin^{-1}\left(\frac{x}{x_+}\right)\right\}, \tag{2.5.148}$$

where we have taken into account the BS quantization condition. We conclude that the wavefunction in region II is given by

$$\psi_{II}(x) \approx \frac{C}{\sqrt{p(x)}} \cos\left[\frac{\pi n}{2} - \frac{E}{\hbar}\left\{\frac{x}{x_+}\sqrt{1 - \frac{x^2}{x_+^2}} + \sin^{-1}\left(\frac{x}{x_+}\right)\right\}\right], \qquad (2.5.149)$$

which is a linear combination of basic WKB wavefunctions. The wavefunction in the region I, where $x < -x_+$, follows from the parity symmetry of the problem. In the case of the harmonic oscillator, the normalization constant in (2.5.140) is simply

$$C = \sqrt{\frac{2}{\pi}}. \qquad (2.5.150)$$

It is instructive to compare the approximate WKB wavefunctions written down above with the exact wavefunctions, as shown in Figure 2.4. The approximate wavefunctions are always divergent at the turning points. However, away from these points, the approximation is extremely good, even for moderate values of the quantum number n.

It is also interesting to consider the uniform WKB approximation to the wavefunctions given by (2.4.82), and in which we set in addition $\phi(x) \approx \phi_0(x)$. This particular approximation, based on an Airy function, is valid near one of the turning points, say $x = x_+$. When $x > x_+$, the approximate, uniform wavefunction is given by

$$\psi_{III}(x) \approx \frac{C_u}{\sqrt{p_1(x)}} \left(\frac{3}{2}\int_{x_+}^{x} p_1(x')dx'\right)^{1/6} \mathrm{Ai}\left(\left(\frac{3}{2\hbar}\int_{x_+}^{x} p_1(x')dx'\right)^{2/3}\right). \qquad (2.5.151)$$

Note that the Bi function in (2.4.82) does not contribute, due to the boundary conditions at $x \to \infty$. For $-x_+ < x < x_+$, we have

$$\psi_{II}(x) \approx \frac{C_u}{\sqrt{p(x)}} \left(\frac{3}{2}\int_{x}^{x_+} p(x')dx'\right)^{1/6} \mathrm{Ai}\left(\left(\frac{3}{2\hbar}\int_{x}^{x_+} p(x')dx'\right)^{2/3}\right). \qquad (2.5.152)$$

In these equations, C_u is a normalization constant given by

$$C_u = \sqrt{2}\,\hbar^{-1/6}. \qquad (2.5.153)$$

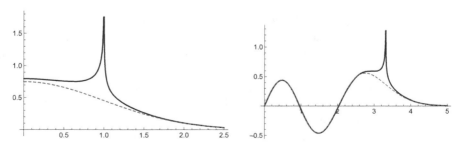

Figure 2.4   The WKB wavefunctions (continuous line) versus the exact wavefunctions (dashed line) for the bound states of the harmonic oscillator, as a function of $x > 0$. (Left) Corresponds to the $n = 0$ wavefunction; (Right) corresponds to $n = 5$. The peak in both figures is the divergence of the WKB wavefunction at the turning point $x = x_+$. We set $m = \hbar = \omega = 1$.

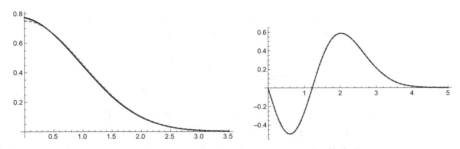

Figure 2.5 The uniform WKB wavefunctions (continuous line) versus the exact wavefunctions (dashed line) for the bound states of the harmonic oscillator, as a function of $x > 0$. (Left) Corresponds to the $n = 0$ wavefunction; (right) corresponds to $n = 3$. We set $m = \omega = \hbar = 1$.

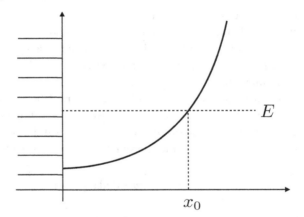

Figure 2.6 A potential with an infinite barrier at $x = 0$.

If we replace the Airy function by its asymptotic behaviors as $x \to \pm\infty$, we recover the wavefunctions (2.5.147) and (2.5.149), respectively. The above forms for the wavefunctions give excellent approximations for $x > 0$, and they are completely smooth at the turning point $x = x_+$, as required. In Figure 2.5 we plot the exact wavefunction versus the uniform approximation in the region $x > 0$, for $n = 0$ and $n = 3$. As we can see, both functions are essentially undistinguishable to the naked eye when $n > 0$. □

Example 2.5.2 *BS quantization condition in the presence of an infinite barrier*. Let us suppose that we have an infinite barrier at, say, $x = 0$, and a potential $V(x)$ satisfying $V(x) \to \infty$ when $x \to \infty$, so that there is a turning point at $x = x_0$, see Figure 2.6. The all-orders quantization condition can be easily generalized to this situation. In addition to having a decaying wavefunction for $x > x_0$, we have to impose that $\psi_n(0) = 0$. This last condition can be implemented with the following trick. Let us extend the potential $V(x)$ symmetrically to $\mathbb{R}$, i.e. $V(-x) = V(x)$. The vanishing of the wavefunction at $x = 0$ is equivalent to considering bound states that are odd under parity in this symmetric potential: $n = 2k + 1$, $k = 0, 1, 2, \ldots$. The BS quantization condition reads

$$\int_0^{x_0} P(x')\mathrm{d}x' = \pi\hbar\left(k + \frac{3}{4}\right), \qquad k = 0, 1, 2, \ldots. \qquad (2.5.154)$$

Note that in the r.h.s. we have a shift of the quantum number k by $3/4$, instead of the usual $1/2$.   □

Example 2.5.3 *BS quantization condition from the uniform WKB method.* It is possible to derive the BS quantization condition by using the uniform WKB approximation, and picking the function (2.3.52). The function $f(\phi)$ has the form (2.3.54), while $\phi(x)$ solves

$$\hbar^2 \left(\phi'(x) \right)^2 \left(2\nu - \phi^2(x) \right) = p^2(x) - \frac{\hbar^2}{2} \{\phi, x\}. \tag{2.5.155}$$

We will perform the approximation in which we neglect the Schwartzian derivative in the r.h.s., so that the approximate solution to (2.5.155) is given by the equation

$$\hbar \int_{\phi(x_0)}^{\phi(x)} \sqrt{2\nu - \phi^2} \, \mathrm{d}\phi = \int_{x_0}^{x} p(x') \mathrm{d}x'. \tag{2.5.156}$$

In order to solve for $\phi(x)$, we will require the wavefunction to be regular at the turning points, in contrast to the usual WKB approximation. Since

$$\hbar^2 \phi'(x)^2 \left(2\nu - \phi^2(x) \right) \approx p^2(x), \tag{2.5.157}$$

and $p(x_{\pm}) = 0$, we deduce that

$$\phi^2(x_{\pm}) = 2\nu. \tag{2.5.158}$$

We then choose

$$\phi(x_-) = -\sqrt{2\nu}, \qquad \phi(x_+) = \sqrt{2\nu}, \tag{2.5.159}$$

and the solution $\phi(x)$ to (2.5.156) has to satisfy

$$\hbar \int_{-\sqrt{2\nu}}^{\phi(x)} \sqrt{2\nu - \phi^2} \, \mathrm{d}\phi = \int_{x_-}^{x} p(x') \mathrm{d}x'. \tag{2.5.160}$$

In addition, the value of ν is related to the value of the energy E by the equation,

$$\hbar \int_{-\sqrt{2\nu}}^{\sqrt{2\nu}} \sqrt{2\nu - \phi^2} \, \mathrm{d}\phi = \int_{x_-}^{x_+} p(x) \mathrm{d}x. \tag{2.5.161}$$

The integral in the l.h.s. is elementary, and we find

$$\int_{x_-}^{x_+} p(x) \mathrm{d}x = \pi \hbar \nu. \tag{2.5.162}$$

This does not fix quantization conditions for the energy. To do that, we have to impose the decay of the wavefunction at infinity. As in (2.3.58), this imposes $\nu = n + 1/2$. Plugging this into (2.5.162) we obtain again the BS quantization condition (2.5.136).   □

So far we have discussed the leading term in the quantization condition (2.5.135). Let us have a first look at the subleading terms of the formal power series in the l.h.s. of (2.5.135). These terms involve the power series for $P(x)$ in (2.2.7),

$$P(x) = \sum_{n \geq 0} Y_{2n}(x) \hbar^{2n}. \tag{2.5.163}$$

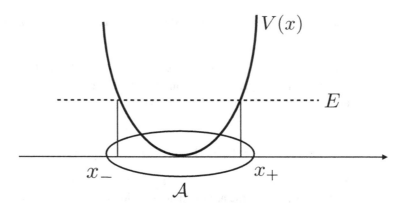

Figure 2.7   A contour $\mathcal{A}$ in the complex plane, encircling the turning points.

By plugging this series into (2.5.135), one finds higher order corrections to the BS quantization condition. We should now recall that the integrals are regularized near the turning points by the procedure described in (2.4.104), where the path C is shown in Figure 2.2. When we have an integral between two turning points, the regularization involves integrating around a contour $\mathcal{A}$ in the complex plane, as shown in Figure 2.7. This contour encircles the interval of allowed classical motion for the corresponding energy, $[x_-, x_+]$. The upper part of the contour, above the real axis, is on the first Riemann sheet, while the lower part of the contour is on the second Riemann sheet. Therefore, with the regularization we have been using, we can write

$$\int_{x_-}^{x_+} P(x')\mathrm{d}x' = \frac{1}{2} \oint_{\mathcal{A}} P(z)\mathrm{d}z. \tag{2.5.164}$$

The all-orders generalization of the BS quantization condition reads then,

$$\sum_{n=0}^{\infty} \hbar^{2n}\mathrm{vol}_n(E) = 2\pi\hbar\left(n + \frac{1}{2}\right), \qquad n = 0, 1, 2, \ldots, \tag{2.5.165}$$

where

$$\mathrm{vol}_n(E) = \oint_{\mathcal{A}} Y_{2n}(z)\mathrm{d}z. \tag{2.5.166}$$

The l.h.s. of this equation can be regarded as a quantum version of the volume of phase space that appeared in (2.5.143). The contour integral appearing in (2.5.142) is the integral of the Liouville differential form $p(x)\mathrm{d}x$ on a cycle of the curve

$$p^2 = 2m(E - V(x)). \tag{2.5.167}$$

In the theory of complex curves, these type of quantities are called *periods*. We will refer to its quantum version to all orders, appearing in (2.5.164), as a *quantum period* or a *WKB period*.

   There is an alternative derivation of (2.5.165) that uses complex analysis and it is very instructive. If $V(x)$ is an analytic function, any function $\psi(x)$ solving the Schrödinger equation is also analytic, by the elementary theory of ODEs in the complex plane. The oscillation theorem asserts that the eigenfunctions $\psi(x)$ corresponding to the nth energy level have n zeros on the real line inside the interval $[x_-, x_+]$. Since $\psi(x)$ is analytic when extended to the full complex plane, the argument theorem says that,

$$\frac{1}{2\pi i} \oint_{\mathcal{A}} \frac{\psi'(z)}{\psi(z)} dz = n, \qquad n = 0, 1, 2, \ldots \qquad (2.5.168)$$

By using the WKB ansatz for the wavefunction (2.2.3), we conclude that

$$\oint_{\mathcal{A}} Y(z) dz = 2\pi n\hbar. \qquad (2.5.169)$$

This is similar to (2.5.165), but in principle it also involves the odd powers of $\hbar$. However, it is easy to see that Y_{odd} only contributes through its first term, since, due to (2.2.10),

$$Y_{\text{odd}}(z) = \frac{i\hbar}{2} \frac{Y_0'(z)}{Y_0(z)} + \sum_{k=1}^{\infty} f_k'(z)\hbar^{2k+1}, \qquad (2.5.170)$$

where $f_k(z)$ is a rational function. The integration of $f_k'(z)$ along the $\mathcal{A}$-cycle gives a vanishing contribution. Let us consider the contribution of the first term. Note that $Y_0(z)$ is not analytic. However,

$$\frac{Y_0'}{Y_0} = \frac{d}{dz} \log Y_0 = \frac{1}{2} \frac{d}{dz} \log\left(2(E - V(z))\right) = \frac{1}{2} \frac{\sigma'(z)}{\sigma(z)}, \qquad (2.5.171)$$

where $\sigma(z) = 2(E - V(z))$, is an analytic function that has two zeros inside $\mathcal{A}$, namely, the two turning points x_-, x_+. By applying the theorem of the argument to $\sigma(z)$, we obtain

$$\frac{i\hbar}{2} \oint_{\mathcal{A}} \frac{Y_0'(z)}{Y_0(z)} dz = \frac{i\hbar}{4} \oint_{\mathcal{A}} \frac{\sigma'(z)}{\sigma(z)} dz = \frac{i\hbar}{4}(4\pi i) = -\pi\hbar. \qquad (2.5.172)$$

We conclude that (2.5.169) is precisely (2.5.165).

Example 2.5.4 Let us present results for the first two corrections in (2.5.165). The expressions for the functions $Y_{2n}(x)$ with $n \geq 1$ can be typically simplified by subtracting total derivatives. For example, we have that

$$Y_2(x) = -\frac{1}{8} \frac{(Y_0'(x))^2}{Y_0^3(x)} + \text{total derivative}. \qquad (2.5.173)$$

The first correction to the BS quantization condition reads then,

$$\oint_{\mathcal{A}} dx\, Y_2(x) = -\frac{1}{8} \oint_{\mathcal{A}} dx \frac{m^2(V'(x))^2}{(2m(E - V(x)))^{5/2}} = \frac{1}{24} \oint_{\mathcal{A}} dx \frac{mV''(x)}{(2m(E - V(x)))^{3/2}}$$
$$= -\frac{1}{24} \frac{d}{dE} \oint_{\mathcal{A}} dx \frac{V''(x)}{(2m(E - V(x)))^{1/2}}. \qquad (2.5.174)$$

In writing the last term in the first line we have used integration by parts. Finally, in order to write the final result, we have decreased the power of the momentum in the denominator by extracting a derivative w.r.t. the energy E. The advantage of having a single power of the momentum in the denominator is that the contour integral can then be written as a convergent integral on the interval $[x_-, x_+]$.

To write down the next correction, we first note that

$$Y_4(x) = \frac{15}{128} \frac{(Y_0'(x))^4}{Y_0^7(x)} - \frac{1}{32} \frac{(Y_0''(x))^2}{Y_0^5(x)} + \text{total derivative}. \tag{2.5.175}$$

Integrating by parts repeatedly and extracting derivatives w.r.t. the energy, we finally obtain,

$$\oint_{\mathcal{A}} Y_4(x)\, dx = \frac{7}{5760m} \frac{d^3}{dE^3} \oint_{\mathcal{A}} \frac{(V''(x))^2}{(2m(E - V(x)))^{1/2}}\, dx$$
$$- \frac{1}{1152m} \frac{d^2}{dE^2} \oint_{\mathcal{A}} \frac{V^{(4)}(x)}{(2m(E - V(x)))^{1/2}}\, dx. \tag{2.5.176}$$

Let us consider the case of the harmonic oscillator, with the conventions of Example 2.5.1. From the general expression (2.5.174) one finds that the first correction is given by

$$\oint_{\mathcal{A}} Y_2(x)\, dx = -\frac{1}{12} \frac{d}{dE} \int_{-x_+}^{x_+} \frac{dx}{(2(E - x^2/2))^{1/2}} = -\frac{1}{12} \frac{d}{dE} \int_{-1}^{1} \frac{1}{\sqrt{1 - u^2}} = 0. \tag{2.5.177}$$

It can be shown that the corrections coming from the terms Y_{2n}, with $n \geq 2$, also vanish. □

Since the l.h.s. of the all-orders quantization condition (2.5.165) involves an infinite power series, how do we implement it in practice? One could hope that this series can be summed to a well-defined function, and this is the case in a handful of examples. However, the formal power series in the l.h.s. of (2.5.165) is, generically, a (doubly) factorially divergent series: for fixed E, one has

$$\text{vol}_n(E) \sim (2n)!\,. \tag{2.5.178}$$

Therefore, this series is asymptotic and it has to be analyzed with the techniques summarized in Appendix A. For example, if $\hbar$ is small, one can use an optimal truncation of the asymptotic series to obtain a reasonable approximation to the exact eigenvalues, by picking the very first corrections to the BS quantization condition. This gives excellent estimates for the energy levels, as we will see in examples in Section 2.6. A more powerful technique is Borel resummation. However, the series appearing in the l.h.s. of (2.5.165) is typically not Borel summable. This means that one should include additional exponentially small effects in $\hbar$ in order to reconstruct the exact eigenvalues. In some cases, the exact connection formulae provide detailed information about these effects. Later on, we will analyze in detail examples of how to compute these nonperturbative corrections.

2.6 Some Examples

Let us now illustrate the results obtained so far with some simple examples. We will focus on *confining potentials*. These potentials are characterized by the behavior

$$V(x) \to \infty, \qquad |x| \to \infty. \tag{2.6.179}$$

Examples of (2.6.179) include polynomial potentials behaving like $V(x) \approx x^{2M}$ at large x, where $M \in \mathbb{Z}_{>0}$. When the potential is confining, the corresponding Hamiltonian

$$H = \frac{p^2}{2m} + V(x), \qquad (2.6.180)$$

has an infinite, discrete, and nondegenerate set of positive eigenvalues E_n, $n = 0, 1, 2, \ldots$. We will use the WKB method to obtain approximate expressions for this spectrum in various examples.

Example 2.6.1 Perhaps the simplest illustration of the method is the family of Hamiltonians with a monic potential

$$H = \frac{p^2}{2m} + gq^{2M}, \qquad M \in \mathbb{Z}_{>0}. \qquad (2.6.181)$$

For $M = 1$, this is the harmonic oscillator. For $M = 2$, we obtain the so called *pure (or monic) quartic oscillator*. Since the potential is confining, it has an infinite tower of bound states with energies $E_k(\hbar, m, g)$. An important property of this Hamiltonian is that, by an appropriate rescaling of the variables in the Schrödinger equation, we can set $m = 1/2$ and $\hbar = g = 1$. Let us denote by $e_k = E_k(1, 1/2, 1)$. Then, one has

$$E_k(\hbar, m, g) = g^{\frac{1}{M+1}} \left(\frac{\hbar^2}{2m} \right)^{\frac{M}{M+1}} e_k. \qquad (2.6.182)$$

This means that there is no adjustable parameter in this problem: the energies of the bound states have a trivial dependence on $\hbar$, g, and m. The only nontrivial dependence is on k, the integer labeling the energy level, and the WKB approximation leads to an expression for E_k that is valid for large k. In view of the scaling (2.6.182), we can consider, without loss of generality, the Hamiltonian

$$H = p^2 + q^{2M}. \qquad (2.6.183)$$

Let us first work out the BS quantization condition. We have

$$\text{vol}_0(e) = 2 \int_{-e^{\frac{1}{2M}}}^{e^{\frac{1}{2M}}} \sqrt{e - q^{2M}} \, dq = \frac{\sqrt{\pi}}{M} \frac{\Gamma\left(\frac{1}{2M}\right)}{\Gamma\left(\frac{3}{2} + \frac{1}{2M}\right)} e^{\frac{1+M}{2M}}. \qquad (2.6.184)$$

We can also compute the next-to-leading order correction by using (2.5.174). One finds,

$$\text{vol}_1(e) = -\frac{M(2M-1)}{6} \frac{d}{de} \int_{-e^{\frac{1}{2M}}}^{e^{\frac{1}{2M}}} \frac{q^{2M-2}}{(e - q^{2M})^{1/2}} \, dq = -\frac{\sqrt{\pi}}{3} \frac{M\Gamma\left(2 - \frac{1}{2M}\right)}{\Gamma\left(\frac{1}{2} - \frac{1}{2M}\right)} e^{-\frac{1+M}{2M}}. \qquad (2.6.185)$$

Let us now compare the numerical results for the spectrum of the Hamiltonian (2.6.183), for $M = 2$ (the pure quartic oscillator), with the results obtained from the BS quantization condition (2.5.143), and with the results obtained from (2.5.169) after including the first correction (2.6.185). We will denote these three quantities by E_n, $E_n^{(0)}$, and $E_n^{(1)}$, respectively. The results are shown in Table 2.1. We observe that the BS approximation gives a reasonable estimate for the first excited state. For the state with $n = 4$, the error is only of 0.2%. The next-to-leading approximation

Table 2.1. Comparison between the numerical values of the energies and the WKB approximations $E_n^{(0)}$, $E_n^{(1)}$ for the pure quartic oscillator. We have set $2m = g = \hbar = 1$.

Level	E_n	$E_n^{(0)}$	$E_n^{(1)}$
0	1.060362091	0.8671453264	0.980766290
1	3.799673029	3.751919923	3.810329951
4	16.261826019	16.233614692	16.261936744

provides a remarkable improvement, and the error for the level $n = 4$ is reduced to one part in 10^5.   □

Example 2.6.2 *The quartic oscillator.* Let us now consider the quartic oscillator, defined by the Hamiltonian

$$H = \frac{p^2}{2m} + \frac{m\omega^2}{2}q^2 + gq^4. \tag{2.6.186}$$

We will set for simplicity $m = \omega = 1$. There are two symmetric turning points in this problem, $\pm a$, determined by the equation

$$a^2 = \frac{\sqrt{16Eg + 1} - 1}{4g}, \tag{2.6.187}$$

where E is the energy. The period (2.5.142) is

$$\text{vol}_0(E) = 2\int_{-a}^{a} \sqrt{2E - x^2 - 2gx^4}\, dx. \tag{2.6.188}$$

This integral can be evaluated in terms of elliptic integrals of the first and second kind, denoted respectively by $K(k)$ and $E(k)$. One finds,

$$\text{vol}_0(E) = \frac{4}{3}(2g)^{\frac{1}{2}}(a^2 + b^2)^{\frac{1}{2}}\left[b^2 K(k) + (a^2 - b^2)E(k)\right], \tag{2.6.189}$$

where

$$b^2 = \frac{\sqrt{16Eg + 1} + 1}{4g}, \qquad k^2 = \frac{a^2}{a^2 + b^2}. \tag{2.6.190}$$

To calculate the next-to-leading correction in (2.5.165), we note that

$$\text{vol}_1(E) = \oint_{\mathcal{A}} Y_2(x)dx = \frac{1}{24}\oint_{\mathcal{A}} \frac{1 + 12gx^2}{(2E - x^2 - 2gx^4)^{3/2}}dx, \tag{2.6.191}$$

where we used (2.5.174). A simple calculation shows that this can be written as

$$\begin{aligned}
\text{vol}_1(E) &= -\left(\frac{g}{2}\frac{d}{dE} + \frac{1 + 48gE}{24}\frac{d^2}{dE^2}\right)\oint_{\mathcal{A}} Y_0(x)dx \\
&= -\left(\frac{g}{2}\frac{d}{dE} + \frac{1 + 48gE}{24}\frac{d^2}{dE^2}\right)\text{vol}_0(E).
\end{aligned} \tag{2.6.192}$$

Table 2.2. Comparison between the numerical values of the energies and the WKB approximations $E_n^{(0)}, E_n^{(1)}$ for the quartic oscillator. We have set $\hbar = m = g = 1$.

Level	E_n	$E_n^{(0)}$	$E_n^{(1)}$
0	0.803770651	0.704201128	0.776387207
1	2.737892268	2.703483166	2.740647416
4	10.963583094	10.945656155	10.963634214

This type of trick, in which the next-to-leading correction is written as a combination of derivatives of the leading-order term w.r.t. the energy E, can be generalized to higher corrections and to many other potentials.

In Table 2.2 we compare the numerical values for the energy levels, E_n, with those obtained using the WKB approximation $E_n^{(0)}$, $E_n^{(1)}$ (at leading and next-to-leading order, respectively), in the case $g = 1$. As in the previous example, the approximation is very good, and improves further as we consider higher energy levels.

The quartic oscillator is often solved with another approximation mechanism, namely stationary perturbation theory in g. Usually, this leads to a power series expansion for E_n in powers of g. Standard techniques give,

$$E_n = \hbar v + \frac{3}{2}\left(v^2 + \frac{1}{4}\right)g\hbar^2 - \frac{1}{4}\left(17v^3 + \frac{67v}{4}\right)g^2\hbar^3$$
$$+ \frac{1}{16}\left(375v^4 + \frac{1707v^2}{2} + \frac{1539}{16}\right)g^3\hbar^4 + \cdots, \tag{2.6.193}$$

where

$$v = n + \frac{1}{2}. \tag{2.6.194}$$

We note that the coefficient of g^k is a polynomial in v of order $k + 1$. Let us now compare (2.6.193) with the WKB result. In the WKB method, we found at each order a function of E, $\text{vol}_k(E)$, which also depends on g. These functions can be expanded around $g = 0$:

$$\text{vol}_0(E) = 2\pi\left(E - \frac{3E^2g}{2} + \frac{35E^3g^2}{4} - \frac{1155E^4g^3}{16} + \cdots\right),$$
$$\text{vol}_1(E) = -\pi\left(\frac{3g}{4} - \frac{85Eg^2}{8} + \frac{2625E^2g^3}{16} + \cdots\right). \tag{2.6.195}$$

By using the all-orders quantization condition (2.5.165), we can solve for E as a function of v, in a series around $g = 0$. The BS quantization condition gives the series

$$E^{(0)}(v, g) = \hbar v + \frac{3v^2}{2}g\hbar^2 - \frac{17v^3}{4}g^2\hbar^3 + \frac{375v^4}{16}g^3\hbar^4 + \cdots. \tag{2.6.196}$$

This captures the coefficient of the highest power in v, for each term in the power series expansion in g (2.6.193). In other words, the leading WKB approximation provides a partial resummation of stationary perturbation theory in the coupling constant g. This resummation captures the leading contribution for large values of the quantum number v. In order to correctly incorporate the subleading corrections for v large, we have to include the corrections in (2.5.165). For example, by using the next-to-leading approximation to the quantization condition,

$$\text{vol}_0(E) + \hbar^2 \text{vol}_1(E) \approx 2\pi\hbar v. \tag{2.6.197}$$

we find

$$
\begin{aligned}
E^{(1)}(v, g) = \hbar v &+ \frac{3}{2}\left(v^2 + \frac{1}{4}\right)\hbar^2 - \frac{1}{4}\left(17v^3 + \frac{67v}{4}\right)g^2\hbar^3 \\
&+ \frac{1}{16}\left(375v^4 + \frac{1707v^2}{2} - \frac{57}{2}\right)g^3\hbar^4 + \cdots.
\end{aligned}
\tag{2.6.198}
$$

This captures correctly the coefficient of the two highest powers in v, at each order in g.

Example 2.6.3 As we mentioned above, the Voros–Silverstone connection formula contains both perturbative and nonperturbative information, i.e. it gives, in principle, exponentially small corrections in $\hbar$ to the perturbative corrections in powers of $\hbar$. These nonperturbative corrections are typically due to tunneling effects. In order to appreciate the importance of these effects, and to understand in detail the role of Borel resummation, let us consider a particle of energy E in the potential

$$
V(x) = \begin{cases} kx, & 0 \leq x \leq x_1, \\ \infty, & x < 0, \ x > x_1. \end{cases}
\tag{2.6.199}
$$

As shown in Figure 2.8, this is a linear potential inside an infinite box. Let us first work out the exact solution to this problem, which is very easy to obtain with the help of Airy functions. As in Example 2.2.1, the wave function is given by a linear combination,

$$\psi(x) = a\text{Ai}\left(\hbar^{-2/3}\phi(x)\right) + b\text{Bi}\left(\hbar^{-2/3}\phi(x)\right),
\tag{2.6.200}$$

where

$$\phi(x) = c(x - x_0), \qquad x_0 = \frac{E}{k}, \qquad c = (2mk)^{1/3}.
\tag{2.6.201}$$

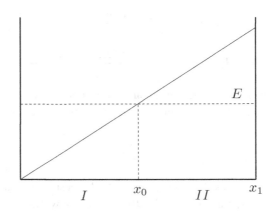

Figure 2.8   The potential of (2.6.199).

The wavefunction must vanish at $x = 0$ and $x = x_1$, and this provides two conditions. One of them fixes the quotient a/b, and the other gives an (exact) quantization condition for the energy, which we can write as

$$\text{Ai}\left(\hbar^{-2/3}\phi(x_1)\right)\text{Bi}\left(\hbar^{-2/3}\phi(0)\right) - \text{Ai}\left(\hbar^{-2/3}\phi(0)\right)\text{Bi}\left(\hbar^{-2/3}\phi(x_1)\right) = 0. \quad (2.6.202)$$

It is possible to solve this condition numerically, and one finds a discrete spectrum of energy levels E_n, $n = 0, 1, \ldots$. Note that, if $x_1 \to \infty$, we must have $b = 0$ in (2.6.200), and the quantization condition reduces to

$$\text{Ai}\left(\hbar^{-2/3}\phi(0)\right) = 0, \quad (2.6.203)$$

i.e. the spectrum is determined by the zeros of the Airy function.

The quantization condition (2.6.202) can be obtained by using the exact WKB method, which provides a semiclassical picture for the quantization. There are two regions to consider, as shown in Figure 2.8. Region I, where $x < x_0$, is a classically allowed region, and the all-orders WKB wavefunction can be written as

$$\psi_I(x) = \frac{1}{\sqrt{P(x)}}\cos\left(\frac{1}{\hbar}\int_x^{x_0} P(x')dx' - \frac{\pi}{4} - \delta\right). \quad (2.6.204)$$

We now use the Voros–Silverstone connection formula to obtain the form of the wavefunction in region II:

$$\psi_{II}(x) = \frac{1}{\sqrt{P_1(x)}}\left\{-\sin\delta\, e^{\frac{1}{\hbar}\int_{x_0}^x P_1(x')dx'} + \frac{1}{2}e^{\mp i\delta}e^{-\frac{1}{\hbar}\int_{x_0}^x P_1(x')dx'}\right\}. \quad (2.6.205)$$

The condition that $\psi_{II}(x)$ vanishes at $x = x_1$ fixes the value of δ to be

$$\cot\delta = 2\exp\left(\frac{2}{\hbar}\int_{x_0}^{x_1} P_1(x')dx'\right) \pm i. \quad (2.6.206)$$

The other boundary condition imposes that

$$\psi_I(0) = 0, \quad (2.6.207)$$

which leads to

$$\frac{1}{\hbar}\int_0^{x_0} P(x')dx' - \delta = \left(n - \frac{1}{4}\right)\pi, \quad n = 1, 2, \ldots. \quad (2.6.208)$$

The structure of this quantization condition is easy to understand. When $\delta = 0$ (which occurs physically in the limit in which $x_1 \to \infty$), we obtain the all-orders quantization condition (2.5.154) corresponding to a potential well with an infinite barrier at $x = 0$. The term δ gives additional, exponentially small corrections due to the barrier penetration effects in the interval $[x_0, x_1]$. For small $\hbar$ we have

$$\delta \approx \frac{1}{2}e^{-\frac{2}{\hbar}\int_{x_0}^{x_1} P_1(x')dx'}. \quad (2.6.209)$$

The equation (2.6.206) seems however to give a complex value for δ. However, this is not so. What happens is that the integral $P_1(x)$ appearing in the r.h.s. of this equation leads to a formal power series in $\hbar^2$ that is not Borel summable, but admits lateral Borel resummations. The lateral Borel resummations of this series leads to an imaginary part that cancels with the term $\pm i$ in the r.h.s. to produce a real value for δ. The choice of lateral resummation is correlated to the sign of the term $\pm i$, in such a way that the cancellation takes place.

Let us now be more explicit about the formal power series appearing in (2.6.208). A calculation similar to the one performed in Example 2.2.1, or in (2.2.44), gives

$$\frac{1}{\hbar} \int_x^{x_0} P(x')\mathrm{d}x' = \zeta(x) + \sum_{n \geq 1} \widetilde{c}_{2n-1} (-1)^n \left(\zeta(x)\right)^{-2n+1}, \qquad 0 \leq x \leq x_0,$$

$$(2.6.210)$$

where

$$\zeta(x) = \frac{2}{3\hbar} \left(-\phi(x)\right)^{3/2}, \qquad (2.6.211)$$

and the coefficients $\widetilde{c}_m$ are defined in (2.2.39). The series (2.6.210) is Borel summable for all $x \in [0, x_0]$. The other formal power series involved in (2.6.208) is

$$\frac{1}{\hbar} \int_{x_0}^x P_1(x')\mathrm{d}x' = \widehat{\zeta}(x) + \sum_{n \geq 1} \widetilde{c}_{2n-1} \left(\widehat{\zeta}(x)\right)^{-2n+1}, \qquad x_0 \leq x \leq x_1, \quad (2.6.212)$$

where

$$\widehat{\zeta}(x) = \frac{2}{3\hbar} \left(\phi(x)\right)^{3/2}. \qquad (2.6.213)$$

This series is not Borel summable, but it has lateral Borel resummations. The choice of lateral resummation has to be correlated with the choice of sign in the r.h.s. of (2.6.206), in such a way that δ is real.

It is possible to show that the asymptotic expansion of the exact WKB quantization condition (2.6.202) agrees with (2.6.208). First of all, we note that (2.6.208) can be written as

$$\tan\left(\frac{1}{\hbar} \int_0^{x_0} P(x')\mathrm{d}x' + \frac{\pi}{4}\right) = \tan \delta. \qquad (2.6.214)$$

By using the expressions in Appendix B.1, one obtains the following asymptotic expansion for $z < 0$,

$$\frac{\mathrm{Ai}(-z)}{\mathrm{Bi}(-z)} \sim \tan\left(\Phi(\zeta) + \frac{\pi}{4}\right), \qquad (2.6.215)$$

where

$$\zeta = \frac{2}{3} z^{3/2}, \qquad \Phi(\zeta) = -\frac{\mathrm{i}}{2} \log\left(\frac{\beta(\mathrm{i}\zeta)}{\beta(-\mathrm{i}\zeta)}\right). \qquad (2.6.216)$$

By comparing these expressions to (2.2.44), we conclude that

$$\frac{\mathrm{Ai}\left(\hbar^{-2/3}\phi(0)\right)}{\mathrm{Bi}\left(\hbar^{-2/3}\phi(0)\right)} \sim \tan\left(\frac{1}{\hbar} \int_0^{x_0} P(x')\mathrm{d}x' + \frac{\pi}{4}\right). \qquad (2.6.217)$$

Next, we note that

$$2 \exp\left(\frac{2}{\hbar} \int_{x_0}^{x_1} P_1(x')\mathrm{d}x'\right) \pm \mathrm{i} = 2\frac{\beta(\widehat{\zeta}(x_1))}{\beta(-\widehat{\zeta}(x_1))} \pm \mathrm{i}. \qquad (2.6.218)$$

It follows from (B.32) and (B.35) that this is the asymptotic expansion of

$$\frac{\mathrm{Bi}\left(\hbar^{-2/3}\phi(x_1)\right)}{\mathrm{Ai}\left(\hbar^{-2/3}\phi(x_1)\right)}. \qquad (2.6.219)$$

Equivalently, we have that

$$\frac{\text{Ai}\left(\hbar^{-2/3}\phi(x_1)\right)}{\text{Bi}\left(\hbar^{-2/3}\phi(x_1)\right)} \sim \tan \delta. \tag{2.6.220}$$

We conclude that the asymptotic expansion of the exact quantization condition (2.6.202) agrees with (2.6.214). Moreover, after appropriate lateral Borel resummations of the right-hand sides, the asymptotic expansions (2.6.215), (2.6.220) become equalities.   □

Example 2.6.3 shows that the Voros–Silverstone connection formula gives quantization conditions that go beyond the all-orders WKB formula (2.5.165). The quantization condition (2.6.208), after appropriate (lateral) Borel resummations of the series, is in fact *exact*, and it reproduces the solution (2.6.202). The perturbative, all-orders WKB method can then be improved into an *exact* WKB method by using Borel resummation techniques and by incorporating exponentially small corrections. We will discuss in detail other examples of exact quantization conditions in Sections 2.8, 2.10, and 5.3.

2.7 Spectral Functions and the WKB Method

In this section we will discuss some additional properties of confining potentials, in the light of the WKB method. We will assume, without loss of generality, that $\inf V(x) = 0$. We note that, with this assumption, the energy eigenvalues are strictly positive.

As we mentioned in Section 2.6, when $V(x)$ is confining, the spectrum of H is a discrete set of nondegenerate eigenvalues. The spectral information contained in this tower of real numbers can be encoded conveniently in so-called *spectral functions*. The first one is the trace of the resolvent (1.3.81), which can be written as

$$G(E) = \sum_{n=0}^{\infty} \frac{1}{E - E_n}. \tag{2.7.221}$$

The second one is the *Fredholm determinant*, or *spectral determinant* $\Xi(E)$,

$$\Xi(E) = \det\left(1 - E\text{H}^{-1}\right) = \prod_{n=0}^{\infty}\left(1 - \frac{E}{E_n}\right). \tag{2.7.222}$$

Finally, we have the canonical partition function $Z(\beta)$, given by

$$Z(\beta) = \sum_{n \geq 0} e^{-\beta E_n}. \tag{2.7.223}$$

These three spectral functions are not independent, and one finds

$$\log \Xi(E) = \int_0^E G(E')dE', \tag{2.7.224}$$

as well as

$$G(E) = - \int_0^\infty Z(\beta) e^{E\beta} d\beta, \qquad E < 0. \tag{2.7.225}$$

It is important to note that the infinite sums and products above converge only if the E_n increase sufficiently fast as n grows. More precisely, one should have

$$\sum_{n=0}^\infty \frac{1}{E_n} < \infty. \tag{2.7.226}$$

When this is the case, the inverse Hamiltonian H^{-1} (which is a compact operator) is called a *trace class operator*. The Fredholm determinant of a trace class operator is an analytic function of E, whose zeros occur precisely for the values of E that belong to the spectrum of H.

When the operator H^{-1} is not trace class, but all higher traces

$$\sum_{n=0}^\infty E_n^{-k}, \qquad k = 2, 3, \ldots \tag{2.7.227}$$

converge, one can define regularized versions of the trace of the resolvent, and of the spectral determinant:

$$G_2(E) = \sum_{n=0}^\infty \left\{ \frac{1}{E - E_n} + \frac{1}{E_n} \right\},$$

$$\Xi_2(E) = \prod_{n=0}^\infty \left(1 - \frac{E}{E_n} \right) e^{E/E_n}. \tag{2.7.228}$$

In this case, one has

$$G_2(E) = - \int_0^\infty Z(\beta) \left(e^{E\beta} - 1 \right) d\beta \tag{2.7.229}$$

and

$$\log \Xi_2(E) = \int_0^E G_2(E') dE'. \tag{2.7.230}$$

We will now study the asymptotic behavior of these functions for *negative* energies $E < 0$. We will set $\hbar = 2m = 1$ for simplicity, and we will make the following choice of the branch cut: $p(x) = -ip_1(x)$, where $p_1(x) > 0$. We want to characterize first the asymptotic behavior of the solutions to the Schrödinger equation. One important property of many confining potentials is that, when $|x| \to \infty$, this behavior is given by a linear combination of the basic WKB solutions

$$\psi(x, E) \sim \frac{A}{\sqrt{p_1(x)}} e^{\int_0^x p_1(x') dx'} + \frac{B}{\sqrt{p_1(x)}} e^{-\int_0^x p_1(x') dx'}, \qquad |x| \to \infty, \tag{2.7.231}$$

where the choice of $x = 0$ in the integration is made for convenience. To justify (2.7.231), we note that any solution to the Schrödinger equation can be written as a linear combination of the form (2.2.20). Let us assume for concreteness that $V(x) \approx x^{2M}$ for $M \in \mathbb{Z}_{>0}$. Then, by using the recursion (2.2.6), we find that

$$Y_n(x) \approx |x|^{-n-(n-1)M}, \qquad n \geq 1, \quad |x| \to \infty, \tag{2.7.232}$$

so that all higher order corrections to $S_0(x)$ in (2.2.20) are suppressed at large x. As in the study of scattering problems, where we introduced the Jost functions, we define two special solutions of the Schrödinger equation, $\psi_\pm(x)$, characterized by their exponential decay as $x \to \pm\infty$, respectively. These functions have the asymptotic form

$$\psi_+(x, E) \sim \frac{1}{\sqrt{p_1(x)}} e^{-\int_0^x p_1(x')dx'}, \qquad x \to \infty,$$

$$\psi_-(x, E) \sim \frac{1}{\sqrt{p_1(x)}} e^{\int_0^x p_1(x')dx'}, \qquad x \to -\infty. \tag{2.7.233}$$

For general values of E, the functions $\psi_\pm(x, E)$ do not decrease necessarily in the other direction, and they involve an increasing exponential

$$\psi_+(x, E) \sim \frac{a_+(E)}{\sqrt{p_1(x)}} e^{-\int_0^x p_1(x')dx'}, \qquad x \to -\infty,$$

$$\psi_-(x, E) \sim \frac{a_-(E)}{\sqrt{p_1(x)}} e^{\int_0^x p_1(x')dx'}, \qquad x \to \infty. \tag{2.7.234}$$

Let us calculate the Wronskian of the two solutions,

$$W(\psi_+, \psi_-) = \psi_+\psi_-' - \psi_+'\psi_-. \tag{2.7.235}$$

This is a constant, so we can evaluate it as $x \to -\infty$ by using the asymptotic form above. We find, as $x \to -\infty$,

$$\psi_+\psi_-' \sim \left(1 - \frac{1}{2}\frac{p_1'}{p_1^2}\right)a_+(E), \qquad \psi_+'\psi_- \sim -\left(1 + \frac{1}{2}\frac{p_1'}{p_1^2}\right)a_+(E), \tag{2.7.236}$$

therefore

$$W(\psi_+, \psi_-) = 2a_+(E). \tag{2.7.237}$$

A similar evaluation as $x \to \infty$ leads to

$$W(\psi_+, \psi_-) = 2a_-(E). \tag{2.7.238}$$

We conclude in particular that

$$a_+(E) = a_-(E) \equiv a(E). \tag{2.7.239}$$

Note that the eigenfunctions of the spectral problem are normalizable wavefunctions, and they are characterized by the condition

$$a(E) = 0, \tag{2.7.240}$$

after analytic continuation of the function $a(E)$ to positive energies. This is similar to the condition $a(k_0) = 0$ for bound states discussed in Section 1.5, which also involves the analytic continuation of $a(k)$ to a neighborhood of the positive imaginary axis. There are other similarities with the theory developed for scattering states. For example, the resolvent of the Schrödinger operator can be explicitly written in terms of the solutions $\psi_\pm(x)$, as follows

$$G(x, x'; E) = -\frac{1}{2a(E)}\Big(\psi_+(x, E)\psi_-(x', E)\theta(x - x')$$

$$+ \psi_+(x', E)\psi_-(x, E)\theta(x' - x)\Big). \tag{2.7.241}$$

This expression can be verified by direct calculation. There is also an analogue of the result (1.3.81), relating the quantity $a(E)$ to the trace of the resolvent, which in particular relates the coefficient $a(E)$ to the spectral determinant of the Schrödinger operator. To derive this relationship, we will follow the same strategy that we used in the scattering case. Let us denote by $\dot{a}$ the derivative w.r.t. E. We have,

$$\frac{\mathrm{d}}{\mathrm{d}E} W(\psi_+, \psi_-) = W(\dot{\psi}_+, \psi_-) + W(\psi_+, \dot{\psi}_-) = 2\dot{a}(E). \tag{2.7.242}$$

On the other hand, if we take a derivative w.r.t. E in the Schrödinger equation for ψ_+, and we multiply the result by ψ_-, we find

$$\psi_- \left(-\frac{\mathrm{d}^2}{\mathrm{d}x^2} + V(x) \right) \dot{\psi}_+ = \left(\psi_+ + E\dot{\psi}_+ \right) \psi_-. \tag{2.7.243}$$

At the same time, we can multiply the Schrödinger equation for ψ_- by $\dot{\psi}_+$, to obtain

$$\dot{\psi}_+ \left(-\frac{\mathrm{d}^2}{\mathrm{d}x^2} + V(x) \right) \psi_- = E\dot{\psi}_+ \psi_-. \tag{2.7.244}$$

Subtracting both equations we find

$$\frac{\mathrm{d}}{\mathrm{d}x} W(\dot{\psi}_+, \psi_-) = \psi_+ \psi_-. \tag{2.7.245}$$

Exchanging + with −, we obtain,

$$-\frac{\mathrm{d}}{\mathrm{d}x} W(\psi_+, \dot{\psi}_-) = \psi_+ \psi_-. \tag{2.7.246}$$

The same argument that led to (1.5.186) leads now to

$$\int_{-A}^{A} \psi_+(x, E)\psi_-(x, E)\mathrm{d}x = W(\dot{\psi}_+, \psi_-)(A) + W(\psi_+, \dot{\psi}_-)(-A) - 2\dot{a}(E). \tag{2.7.247}$$

The Wronskians as $A \to \infty$ can be evaluated by again using the asymptotic expressions for the wavefunctions. We find,

$$W(\dot{\psi}_+, \psi_-)(A) \sim a(E) \left\{ -2 \int_0^A \dot{p}_1(x)\mathrm{d}x + \frac{1}{2p_1(A)} \left(\frac{\dot{p}_1}{p_1} \right)' (A) \right\},$$
$$W(\psi_+, \dot{\psi}_-)(-A) \sim a(E) \left\{ 2 \int_0^{-A} \dot{p}_1(x)\mathrm{d}x - \frac{1}{2p_1(-A)} \left(\frac{\dot{p}_1}{p_1} \right)' (-A) \right\}. \tag{2.7.248}$$

Note that

$$\frac{1}{2p_1(x)} \left(\frac{\dot{p}_1}{p_1} \right)' (x) \propto \frac{V'(x)}{(V(x) - E)^{5/2}}, \tag{2.7.249}$$

and we will assume that this vanishes as $|x| \to \infty$ (this is the case for polynomial potentials, for example). We then obtain,

$$\int_{-\infty}^{\infty} \psi_+(x, E)\psi_-(x, E)\mathrm{d}x = a(E) \int_{-\infty}^{\infty} \frac{\mathrm{d}x}{\sqrt{V(x) - E}} - 2\dot{a}(E). \tag{2.7.250}$$

On the other hand, by using the explicit expression for the resolvent (2.7.241), we find,

$$\mathrm{Tr}\, G(E) = -\frac{1}{2a(E)} \int_{-\infty}^{\infty} \psi_+(x, E)\psi_-(x, E)\mathrm{d}x, \tag{2.7.251}$$

and we conclude that

$$G(E) = \frac{d}{dE} \log a(E) - \frac{1}{\hbar}T(E), \qquad (2.7.252)$$

where

$$T(E) = m \int_{-\infty}^{\infty} \frac{dx}{\sqrt{2m(V(x) - E)}}, \qquad (2.7.253)$$

and we have restored the dependence on m and $\hbar$.

We can now use the all-orders WKB method to write down a complete asymptotic expansion for $\psi_+(x, E)$. Indeed, let us consider

$$\psi_+(x, E) = \frac{1}{\sqrt{P(x)}} \exp\left\{-i\int_0^x p(x')dx' - i\int_\infty^x (P(x') - p(x'))\, dx'\right\}, \qquad (2.7.254)$$

where $P(x)$ is the formal power series in (2.2.7). It is of the form (2.2.20) (i.e. it solves the Schrödinger equation at all orders in $\hbar$) and it has the asymptotic behavior indicated at (2.7.233). By looking at its behavior as $x \to -\infty$, we deduce that

$$a(E) = \exp\left\{\frac{i}{\hbar}\int_{-\infty}^{\infty} (P(x) - p(x))\, dx\right\} = \exp\left\{i\sum_{n\geq 1} \hbar^{2n-1} \int_{-\infty}^{\infty} Y_{2n}(x)dx\right\}, \qquad (2.7.255)$$

where we have restored the $\hbar$ factors. This is an asymptotic expansion for $\hbar$ small or E negative and large. From this expression, it follows that

$$\lim_{E \to -\infty} a(E) = 1, \qquad (2.7.256)$$

which can be regarded as a boundary condition for $a(E)$. We can now integrate (2.7.252) to obtain

$$a(E) = \exp\left\{\int_{-\infty}^{E} \left(G(E') + \frac{1}{\hbar}T(E')\right)dE'\right\}. \qquad (2.7.257)$$

Finally, by using (2.7.224), we obtain the following expression for the Fredholm determinant

$$\Xi(E) = a(E) \exp\left\{-\frac{1}{\hbar}\int_0^E T(E')dE' - \int_{-\infty}^0 \left(G(E') + \frac{1}{\hbar}T(E')\right)dE'\right\}. \qquad (2.7.258)$$

Example 2.7.1 *Spectral functions for the harmonic oscillator.* In the case of the harmonic oscillator, we can compute the function $a(E)$ explicitly. We will use a normalization in which $m = \hbar = 1$ and $V(x) = x^2/2$. In this case,

$$\int_0^x p_1(x')dx' = \frac{1}{2}x\sqrt{x^2 - 2E} - E\log\left(x + \sqrt{x^2 - 2E}\right) + E\log\left(\sqrt{-2E}\right). \qquad (2.7.259)$$

We recall that we take $E < 0$ so that there are no real turning points on the real axis. Then, the function ψ_+ is defined by the asymptotic behavior

$$\psi_+(x, E) \approx \left(-\frac{E}{2e}\right)^{-E/2} x^{E-1/2}e^{-x^2/2}, \qquad x \to \infty. \qquad (2.7.260)$$

On the other hand, the Schrödinger equation in a quadratic potential can be solved in terms of parabolic cylinder functions, as in (2.3.54), with

$$\nu = E. \tag{2.7.261}$$

By comparing the asymptotics of $\psi_+(x, E)$ with those of $D_{\nu-1/2}(x)$ in (B.43), we deduce that

$$\psi_+(x, E) = 2^{1/4}\left(-\frac{E}{e}\right)^{-E/2} D_{\nu-1/2}(\sqrt{2}x). \tag{2.7.262}$$

On the other hand, from (2.7.259) and (2.7.234) we obtain the asymptotics as $x \to -\infty$:

$$\psi_+(x, E) \approx a(E)\left(\frac{E}{2e}\right)^{E/2} (-x)^{-E-1/2}e^{x^2/2}, \qquad x \to -\infty. \tag{2.7.263}$$

By comparing this expression with the asymptotics of (2.7.262), determined by (B.44), we deduce that

$$a(E) = \sqrt{2\pi}\left(-\frac{E}{e}\right)^{-E} \frac{1}{\Gamma\left(\frac{1}{2} - E\right)}. \tag{2.7.264}$$

This can be analytically continued to positive energies, and the zeros of $a(E)$ on the positive real axis give the spectrum of the system. Indeed, these zeros occur at

$$\frac{1}{2} - E = -n, \qquad n = 0, 1, 2, \ldots, \tag{2.7.265}$$

which is the well-known energy spectrum of the harmonic oscillator. For $E = -u < 0$, the expression (2.7.264) has a large u expansion of the form

$$\log a(-u) = -\sum_{n\geq 1}\left(2^{1-2n} - 1\right)\frac{B_{2n}}{2n(2n-1)}u^{1-2n}$$
$$= \frac{1}{24u} - \frac{7}{2880u^3} + \cdots . \tag{2.7.266}$$

In the first line of (2.7.266), B_{2n} are Bernoulli numbers. This is precisely the result expected from the WKB expansion (2.7.255). A simple analysis of $Y_{2n}(x)$ shows that it is a homogeneous function of $u^{1/2}$ and x of degree $-4n + 1$ (as prescribed by (2.7.232) for $M = 1$), and from this it follows that $\log a(-u)$ has the asymptotic expansion

$$\log a(-u) = \sum_{n\geq 1} \alpha_n \left(\frac{\hbar}{u}\right)^{2n-1}. \tag{2.7.267}$$

The coefficients α_n can be read from (2.7.266), and they can be verified at low degrees from the explicit computation of $Y_{2n}(x)$ and their integrals. For example, we have

$$Y_2(x) = i\frac{3x^2 - 4u}{8(x^2 + 2u)^{5/2}}, \tag{2.7.268}$$

where we take into account the choice of branch cut for $p(x)$, and

$$i \int_{-\infty}^{\infty} Y_2(x) dx = \frac{1}{24u},$$
(2.7.269)

in agreement with the result in (2.7.266).

A different way to compute $a(E)$ is based on the identity (2.7.252). However, in the case of the harmonic oscillator, the trace of the resolvent is not convergent. This is because H^{-1} is not a trace class operator. Therefore, (2.7.252) is not meaningful as it stands. This is easily remedied if we take a further derivative w.r.t. E in both sides. We find in this way,

$$-\mathrm{Tr} \frac{1}{(E-H)^2} = \frac{\mathrm{d}^2}{\mathrm{d}E^2} \log a(E) - \int_{-\infty}^{\infty} \frac{\mathrm{d}x}{(2(V(x)-E))^{3/2}}.$$
(2.7.270)

The trace in the l.h.s. can be computed explicitly:

$$-\mathrm{Tr} \frac{1}{(E-H)^2} = -\sum_{n=0}^{\infty} \frac{1}{(n+1/2-E)^2} = -\psi'\left(\frac{1}{2}-E\right),$$
(2.7.271)

where $\psi(z)$ is the digamma function, i.e. the logarithmic derivative of the Gamma function. We can also compute the integral,

$$\int_{-\infty}^{\infty} \frac{\mathrm{d}x}{(x^2-2E)^{3/2}} = -\frac{1}{E},$$
(2.7.272)

where we have assumed that $E < 0$. We conclude that

$$\frac{\mathrm{d}^2}{\mathrm{d}E^2} \log a(E) = -\psi'\left(\frac{1}{2}-E\right) - \frac{1}{E}.$$
(2.7.273)

This can be integrated to obtain,

$$a(E) = \exp(c_0 + c_1 E) \frac{(-E)^{-E}}{\Gamma\left(\frac{1}{2}-E\right)},$$
(2.7.274)

where c_0, c_1 are integration constants. These can be easily fixed by the boundary condition (2.7.256). In this way, we recover the result (2.7.264).

We can also determine the regularized versions of the resolvent and the spectral determinant (2.7.228) by using (2.7.229). We obtain,

$$G_2(E) = -\int_0^{\infty} \frac{\mathrm{e}^{E\beta}-1}{2\sinh\left(\frac{\beta}{2}\right)} \mathrm{d}\beta = \psi\left(\frac{1}{2}-E\right) + \gamma_E + 2\log(2),$$
(2.7.275)

where γ_E is the Euler–Mascheroni constant. As expected, this is an integral of (2.7.271) w.r.t. E. The integration constant is fixed by our choice of regularization. In addition, we find for the spectral determinant

$$\Xi_2(E) = \frac{\sqrt{\pi} \mathrm{e}^{(\gamma_E + 2\log(2))E}}{\Gamma\left(\frac{1}{2}-E\right)}.$$
(2.7.276)

$\square$

Example 2.7.2 Let us consider the Hamiltonian

$$H = p^2 + x^4. \tag{2.7.277}$$

When $E = 0$, the functions $\psi_\pm(x, 0)$ can be computed explicitly in terms of Bessel functions,

$$\psi_\pm(x, 0) = \left(\pm \frac{2x}{3\pi}\right)^{1/2} K_{1/6}\left(\pm \frac{x^3}{3}\right). \tag{2.7.278}$$

In this case,

$$p_1(x) = x^2, \qquad \int_0^x p_1(x')\mathrm{d}x' = \frac{x^3}{3}, \tag{2.7.279}$$

and we can verify that

$$\psi_+(x, 0) \sim \frac{1}{x}\mathrm{e}^{-x^3/3}, \qquad x \to \infty, \tag{2.7.280}$$

in agreement with (2.7.233). In addition,

$$\psi_+(x, 0) \sim -\frac{2}{x}\mathrm{e}^{-x^3/3}, \qquad x \to -\infty, \tag{2.7.281}$$

and we find

$$a(0) = 2. \tag{2.7.282}$$

□

2.8 Nonperturbative Effects and the Double-Well Potential

One of the most interesting applications of the Voros–Silverstone connection formula is the derivation of *exact* quantization conditions (EQCs) for confining potentials, which incorporate exponentially small corrections. A striking example occurs in the case of a symmetric double-well potential with two degenerate vacua, shown in Figure 2.9. In the classical theory, one has two degenerate vacua corresponding to the minima of the potential. This classical degeneracy is inherited by conventional

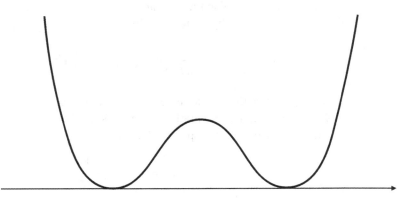

Figure 2.9   A symmetric double-well.

perturbation theory, and each energy level is doubly degenerate. However, according to the general results for confining potentials in quantum mechanics that we mentioned earlier in this chapter, there should be an infinite tower of nondegenerate energy eigenstates. Therefore, there must be a nonperturbative effect that is not visible in perturbation theory and removes the degeneracy. Indeed, there does exist such an effect, due to tunnelling across the energy barrier separating the two minima. The EQCs we will derive incorporate these tunneling effects in a very transparent way. They also allow a precise calculation of the energy levels by using techniques of Borel resummation.

We will first consider a slightly more general situation: a quartic potential that leads to an asymmetric double well with two classical vacua (not necessarily degenerate), and with four real turning points, see Figure 2.10. Let us study the spectral problem for this potential by using the WKB method. In regions V and I, the all-orders WKB wavefunctions have to decay at infinity, therefore they have the form,

$$\psi_I(x) = \frac{A}{\sqrt{P_1(x)}} \exp\left(-\frac{1}{\hbar} \int_x^a P_1(x')\mathrm{d}x'\right),$$
$$\psi_V(x) = \frac{B}{\sqrt{P_1(x)}} \exp\left(-\frac{1}{\hbar} \int_d^x P_1(x')\mathrm{d}x'\right),$$

$$(2.8.283)$$

where A, B are constants. Let us first connect region I to region III. By using the connection formula (2.4.126) around the turning point a, we obtain in region II,

$$\psi_{II}(x) = \frac{2A}{\sqrt{P(x)}} \cos\left(\frac{1}{\hbar} \int_a^x P(x')\mathrm{d}x' - \frac{\pi}{4}\right).$$

$$(2.8.284)$$

This can be written as

$$\psi_{II}(x) = \frac{2A}{\sqrt{P}} \left\{ \sin\left(\frac{1}{\hbar} \int_a^b P(x')\mathrm{d}x'\right) \cos\left(\frac{1}{\hbar} \int_x^b P(x')\mathrm{d}x' - \frac{\pi}{4}\right) \right.$$
$$\left. - \cos\left(\frac{1}{\hbar} \int_a^b P(x')\mathrm{d}x'\right) \cos\left(\frac{1}{\hbar} \int_x^b P(x')\mathrm{d}x' - \frac{\pi}{4} - \frac{\pi}{2}\right) \right\}.$$

$$(2.8.285)$$

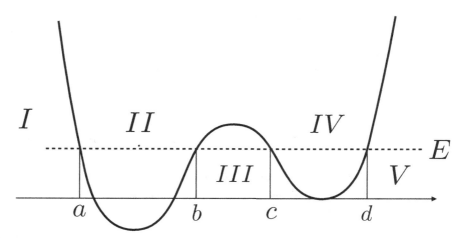

Figure 2.10 An asymmetric double-well.

Applying the connection formula (2.4.127) to move from region II to region III, we find

$$\widetilde{\psi}_{III}(x) = \pm \, \mathrm{i} \frac{A}{\sqrt{P_1(x)}} \exp\left(\mp\frac{\mathrm{i}}{\hbar} \int_a^b P(x')\mathrm{d}x'\right) \exp\left(-\frac{1}{\hbar}\int_b^x P_1(x')\mathrm{d}x'\right)$$
$$+ \frac{2A}{\sqrt{P_1(x)}} \cos\left(\frac{1}{\hbar}\int_a^b P(x')\mathrm{d}x'\right) \exp\left(\frac{1}{\hbar}\int_b^x P_1(x')\mathrm{d}x'\right). \tag{2.8.286}$$

The $\pm$ signs correspond to the two choices of Borel lateral resummation. A similar procedure, coming this time from region V, gives the wavefunction

$$\psi_{III}(x) = \pm \, \mathrm{i} \frac{B}{\sqrt{P_1(x)}} \exp\left(\mp\frac{\mathrm{i}}{\hbar} \int_c^d P(x')\mathrm{d}x'\right) \exp\left(-\frac{1}{\hbar}\int_x^c P_1(x')\mathrm{d}x'\right)$$
$$+ \frac{2B}{\sqrt{P_1(x)}} \cos\left(\frac{1}{\hbar}\int_c^d P(x')\mathrm{d}x'\right) \exp\left(\frac{1}{\hbar}\int_x^c P_1(x')\mathrm{d}x'\right). \tag{2.8.287}$$

Let us know write (2.8.286) as

$$\widetilde{\psi}_{III}(x) = \pm\mathrm{i}\frac{A}{\sqrt{P_1(x)}} \exp\left(\mp\frac{\mathrm{i}}{\hbar}\int_a^b P(x')\mathrm{d}x'\right)$$
$$\times \exp\left(-\frac{1}{\hbar}\int_b^c P_1(x')\mathrm{d}x'\right) \exp\left(\frac{1}{\hbar}\int_x^c P_1(x')\mathrm{d}x'\right)$$
$$+ \frac{2A}{\sqrt{P_1(x)}} \cos\left(\frac{1}{\hbar}\int_a^b P(x')\mathrm{d}x'\right) \exp\left(\frac{1}{\hbar}\int_b^c P_1(x')\mathrm{d}x'\right)$$
$$\times \exp\left(-\frac{1}{\hbar}\int_x^c P_1(x')\mathrm{d}x'\right). \tag{2.8.288}$$

We can now compare it to (2.8.287), and by identifying the coefficients of the leading and subleading exponentials, we find two equations

$$\pm \mathrm{i}A \exp\left(\mp\frac{\mathrm{i}}{\hbar}\int_a^b P(x)\mathrm{d}x\right)\exp\left(-\frac{1}{\hbar}\int_b^c P_1(x)\mathrm{d}x\right) = 2B\cos\left(\frac{1}{\hbar}\int_c^d P(x)\mathrm{d}x\right),$$
$$2A \cos\left(\frac{1}{\hbar}\int_a^b P(x)\mathrm{d}x\right)\exp\left(\frac{1}{\hbar}\int_b^c P_1(x)\mathrm{d}x\right) = \pm\mathrm{i}B \exp\left(\mp\frac{\mathrm{i}}{\hbar}\int_c^d P(x)\mathrm{d}x\right).$$
$$\tag{2.8.289}$$

After dividing one equation by the other and simplifying, we finally obtain

$$\left(1 + \exp\left(\pm\frac{2\mathrm{i}}{\hbar}\int_a^b P(x)\mathrm{d}x\right)\right)\left(1 + \exp\left(\pm\frac{2\mathrm{i}}{\hbar}\int_c^d P(x)\mathrm{d}x\right)\right)$$
$$+ \exp\left(-\frac{2}{\hbar}\int_b^c P_1(x)\mathrm{d}x\right) = 0. \tag{2.8.290}$$

This is the EQC for the quartic potential shown in Figure 2.10.

In the case of the symmetric double-well potential shown in Figure 2.9, we can simplify the EQC (2.8.290) to

$$4\cos^2\left(\frac{1}{\hbar}\int_a^b P(x)\mathrm{d}x\right) + \exp\left(-\frac{2}{\hbar}\int_b^c P_1(x)\mathrm{d}x \mp \frac{2\mathrm{i}}{\hbar}\int_a^b P(x)\mathrm{d}x\right) = 0. \tag{2.8.291}$$

Extracting the square root, we find

$$2\cos\left(\frac{1}{\hbar}\int_a^b P(x)\mathrm{d}x\right) = \sigma\mathrm{i}\exp\left(-\frac{1}{\hbar}\int_b^c P_1(x)\mathrm{d}x \mp \frac{\mathrm{i}}{\hbar}\int_a^b P(x)\mathrm{d}x\right), \quad (2.8.292)$$

where $\sigma = \pm 1$. This sign is related to the parity of the states ϵ (where $\epsilon = \pm 1$ for even, respectively odd states), and to the choice of lateral resummation. It is easy to see from the explicit expression for $\psi_{III}(0)$, $\psi'_{III}(0)$ that $\sigma = \pm\epsilon$, where the $\pm$ sign corresponds to the choice of lateral resummation. Our final expression for the EQC in the symmetric case is

$$1 + \mathrm{e}^{\pm 2\pi\mathrm{i}\nu} = \pm\epsilon\mathrm{i}f(\nu), \quad (2.8.293)$$

where

$$\nu = \frac{1}{\pi\hbar}\int_a^b P(x)\mathrm{d}x, \quad (2.8.294)$$

and

$$f(\nu) = \exp\left(-\frac{1}{\hbar}\int_b^c P_1(x)\mathrm{d}x\right). \quad (2.8.295)$$

Let us explain in detail the meaning of these equations. We first recall that the all-orders quantization condition (2.5.165) involves what we called a quantum period, i.e. a contour integral of $P(x)$, where the contour goes around the minimum of the potential. In the case of the symmetric double-well potential, the analogue of this quantum period is (2.8.294), and the corresponding contour $\mathcal{A}$ encircles the interval $[a, b]$. More precisely, we can write

$$\nu = \frac{1}{2\pi\hbar}\Pi_{\mathcal{A}}, \qquad \Pi_{\mathcal{A}} = 2\int_a^b P(x)\mathrm{d}x, \quad (2.8.296)$$

and we will refer to $\Pi_{\mathcal{A}}$ as the *perturbative quantum period*. It depends on the energy E and on the parameters of the potential. Conversely, we can use (2.8.294) to obtain the energy as a function of ν, so that (2.8.294) defines

$$E = E(\nu), \quad (2.8.297)$$

as a formal power series in $\hbar$. The quantity $f(\nu)$ in (2.8.295) is the exponential of another contour integral. The contour, which we will denote by $\mathcal{B}$, goes around the interval $[b, c]$ associated to tunneling. We will write

$$f(\nu) = \exp\left(-\frac{1}{2\hbar}\Pi_{\mathcal{B}}\right), \qquad \Pi_{\mathcal{B}} = 2\int_b^c P_1(x)\mathrm{d}x, \quad (2.8.298)$$

and we will call $\Pi_{\mathcal{B}}$ the *nonperturbative quantum period*. As our notation indicates, it is useful to think as $f(\nu)$ as a function of ν, instead of the energy. This function is exponentially small in $\hbar$ and quantifies the effects of tunneling.

If we neglect $f(\nu)$, the quantization condition (2.8.293) reduces to

$$\nu = n + \frac{1}{2}, \qquad n \in \mathbb{Z}_{\geq 0}. \quad (2.8.299)$$

This is the all-orders quantization condition (2.5.165), which we will call the *perturbative* quantization condition. The first thing we should note is that ν is an infinite formal power series in $\hbar^2$, so in order to make sense of (2.8.299) we have to explain how to sum this series. As we mentioned in (2.5.178), its coefficients grow like $(2k)!$, so the series has zero radius of convergence. One way to make sense of it is by using an appropriate truncation. For example, if we just keep the first term, we obtain the BS quantization condition. One can improve on this and use optimal truncation, a method explained in Appendix A. By using the truncated series we find only an approximate spectrum of energies. However, there is an deeper problem with this procedure: any solution of the truncated perturbative quantization condition, labeled by a choice of quantum number n, is doubly degenerate, since it is the same for the two different choices of the parity ϵ appearing in (2.8.293). In other words, when using a truncated version of the quantization condition (2.8.299), for each choice of n we have a symmetric and an antisymmetric wavefunction with the same energy. Each energy level obtained in this scheme is doubly degenerate. As we mentioned at the beginning of this section, this is not possible in the exact quantum theory, since parity symmetry cannot be spontaneously broken, and it is an artefact of the truncated version of the perturbative quantization condition.

One can then try to solve this problem by going beyond the truncated version of the quantum period and make sense of ν by using the Borel resummation method explained in Appendix A. This is in fact what we should do, since the Voros–Silverstone connection formula requires the Borel resummation of the $\hbar$ expansion. However, it turns out that ν is not Borel summable, and one has to use lateral Borel resummations. In other words, from the point of Borel resummation, the perturbative quantization condition is ambiguous.

In order to solve these problems, and to obtain the right picture of the spectrum, we have to appeal to the exponentially small corrections in $\hbar$ appearing in the EQC (2.8.293). These corrections are due to the quantum period Π_B around the tunneling contour, which is in the exponent of $f(\nu)$. This quantum period is also given by a factorially divergent power series in $\hbar^2$, but this time the series is Borel resummable. How should we use the EQC (2.8.293)? We first note that (2.8.293) gives two different equations, corresponding to the two choices of lateral Borel resummation of ν in the l.h.s. In the r.h.s., one uses the the conventional Borel resummation of Π_B. The solution to these equations is an infinite discrete set of values for the energy. The two equations turn out to be equivalent, in the sense that they lead to the same set of energies. This is of course required by physical consistency: to implement the EQC (2.8.293) we have to make a nonphysical choice of lateral Borel resummation, but the physical spectrum of energies does not depend on this choice. Note that the resulting spectrum does depend on the parity of the state, which appears explicitly in the r.h.s.: thanks to the inclusion of the exponentially small corrections, the spectrum that is obtained from the EQC is no longer degenerate. Moreover, it turns out to be the *exact* spectrum of the Hamiltonian.

In order to understand more precisely the effect of $f(\nu)$, let us calculate the energy levels from the exact quantization condition (2.8.293), by regarding $f(\nu)$ as a small perturbation. First of all, we can regard (2.8.294) as establishing a relationship between ν and the energy, as we remarked before. In particular, it defines a function $E = E(\nu)$. Let us then write the solution to (2.8.293) in the form

$$\widehat{\nu} = \nu + \Delta\nu, \qquad (2.8.300)$$

where v is given by (2.8.299) and Δv satisfies the equation

$$\Delta v = \frac{i}{2\pi} \log\left(1 + i\epsilon\lambda f(v + \Delta v)\right). \qquad (2.8.301)$$

We have introduced a parameter λ to keep track of the powers of $f(v)$, and we will set it to 1 at the end of the calculation. We now solve for Δv as a formal series in λ,

$$\Delta v = \sum_{k \geq 1} \Delta v^{(k)} \lambda^k. \qquad (2.8.302)$$

From (2.8.301) one easily obtains,

$$\Delta v^{(1)} = -\frac{\epsilon}{2\pi} f(v),$$
$$\Delta v^{(2)} = \frac{i}{4\pi} f^2(v) + \frac{1}{4\pi^2} f'(v) f(v), \qquad (2.8.303)$$

and so on. The series for $\widehat{v}$ can now be plugged in the power series for the energy. In this way we obtain,

$$E_{n,\epsilon} = E(\widehat{v}) = E(v) + \Delta v \frac{\partial E}{\partial v} + \frac{1}{2}(\Delta v)^2 \frac{\partial^2 E}{\partial v^2} + \cdots \qquad (2.8.304)$$

This can be written in turn as a series in λ,

$$E_{n,\epsilon} = \sum_{k \geq 0} E^{(k)}(v) \lambda^k, \qquad (2.8.305)$$

and one finds, for the very first orders,

$$E^{(0)}(v) = E(v),$$

$$E^{(1)}(v) = -\frac{\epsilon}{2\pi} f(v) \frac{\partial E}{\partial v},$$

$$E^{(2)}(v) = \frac{i}{4\pi} f^2(v) \frac{\partial E}{\partial v} + \frac{\hbar^2}{8\pi^2} \frac{\partial}{\partial v}\left(f^2(v) \frac{\partial E}{\partial v}\right),$$

$$E^{(3)}(v) = \frac{\epsilon}{6\pi} f^3(v) \frac{\partial E}{\partial v} - \frac{i\epsilon}{8\pi^2} \frac{\partial}{\partial v}\left(f^3(v) \frac{\partial E}{\partial v}\right) - \frac{\epsilon}{48\pi^3} \frac{\partial^2}{\partial v^2}\left(f^3(v) \frac{\partial E}{\partial v}\right).$$
$$(2.8.306)$$

We see that the corrections to the perturbative energy $E^{(0)}(v)$ depend on the parity ϵ, and the degeneracy is removed by $f(v)$. In particular, there is a splitting of the energy levels according to their parity. The energy gap between the levels of opposite parity is

$$E_{n,+} - E_{n,-} = -\frac{1}{\pi} f(v) \frac{\partial E}{\partial v} + \cdots, \qquad (2.8.307)$$

which is exponentially small in $\hbar$ and is due to tunneling across the forbidden region between the two wells.

Let us now obtain concrete values for the functions $E(v)$ and $f(v)$. The symmetric double-well potential can be taken to be

$$V(x) = \frac{1}{2}x^2(1 + gx)^2. \qquad (2.8.308)$$

This form of the potential makes it manifest that it is a perturbation of the harmonic oscillator. The potential has two degenerate minima at $x = 0$ and $x = -1/g$, and it is symmetric w.r.t. $x = -1/(2g)$. We consider a positive energy smaller than the height of the barrier $32g^2E < 1$, so that there are four turning points. The WKB expansion gives

$$E(v) = \sum_{k \geq 0} E_{(k)}(v)\hbar^{2k}. \tag{2.8.309}$$

(The series $E_{(k)}(v)$ appearing here shouldn't be confused with the functions $E^{(k)}(v)$ introduced above.) This expansion is obtained by inverting the relation (2.8.294). At leading order in $\hbar$, (2.8.294) reads

$$\hbar v = \frac{1}{g^2} \frac{\sqrt{1 + \sqrt{32\xi}}}{12\pi} \left(E(k(\xi)) + \left(\sqrt{32\xi} - 1 \right) K(k(\xi)) \right) + \mathcal{O}(\hbar^2), \tag{2.8.310}$$

where we have introduced

$$\xi = g^2 E, \tag{2.8.311}$$

and the argument of the elliptic functions $E(k)$, $K(k)$ in (2.8.310) is

$$k^2(E) = 2 - \frac{2}{1 + \sqrt{32\xi}}. \tag{2.8.312}$$

The function in the r.h.s. of (2.8.310) can be expanded as a series in ξ around $\xi = 0$, and we obtain

$$g^2\hbar v = \xi + 3\xi^2 + 35\xi^3 + \frac{1155}{2}\xi^4 + \frac{45045}{4}\xi^5 + \mathcal{O}(\xi^6, \hbar^2). \tag{2.8.313}$$

This expansion is equivalent to a perturbative expansion in the coupling constant g^2. In fact, by inverting this series, we find

$$E_{(0)}(v) = \hbar v - 3g^2(\hbar v)^2 - 17g^4(\hbar v)^3 - \frac{375}{2}g^6(\hbar v)^4 - \frac{10689}{4}g^8(\hbar v)^5 + \mathcal{O}(v^6). \tag{2.8.314}$$

This is the leading approximation to the standard perturbative series expansion for the energy, as we discussed in the case of the quartic oscillator in Example 2.6.2. Indeed, by doing standard perturbation theory in the coupling constant g^2, one finds

$$E(v) = \hbar v - \left(3v^2 + \frac{1}{4}\right)\hbar^2 g^2 - \left(17v^3 + \frac{19v}{4}\right)g^4\hbar^3$$
$$- \left(\frac{375v^4}{2} + \frac{459v^2}{4} + \frac{131}{32}\right)g^6\hbar^4 \tag{2.8.315}$$
$$- \left(\frac{10689v^5}{4} + \frac{23405v^3}{8} + \frac{22709v}{64}\right)g^8\hbar^5 + \mathcal{O}(g^{10}).$$

One can do a similar calculation for the exponent of (2.8.295). However, for this quantity, a new phenomenon appears: when written in terms of v, all $\hbar^2$ corrections to the nonperturbative quantum period contribute to the leading-order term. Indeed, a straightforward but somewhat tedious calculation by using elliptic functions leads to the result

$$
\frac{2}{\hbar} \int_b^c P_1(x)\,\mathrm{d}x
$$

$$
= \frac{1}{\hbar} \left(2v\hbar \log\left(\frac{vg^2\hbar}{2}\right) - 2v\hbar + \frac{1}{3g^2} + 17v^2g^2\hbar^2 + \cdots \right)
$$

$$
+ \hbar \left(-\frac{1}{12v\hbar} + \frac{19g^2}{12} + \frac{153v\hbar g^4}{4} + \frac{23405v^2\hbar^2 g^6}{24} + \cdots \right)
$$

$$
+ \hbar^3 \left(\frac{7}{1440v^3\hbar^3} + \frac{22709}{576}g^6 + \frac{217663v\hbar g^8}{64} + \frac{61936297v^2\hbar^2 g^{10}}{22320} + \cdots \right)
$$

$$
+ \mathcal{O}(\hbar^5).
$$

(2.8.316)

Therefore, in order to make progress, we have to resum the series:

$$
2v \log\left(\frac{vg^2\hbar}{2}\right) - 2v - \frac{1}{12v} + \frac{7}{1440v^3} + \mathcal{O}(v^{-5}).
$$

(2.8.317)

Fortunately, this can be done. To see this, and to verify other aspects of the above analysis, it is useful to re-derive the quantization condition for the symmetric double-well potential (2.8.308) by using the uniform WKB method. In this calculation we will set $m = \hbar = 1$.

We first change coordinates in the Schrödinger equation,

$$
z = gx,
$$

(2.8.318)

so that it reads

$$
-\frac{g^4}{2}\psi''(z) + \frac{1}{2}z^2(1+z)^2\psi(z) = g^2 E\psi(z).
$$

(2.8.319)

Since the potential is quadratic around the minima, at leading order, it is natural to use a uniform WKB approximation similar to (2.3.54). We then write,

$$
\psi(z) = \frac{1}{\sqrt{u'(z)}} D_{v-1/2}\left(\frac{\sqrt{2}}{g}u(z)\right).
$$

(2.8.320)

This is the uniform WKB approximation used in (2.3.54), with $\xi = g^2$. The function $u(z)$ satisfies the analogue of equation (2.3.66), namely,

$$
u'(z)^2 u^2(z) = z^2(1+z)^2 - 2g^2\left(E - vu'(z)^2\right) + \frac{g^4}{2}\{u, z\}.
$$

(2.8.321)

We solve this equation as a formal power series in g^2, so we set

$$
u(z) = \sum_{n\geq 0} u_n(z)g^{2n}, \qquad E = \sum_{n\geq 0} E_n(v)g^{2n},
$$

(2.8.322)

and one finds at leading order

$$
u_0(z) = z\left(1 + \frac{2z}{3}\right)^{1/2}.
$$

(2.8.323)

As in Example 2.3.1, the behavior $u_0(z) \sim z$ near $z = 0$ follows from (2.8.321). The next orders can be solved recursively. The values of $E_n(v)$ can be found by requiring regularity of the solution $u(z)$ at $z = 0$, as in Example 2.3.1. Let us consider for example the next-to-leading order. The equation to solve is (2.3.75), which is immediately integrated to

$$u_0(z)u_1(z) = v \log u_0(z) - E_0(v) (\log z - \log(1 + z)) + C, \qquad (2.8.324)$$

where we have used (2.8.323) and C is an integration constant. Regularity at $z = 0$ requires $C = 0$ and

$$E_0(v) = v. \qquad (2.8.325)$$

The solution is then,

$$u_1(z) = \frac{v}{2} \frac{\log\left[(1 + z)^2 (1 + 2z/3)\right]}{z(1 + 2z/3)^{1/2}}. \qquad (2.8.326)$$

Higher orders can be computed similarly. The resulting series for $E(v)$ is identical to the perturbative series obtained from standard Rayleigh–Schrödinger stationary perturbation theory, where v is related to the quantum number through the equation,

$$v = n + \frac{1}{2}. \qquad (2.8.327)$$

Let us now find the quantization conditions. The ansatz for the wavefunction (2.8.320) already has the right behavior as $z \to +\infty$, since $u(z) \to +\infty$. Since the potential is even w.r.t. $z = -1/2$, we can then impose the solutions to be even or odd in $z + 1/2$. This gives the condition

$$\psi'(-1/2) = 0 \qquad (2.8.328)$$

for even solutions, and

$$\psi(-1/2) = 0 \qquad (2.8.329)$$

for odd solutions. Let us first look at the odd solutions, which are easier to analyze. Since $u_0(z)$ is negative for $z = -1/2$, we have to study the parabolic cylinder equation in a Stokes direction. We then use the asymptotic expansion (B.44), where the $\pm$ signs correspond to different choices of lateral resummation. We will denote for simplicity

$$F_1(z^2) = {}_2F_0\left(\frac{1}{4} + \frac{v}{2}, \frac{3}{4} - \frac{v}{2}; ; 2z^{-2}\right), \qquad F_2(z^2) = {}_2F_0\left(\frac{1}{4} - \frac{v}{2}, \frac{3}{4} + \frac{v}{2}; ; -2z^{-2}\right).$$
$$(2.8.330)$$

If we use the reflection formula for the Gamma function,

$$\Gamma\left(\frac{1}{2} - v\right)\Gamma\left(\frac{1}{2} + v\right) = \frac{\pi}{\cos(\pi v)}, \qquad (2.8.331)$$

the condition that $\psi(z)$ vanishes at $z = -1/2$ becomes

$$\mp i f(v) = 1 + e^{\pm 2\pi i v}, \qquad (2.8.332)$$

where the $\pm$ sign here corresponds to the $\pm$ sign in (B.44), and

$$f(v) = \frac{\sqrt{2\pi}}{\Gamma\left(\frac{1}{2} + v\right)} \exp\left[-\frac{u^2(-1/2)}{g^2}\right] \left(\frac{2u^2(-1/2)}{g^2}\right)^v \frac{F_2(2u^2(-1/2)/g^2)}{F_1(2u^2(-1/2)/g^2)}. \qquad (2.8.333)$$

If we use the explicit values of $u_0(z)$, $u_1(z)$ obtained above, it is easy to see that $f(v)$ can be written as

$$f(v) = \frac{\sqrt{2\pi}}{\Gamma\left(\frac{1}{2}+v\right)}\left(\frac{2}{g^2}\right)^v e^{-A(v)/2}, \tag{2.8.334}$$

where $A(v)$ has the form

$$A(v) = \frac{1}{3g^2} + \sum_{k \geq 1} c^{(k)}(v) g^{2k}. \tag{2.8.335}$$

Although it is not manifest from the above calculation, the coefficients $c^{(k)}(v)$ are independent of z. This is required for consistency of the whole procedure, since $f(v)$ enters into the quantization condition (2.8.332) and cannot depend on z. The coefficients $c^{(k)}(v)$ can be obtained in a systematic way from the solution for the $u_n(z)$, and the explicit formula (2.8.333), and they are polynomials in v with rational coefficients. We list here the very first values,

$$c^{(1)}(v) = \frac{19}{12} + 17v^2,$$

$$c^{(2)}(v) = \frac{153v}{4} + 125v^3, \tag{2.8.336}$$

$$c^{(3)}(v) = \frac{22709}{576} + \frac{23405v^2}{24} + \frac{17815v^4}{12}.$$

A similar structure is found in the uniform WKB analysis of the cubic oscillator, which is presented in Section 5.4, and in particular in Example (5.4.1).

Let us now pause to compare these results with the ones obtained with the Voros–Silverstone connection formula. First of all, since

$$-2\log\left(\frac{\sqrt{2\pi}}{\Gamma\left(v+\frac{1}{2}\right)}\right) \sim 2v\left(\log(v)-1\right) - \frac{1}{12v} + \frac{7}{1440v^3} + \mathcal{O}\left(v^{-5}\right), \qquad v \gg 1, \tag{2.8.337}$$

we see that the expression (2.8.334) for $f(v)$ successfully resums the series (2.8.317). In addition, the values obtained for (2.8.336) agree with the results in (2.8.316). The quantization condition (2.8.332) is precisely (2.8.293) for odd states $\epsilon = -1$, where the choice of lateral resummation in the Voros–Silverstone connection formula corresponds to the choice of lateral resummation in the asymptotic expansion (B.44). It can easily be shown that the quantization condition for even states leads to the result (2.8.332) with an additional minus sign in $f(v)$, in complete agreement with (2.8.293).

Let us note that, although the WKB calculation is naturally organized as a series in $\hbar^2$, in the calculation that uses the uniform WKB method it is natural to perform an expansion in the coupling constant g^2. This expansion is used in both the perturbative energy $E(v)$ and in the exponent of the nonperturbative correction, $A(v)$. Clearly, we can always obtain the expansions in g^2 of these functions by using the WKB expansions in $\hbar^2$, up to the required order, and re-expanding them in g^2.

The determination of the coefficients $c^{(k)}(v)$, which enter into the *nonperturbative* quantity $f(v)$, involves a relatively complex procedure. However, it has been noted that the $c^{(k)}(\nu)$ are related in a simple way to the coefficients of the *perturbative* ground state energy (2.8.315). This relation can be written as

$$\frac{\partial E}{\partial \nu} = -6g^2 \nu - 3g^4 \frac{\partial A}{\partial g^2}, \tag{2.8.338}$$

after setting $\hbar = 1$. This is sometimes called in the literature a "P/NP" relation, since it connects a perturbative quantity (the perturbative expansion of the ground state energy appearing in the l.h.s. of (2.8.338)) to a non-perturbative quantity (the exponent in (2.8.334) appearing in the r.h.s. of (2.8.338)).

As a first application of the above results, we can use the expression (2.8.334), together with (2.8.307), to give the energy gap between the ground state and the first excited state,

$$E_{0,+} - E_{0,-} \approx -\frac{2}{\sqrt{\pi} g} e^{-\frac{1}{6g^2}}, \qquad g \to 0. \tag{2.8.339}$$

In addition, we can use these results to perform precision calculations of the energy levels in the symmetric double-well potential, including both perturbative and nonperturbative corrections. Let us study in some detail the ground state, which corresponds to $\nu = 1/2$ and $\epsilon = 1$. The perturbative series can be obtained from (2.8.315), and it is given by

$$E^{(0)}\left(\frac{1}{2}\right) = \frac{1}{2} - g^2 - \frac{9g^4}{2} - \frac{89g^6}{2} - \frac{5013g^8}{8} - \frac{88251g^{10}}{8} + \mathcal{O}\left(g^{12}\right). \tag{2.8.340}$$

The calculation of this series to very high order can be done efficiently with a symbolic program such as the *Bender Wu* code of Sulejmanpasic and Ünsal (2018). The first non-perturbative correction to the ground state energy is given by

$$E^{(1)}\left(\frac{1}{2}\right) = -\frac{1}{\sqrt{\pi} g} e^{-\frac{1}{6g^2}} \left(\frac{\partial E^{(0)}}{\partial \nu}\right)_{\nu=\frac{1}{2}} \exp\left(-\frac{1}{2}\sum_{k \geq 1} c^{(k)}(1/2)g^{2k}\right)$$

$$= -\frac{1}{\sqrt{\pi} g} e^{-\frac{1}{6g^2}} \left(1 - \frac{71g^2}{12} - \frac{6299g^4}{288} - \frac{2691107g^6}{10368}\right. \tag{2.8.341}$$

$$\left. - \frac{2125346615g^8}{497664} + \mathcal{O}\left(g^{10}\right)\right).$$

We note that the coefficients $c^{(k)}(\nu)$ can be calculated to high order by using the P/NP relation (2.8.338). How do we extract actual values for the energies from these series? We should consider lateral Borel resummations of *all* the series involved. The direction in which we perform the lateral Borel resummation should be correlated with the choice of sign in the quantization condition. In this case, since the final answer for the energies should be real, we make a choice that guarantees this reality condition.

To see how the procedure works, let us pick a particular value of g^2, say

$$g^2 = \frac{7}{1000}. \tag{2.8.342}$$

The lateral resummation of $E^{(0)}(1/2)$ above the real axis gives the following value

$$E^{(0)}(1/2) = 0.4927625141294221339923438509\ldots - 2.792818158\ldots \cdot 10^{-19}\mathrm{i}, \tag{2.8.343}$$

which can be obtained by Borel–Padé resummation. Note that it has a small imaginary part. The lateral Borel resummation of the first nonperturbative correction above the real axis gives,

$$E^{(1)}(1/2) = \pm 2.9486924646... \cdot 10^{-10} \mp 1.554... \cdot 10^{-27}\mathrm{i}, \qquad (2.8.344)$$

where the sign $\pm$ corresponds to odd/even states. We see that

$$\mathrm{Re}\left(E^{(0)}(1/2) - E^{(1)}(1/2)\right) = 0.492762513834552888..., \qquad (2.8.345)$$

which is the correct ground state energy at this precision. What happens to the imaginary contributions of the lateral Borel resummations? They turn out to *cancel* against the resummation of the formally imaginary terms in the higher nonperturbative corrections. Let us consider a concrete example of this cancellation. The formally imaginary piece of the second nonperturbative correction $E^{(2)}(\nu)$ is given by

$$\frac{\mathrm{i}}{4\pi}f^2(\nu)\frac{\partial E}{\partial \nu}. \qquad (2.8.346)$$

For $\nu = 1/2$, this gives the following power series in g^2:

$$\frac{\mathrm{i}}{g^2}\mathrm{e}^{-\frac{1}{3g^2}}\left(1 - \frac{53g^2}{6} - \frac{1277g^4}{72} - \frac{336437g^6}{1296} - \frac{141158555g^8}{31104} + \mathcal{O}\left(g^{10}\right)\right).$$
$$(2.8.347)$$

The imaginary part of the lateral Borel resummation of this series is

$$2.7928181575... \cdot 10^{-19}\mathrm{i}, \qquad (2.8.348)$$

which cancels the imaginary part of (2.8.343) up to order 10^{-27}. At this point, one has to consider the imaginary contribution of (2.8.344), as well as contributions coming from higher nonperturbative corrections $E^{(k)}(\nu)$ in (2.8.305), with $k \geq 3$. Proceeding in this way, one obtains a real result that agrees with the exact energy eigenvalue.

The procedure we have sketched to obtain the exact spectrum uses the EQC (2.8.293) but it is, in practice, based on power series in the coupling constant g and their resummations. It is possible to obtain the spectrum from EQC by using (lateral) Borel resummations of the quantum periods, understood as series in $\hbar^2$, as we explained in the discussion around (2.5.178) (and in the original spirit of the exact connection formulae). Examples of this procedure can be found in the bibliographical references in Section 2.13.

2.9 Nonperturbative Effects and Large-Order Behavior

One important application of the calculation of nonperturbative corrections is the determination of the large-order behavior of the original, perturbative series. Let us present the general argument for this connection, which we will then make concrete in the case of the symmetric double-well potential. We will present other applications of this general principle in Chapter 5. As in Sections 2.4 and 2.8, we will use results from the theory of Borel resummation presented in Appendix A.

Let us consider a formal power series of the form

$$\varphi(z) = \sum_{n=0}^{\infty} a_n z^n. \qquad (2.9.349)$$

Let us assume that its Borel transform $\widehat{\varphi}(\zeta)$, defined in (A.15), has a single singularity at $\zeta = A > 0$, which leads to a discontinuity of the form

$$\mathrm{disc}(\varphi)(z) = \mathrm{i}\, e^{-A/z} z^{-b} \sum_{n=0}^{\infty} c_n z^n. \qquad (2.9.350)$$

The coefficients of the Borel transform are given by the Cauchy formula

$$\frac{a_k}{k!} = \frac{1}{2\pi\mathrm{i}} \oint_{\mathcal{C}_0} \frac{\widehat{\varphi}(\zeta)}{\zeta^{k+1}} d\zeta, \qquad (2.9.351)$$

where $\mathcal{C}_0$ is a small circle around $\zeta = 0$. Let us choose $\delta > 0$. We now enlarge the contour $\mathcal{C}_0$ to a contour $\mathcal{C}_{A+\delta} \cup \mathcal{H}_A$, where $\mathcal{C}_{A+\delta}$ is a circle of radius $A + \delta$, minus an arc, and $\mathcal{H}_A$ is a Hankel contour centered around A (see Figure 2.11). By deforming the contour we find

$$\frac{a_k}{k!} = \frac{1}{2\pi\mathrm{i}} \oint_{\mathcal{C}_{A+\delta}} \frac{\widehat{\varphi}(\zeta)}{\zeta^{k+1}} d\zeta + \frac{1}{2\pi\mathrm{i}} \oint_{\mathcal{H}_A} \frac{\widehat{\varphi}(\zeta)}{\zeta^{k+1}} d\zeta. \qquad (2.9.352)$$

The first integral can be estimated to be of order $\mathcal{O}((A + \delta)^{-k})$. Since, as we will now show, the leading large k asymptotics go like A^{-k}, and $A + \delta > A$, this is a subleading, exponentially small correction as k grows large, and it does not contribute to the leading $1/k$ asymptotics. On the other hand, the integral around the contour $\mathcal{H}_A$ can be evaluated by using the behavior of the Borel transform near the singularity. As explained in Appendix A, it follows from the discontinuity (2.9.350) that this behavior is of the form

$$\widehat{\varphi}(A + \xi) = (-\xi)^{-b} \sum_{n\geq 0} \frac{c_n}{2\sin(\pi b)\Gamma(n+1-b)} \xi^n + \cdots, \qquad (2.9.353)$$

and we obtain

$$\frac{1}{2\pi\mathrm{i}} \oint_{\mathcal{H}_A} \frac{\widehat{\varphi}(\zeta)}{\zeta^{k+1}} d\zeta = \frac{1}{2\pi\Gamma(n+1-b)} \sum_{n\geq 0} c_n \int_0^\delta \frac{\xi^{n-b}}{(A+\xi)^{k+1}} d\xi, \qquad (2.9.354)$$

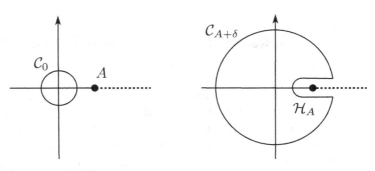

Figure 2.11 Contour deformation in (2.9.352).

where we have set $\zeta = A + \xi$ and the calculation is similar to the one in (A.37). On the other hand, an easy estimate shows that, at fixed n,

$$\int_0^\delta \frac{\xi^{n-b}}{(A+\xi)^{k+1}} \mathrm{d}\xi = \int_0^\infty \frac{\xi^{n-b}}{(A+\xi)^{k+1}} \mathrm{d}\xi + \mathcal{O}\left((A+\delta)^{-k}\right)$$

$$= A^{n-b-k} \frac{\Gamma(k+b-n)\Gamma(n-b+1)}{\Gamma(k+1)} + \mathcal{O}\left((A+\delta)^{-k}\right)$$

(2.9.355)

We conclude that the asymptotic expansion of a_k can be expressed as

$$a_k \sim \frac{1}{2\pi} \sum_{n\geq 0} A^{n-b-k} c_n \Gamma(k+b-n).$$

(2.9.356)

The above derivation can be simplified if the Borel transform $\widehat{\varphi}(z)$ decays sufficiently fast at infinity, so that we can take $\delta \to \infty$. If there are further singularities along the real axis, one can include them systematically if one knows the singular behavior of the Borel transform in their neighborhood. They lead to subleading, exponentially small corrections to the asymptotics that we have obtained.

The expression (2.9.356) should be regarded as an asymptotic expansion of a_k at large k. The very first terms are

$$a_k \sim \frac{1}{2\pi} A^{-b-k} \Gamma(k+b) \left[c_0 + \frac{c_1 A}{k+b-1} + \frac{c_2 A^2}{(k+b-2)(k+b-1)} + \cdots \right].$$

(2.9.357)

It can be easily seen that, when $b = 0$, this formula gives the result that is obtained for a Borel transform with a logarithmic singularity. If the singularity occurs on the negative real axis, $\zeta = -A$ with $A > 0$, as occurs in many interesting cases, the above derivation can be repeated verbatim, by using the discontinuity along the direction $\theta = \pi$. Let us write this discontinuity as

$$-\mathrm{disc}(\varphi)(-z) = \mathrm{i}e^{-A/z} z^{-b} \sum_{n=0}^\infty c_n z^n, \qquad z > 0.$$

(2.9.358)

The overall minus sign is a convenient convention due to the fact that, along the negative real axis, the discontinuity, as we defined it in (A.34), computes the difference between the functions below the axis and above the axis. Then, the large-order behavior is given by

$$a_k \sim \frac{1}{2\pi} (-1)^{k+1} A^{-b-k} \Gamma(k+b) \left[c_0 - \frac{c_1 A}{k+b-1} + \frac{c_2 A^2}{(k+b-2)(k+b-1)} + \cdots \right],$$

(2.9.359)

and the original series is alternating.

Example 2.9.1 Let us illustrate these results with the quartic integral (5.6.149). The leading-order behavior of the discontinuity is given in (5.6.160). The singularity of the Borel transform occurs at $\zeta = -1/4$. The result (2.9.359) gives, in this case,

$$a_k \sim \frac{1}{\pi\sqrt{2}} (-1)^k 4^k \Gamma(k),$$

(2.9.360)

which is indeed the asymptotic behavior found in (A.12). $\qquad\qquad\square$

Let us now present an application of this general result to the perturbative series of the double-well potential. We focus again on the ground state energy, although higher energy levels can be analyzed in the same way. The perturbative series is a formal power series in g^2 given by

$$E^{(0)}\left(\frac{1}{2}\right) = \sum_{k \geq 0} a_k g^{2k}. \qquad (2.9.361)$$

As we noted in Section 2.8, the imaginary part of the Borel resummation of this perturbative series is cancelled at leading order by the imaginary part of the non-perturbative correction $E^{(2)}(\nu)$, which is given in (2.8.347). It follows that the discontinuity (2.9.350) is given by the series (2.8.347) multiplied by a factor of -2. We can now apply the general formula (2.9.357) to obtain the large-order behavior of the coefficients a_k in (2.9.361). In this case, we have

$$A = \frac{1}{3}, \qquad b = 1, \qquad c_0 = -1, \qquad c_1 = \frac{53}{6}. \qquad (2.9.362)$$

Therefore, we obtain

$$a_k \sim -\frac{1}{\pi} k! \, 3^{k+1} \left(1 - \frac{53}{18k} + \mathcal{O}(k^{-2})\right), \qquad k \gg 1. \qquad (2.9.363)$$

This prediction for the asymptotic behavior can be tested in detail as follows. Let us consider the sequences

$$s_k^{(1)} = -\pi \frac{a_k}{k! \, 3^{k+1}}, \qquad s_k^{(2)} = k\left(s_k^{(1)} - 1\right). \qquad (2.9.364)$$

According to (2.9.363), the first sequence tends to 1 as $k \to \infty$, while the second sequence tends to $-53/18$. However, due to the tails in $1/k$, the convergence is relatively slow. One way to accelerate the convergence is to do a Richardson transform of the original sequence. Given a sequence with the asymptotic behavior

$$s_k \sim \sum_{n=0}^{\infty} \frac{\sigma_n}{k^n}, \qquad k \gg 1, \qquad (2.9.365)$$

its Nth Richardson transform is defined by

$$s_k^{(N)} = \sum_{\ell=0}^{N} \frac{s_{k+\ell}(k+\ell)^N (-1)^{\ell+N}}{\ell! \, (N-\ell)!}. \qquad (2.9.366)$$

The effect of this transformation is to remove the first $N-1$ subleading tails in (2.9.365), thereby resulting in a sequence that convergences to σ_0 much faster than the original one. By using Richardson transforms of the sequences $s_k^{(1)}$, $s_k^{(2)}$ to accelerate the convergence, we can indeed test their expected behavior with high precision, as shown in Figure 2.12.

Example 2.9.2 *Large-order behavior of the first nonperturbative correction.* The corrections $E^{(k)}(\nu)$, $k \geq 1$, in (2.8.306) are given by an exponential factor, times a series in g^2 and $\log(g^2)$. It turns out that the large-order behavior of the coefficients appearing in these series can be also obtained by the arguments presented above. Let us consider the first nonperturbative correction to the energy, $E^{(1)}(\nu)$. We can see in (2.8.344) that its lateral Borel resummation has a tiny imaginary part,

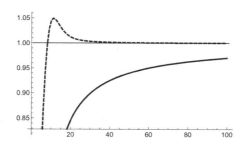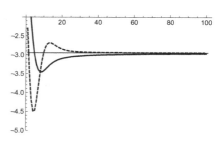

Figure 2.12 In these figures we show the sequences (left) $s_k^{(1)}$ and (right) $s_k^{(2)}$. For visual convenience we have drawn them as continuous, joined lines. We also show their first Richardson transform, as dashed, joined lines. The expected asymptotic limits (1 and $-53/18$, respectively) are shown as thin horizontal lines.

which is cancelled by the formally imaginary part of the third nonperturbative correction $E^{(3)}(\nu)$:

$$\frac{1}{8\pi^2}\frac{\partial}{\partial \nu}\left(f^3(\nu)\frac{\partial E}{\partial \nu}\right). \tag{2.9.367}$$

Let us obtain an explicit expression for (2.9.367). We write

$$f^3(\nu)\frac{\partial E}{\partial \nu} = \xi^3(\nu)b_3(\nu), \tag{2.9.368}$$

where

$$\xi(\nu) = \frac{\sqrt{2\pi}}{\Gamma\left(\frac{1}{2}+\nu\right)}\left(\frac{2}{g^2}\right)^\nu e^{-1/(6g^2)}, \tag{2.9.369}$$

and $b_3(\nu)$ can be easily calculated to be

$$b_3(\nu) = 1 - \left(\frac{19}{8} + 6\nu + \frac{51\nu^2}{2}\right)g^2 + \cdots. \tag{2.9.370}$$

In terms of these quantities, we have

$$\frac{\partial}{\partial \nu}\left(f^3(\nu)\frac{\partial E}{\partial \nu}\right) = \left\{b_3'(\nu) - 3\left(\psi\left(\nu + \frac{1}{2}\right) + \log\left(\frac{g^2}{2}\right)\right)b_3(\nu)\right\}\xi^3(\nu). \tag{2.9.371}$$

Let us now focus on the ground state. The series (2.9.367) for $\nu = 1/2$ is given by

$$\frac{e^{-\frac{1}{2g^2}}}{\pi^{1/2}g^3}\left\{3\left(\gamma_E + \log\left(\frac{2}{g^2}\right)\right) - \frac{63}{2}g^2 - 3\left(\gamma_E + \log\left(\frac{2}{g^2}\right)\right)\frac{47}{4}g^2 + \cdots\right\}. \tag{2.9.372}$$

Let us now write

$$E_0^{(1)}(g^2) = \frac{1}{\sqrt{\pi g^2}}e^{-\frac{1}{6g^2}}\sum_{k\geq 0}a_k^{(1)}g^{2k}. \tag{2.9.373}$$

We want to find the large-order behavior of the coefficients $a_k^{(1)}$. The discontinuity is now given by (2.9.372). However, since (2.9.372) involves logarithmic terms, which were not present in (2.9.350), we cannot use (2.9.357). However, we can go back to the general argument, which involves the integral of the discontinuity, as in (2.9.354). In our case this gives:

$$a_k^{(1)} \sim \frac{1}{\pi}\int_0^\infty \mathrm{Im}\, E_0^{(3)}e^{\frac{1}{6z}}\sqrt{\pi z}\frac{dz}{z^{k+1}}, \tag{2.9.374}$$

where $z = g^2$. To calculate the large k asymptotics, we use the integrals

$$\int_0^\infty e^{-A/z} z^{-k-2} dz = A^{-1-k} \Gamma(1+k),$$

$$\int_0^\infty e^{-A/z} \log(z) z^{-k-2} dz = A^{-k-1} \Gamma(k+1) \left(\log(A) - H_k + \gamma_E \right),$$

(2.9.375)

where H_k is the harmonic number, and we have the following asymptotics at large k,

$$-H_k + \gamma_E = -\log(k) - \frac{1}{2k} + \frac{1}{12k^2} + \mathcal{O}(k^{-3}).$$

(2.9.376)

We then find

$$-\frac{\pi a_k^{(1)}}{3^{k+2} \Gamma(k+1)} = \left\{ 1 - \frac{47}{12k} + \mathcal{O}(k^{-2}) \right\} \log k + \gamma_E + \log(6) + \frac{\mu_1}{k} + \mathcal{O}(k^{-2}),$$

(2.9.377)

where

$$\mu_1 = \frac{1}{2} - \left(\frac{7}{2} + \frac{47}{12} \gamma_E \right) - \log(6) \frac{47}{12}.$$

(2.9.378)

The large k asymptotic expansion (2.9.377) can be verified numerically to very high precision by using Richardson transforms.   □

2.10 Nonperturbative Effects and the Complex WKB Method

We have seen in previous examples that the WKB method gives perturbative as well as nonperturbative contributions to the exact spectrum of quantum systems. The non-perturbative corrections we have seen so far are due to tunneling effects, as in Example 2.6.3 and in the double-well potential presented in Section 2.9. It turns out that, in the exact WKB method, and in order to find the exact spectrum, one also has to consider tunneling effects in the complex plane. More generally, one has to consider period integrals for complex turning points. This extension of the exact WKB method to the full complex realm is sometimes called the *complex WKB method*. We would like to emphasize that the exact WKB method, when taken seriously, inevitably becomes the complex WKB method.

Perhaps the simplest example of the complex WKB method is the pure quartic oscillator, given by the Hamiltonian (2.6.183) with $M = 2$. The all-orders WKB method gives a perturbative period

$$\frac{1}{\hbar} \Pi_p = \sum_{n \geq 0} b_n \sigma^{1-2n},$$

(2.10.379)

where

$$\sigma = \frac{\Gamma(1/4)^2}{3} \sqrt{\frac{2}{\pi}} \frac{E^{3/4}}{\hbar}.$$

(2.10.380)

The coefficients b_n can be computed recursively, and b_0, b_1 can be obtained from the expressions (2.6.184), (2.6.185), with $M = 2$. The normalization in (2.10.380) is such that $b_0 = 1$. The formal power series (2.10.379) diverges doubly-factorially, as we noted in general in (2.5.178). We can try to perform a Borel resummation in order to obtain an EQC for the energy levels. The first problem one finds is that this series is not Borel summable. If we write

$$\frac{1}{\hbar}\Pi_{\mathrm{p}} = \sigma + \frac{1}{\sigma}\varphi(\sigma), \qquad \varphi(\sigma) = \sum_{n \geq 0} b_{n+1}\sigma^{-2n}, \qquad (2.10.381)$$

and we define the Borel transform as

$$\widehat{\phi}(\zeta) = \sum_{n \geq 0} \frac{b_{n+1}}{(2n)!}\zeta^{2n}, \qquad (2.10.382)$$

we find a singularity in the real axis at $\zeta = 1$. Therefore, we have to perform lateral Borel resummations, which will lead to an imaginary, nonperturbative ambiguity, of order $e^{-\sigma}$. In the example of the double-well potential, nonperturbative effects are due to tunneling phenomena, but in this purely quartic potential there is no tunneling on the real axis. This is the first puzzling aspect of this problem.

There is an additional puzzle. We could try to use the real part of the Borel resummation in order to find an approximation to the energy spectrum. Since there is a non-perturbative ambiguity due to the singularity in the real axis, we would expect this to leads to an error of at most order $e^{-\sigma}$. However, as noted by Balian, Parisi, and Voros, the disagreement obtained in this way is larger, of order $e^{-\sigma/2}$. What is the source of this nonperturbative correction?

One way to address this problem is to start with the double-well potential, where non-perturbative effects can be understood using the conventional WKB method, and deform it smoothly to a pure quartic oscillator by changing the coefficient of the x^2 term. As we do this, the turning points b and c in Figure 2.10, corresponding to a fixed energy E, approach each other, collapse, and then move into the imaginary axis. When we reach the pure quartic potential, the turning points a and d are given by $\mp E^{1/4}$, while the points b, c become $\pm iE^{1/4}$. Therefore, the tunneling cycle has *not* disappeared, but has moved into the complex plane. We end up with two different cycles in the complexified x-plane: one of them goes around the real turning points $\pm E^{1/4}$, while the other cycle goes around the complex turning points $\pm iE^{1/4}$, as shown in Figure 2.13.

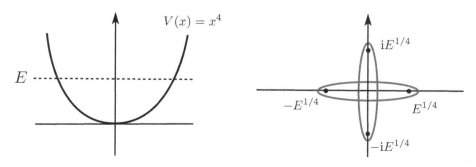

Figure 2.13 (Left) The pure quartic potential leads to (Right) two cycles in the complex x-plane: the "perturbative" one encircles the real turning points $\mp E^{1/4}$, while the "non-perturbative" one encircles the complex turning points $\mp i E^{1/4}$.

In order to have a complete description of the system with the WKB method, we should consider the quantum periods associated with both cycles. The first one is Π_p, as in (2.10.379), while the second one will be denoted as Π_{np}. It is given by a formal power series in which each term is an integral around the cycle encircling the complex turning points. The integral at order n, (2.5.166), can be computed by simply changing $x \to ix$. It is easy to see that this gives the same integral, computed around the *real* turning points, times a factor $i^{2n+1} = i(-1)^n$. The "nonperturbative" WKB period is then given by the formal power series

$$\frac{1}{\hbar}\Pi_{np} = \sum_{n \ge 0}(-1)^n b_n \sigma^{1-2n}. \tag{2.10.383}$$

It is possible to show that Π_{np} is Borel summable. As we deform the double-well potential into the pure quartic potential, the quantum periods $\Pi_{A,B}$ of the symmetric double-well potential, which we defined in (2.8.296) and (2.8.298), will become linear combinations of the periods $\Pi_{p,np}$ of the pure quartic oscillator. The precise relations can be obtained by using a generalization of the Voros–Silverstone connection formula to the complex plane, and we refer the reader to the references at the end of this chapter for a derivation. They are given by:

$$\Pi_A^{\pm} \to \frac{1}{2}\Pi_p^{\pm} \mp \frac{i}{2}\Pi_{np}, \qquad \Pi_B \to -\Pi_{np}. \tag{2.10.384}$$

In the first relation, the $\pm$ superscripts refer to the direction of lateral resummation for Π_A and Π_p. However, it is possible to understand these relations heuristically by looking at the evolution of the cycles in Figure 2.14. The configuration shown in this figure is relevant for the positive lateral resummation of the period, since the deformation of the cycles is such that the singularity in the Borel plane does not cross the integration contour $\mathcal{C}_+$. Then, at the classical level, we have that

$$\int_{-E^{1/4}}^{-iE^{1/4}} \sqrt{E - q^4}\,dq = \int_{-E^{1/4}}^{0} \sqrt{E - q^4}\,dq - i\int_{-E^{1/4}}^{0} \sqrt{E - q^4}\,dq, \tag{2.10.385}$$

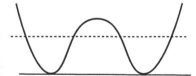

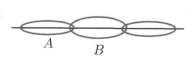
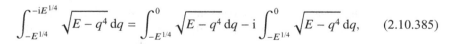
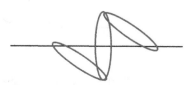

A　　B

Figure 2.14　　The double-well potential can be deformed into a pure quartic potential. In this figure we show one possible deformation path for the cycles, appropriate for one of the choices of lateral resummation.

which leads to the first relation in (2.10.384) (for the positive lateral resummation). We also have that

$$\int_{-iE^{1/4}}^{iE^{1/4}} \sqrt{q^4 - E}\, dq = -\int_{-E^{1/4}}^{E^{1/4}} \sqrt{E - q^4}\, dq, \tag{2.10.386}$$

where we have made the appropriate choice of branch cut in the square root. This leads to the second relation in (2.10.384). The extension of the Voros–Silverstone formula to the complex realm essentially provides a systematic way of keeping track of the signs appearing in the process.

The quantization condition for the pure quartic potential can now be obtained from that of the double-well potential (2.8.293) by implementing the transformation (2.10.384). We find in this way,

$$1 + \exp\left(\pm\frac{i}{2\hbar}\Pi_p^\pm + \frac{1}{2\hbar}\Pi_{np}\right) = \pm i\epsilon \, \exp\left(\frac{1}{2\hbar}\Pi_{np}\right). \tag{2.10.387}$$

This can be written in a more illuminating form by adding the two lateral Borel resummations. We find,

$$\frac{\Pi_p^+ + \Pi_p^-}{2\hbar} - 2\epsilon \tan^{-1}\left[\exp\left(-\frac{1}{2\hbar}\Pi_{np}\right)\right] = 2\pi\left(n + \frac{1}{2}\right), \qquad n \in \mathbb{Z}_{\geq 0}. \tag{2.10.388}$$

Here $\epsilon = (-1)^n$ is the parity of the state. This EQC reduces to the all-orders BS quantization condition (2.5.165) when we neglect the exponentially small correction involving Π_{np}. In addition, we see that this correction is of order

$$\exp\left(-\frac{1}{2\hbar}\Pi_{np}\right) \approx e^{-\sigma/2}, \tag{2.10.389}$$

in accordance with the observation mentioned at the beginning of this section. Finally, by calculating the Borel resummation of the quantum periods Π_p, Π_{np}, one can obtain the spectrum of the pure quartic Hamiltonian (2.6.183) for $M = 2$ by solving the EQC (2.10.388).

As this example should make clear, for generic energies and potentials, we have to use the complex WKB method. For example, if we consider the double-well potential but the energy is above the barrier in the middle, the $\mathcal{B}$ cycle will also move to the complex plane and we will have to again use this procedure [in fact, the quantization condition obtained in this case is again (2.10.388)]. More generally, for any given potential, one has to consider all turning points, both real and complex, as well as the corresponding quantum periods, in order to determine the spectrum.

2.11 The WKB Approximation for Integrable Systems

So far, our analysis of the WKB method has focused on one-dimensional problems. What happens in higher dimensions? It turns out that many aspects of the WKB method can be generalized as long as the underlying mechanical system is completely integrable. In this case, we can write down generalizations of the WKB basic solutions (2.2.22) and of the BS quantization conditions.

Let us consider a classical mechanical system with a phase space $\mathcal{M}$ of dimension $2n$. The canonical coordinates and momenta of $\mathcal{M}$ will be denoted by the n-component vectors $\boldsymbol{q}$ and $\boldsymbol{p}$, respectively. We recall that the canonical symplectic form on $\mathcal{M}$ is given by

$$\omega = \sum_{i=1}^{n} \mathrm{d}p_i \wedge \mathrm{d}q_i. \tag{2.11.390}$$

Let $H(\boldsymbol{q}, \boldsymbol{p})$ be the Hamiltonian. The system is *integrable* (in the sense of Liouville) if it has n independent integrals of motion $F_i(\boldsymbol{q}, \boldsymbol{p})$ in involution, $i = 1, \ldots, n$, where $F_1 = H$. This means that their Poisson brackets vanish:

$$\{F_i, F_j\} = 0, \qquad i, j = 1, \ldots, n. \tag{2.11.391}$$

The condition of independence means that the n one-forms $\mathrm{d}F_i$ are linearly independent at almost all points of $\mathcal{M}$.

In an integrable system, the EOM can be completely solved in terms of quadratures. To see this, one first looks at the level set of the functions F_i:

$$M_{\mathbf{f}} = \{F_i(\boldsymbol{q}, \boldsymbol{p}) = f_i, i = 1, \ldots, n\}, \tag{2.11.392}$$

where $f_i, i = 1, \ldots, n$, are constants, and $\mathbf{f} = (f_1, \ldots, f_n)$. One important property of this level set is that it is a *Lagrangian* submanifold, i.e. the the canonical symplectic form, when restricted to $M_{\mathbf{f}}$, vanishes:

$$\omega|_{M_{\mathbf{f}}} = 0. \tag{2.11.393}$$

This is easily proved as follows. Let us consider the Hamiltonian vector fields X_{F_i}, with components

$$X_{F_i} = \sum_{k=1}^{n} \left(\frac{\partial F_i}{\partial p_k} \frac{\partial}{\partial q_k} - \frac{\partial F_i}{\partial q_k} \frac{\partial}{\partial p_k} \right). \tag{2.11.394}$$

These vector fields are independent, since the $\mathrm{d}F_i$ are independent. They are tangent vectors in $M_{\mathbf{f}}$, since

$$\nabla F_j \cdot X_{F_i} = -\{F_i, F_j\} = 0, \qquad i, j = 1, \ldots, n, \tag{2.11.395}$$

where our convention for Poisson brackets is

$$\{A, B\} = \sum_{k=1}^{n} \left(\frac{\partial A}{\partial q_k} \frac{\partial B}{\partial p_k} - \frac{\partial B}{\partial q_k} \frac{\partial A}{\partial p_k} \right). \tag{2.11.396}$$

Therefore, the vector fields X_{F_i} form a basis for the tangent space to $M_{\mathbf{f}}$. It is now easy to check that,

$$\omega(X_{F_i}, X_{F_j}) = \{F_j, F_i\} = 0, \qquad i, j = 1, \ldots, n, \tag{2.11.397}$$

which is equivalent to (2.11.393). A famous theorem, known as the Liouville–Arnold theorem, states that if $M_{\mathbf{f}}$ is compact and connected, then it is diffeomorphic to an n-dimensional torus, i.e.

$$M_{\mathbf{f}} \simeq \mathbb{T}^n = \mathbb{S}^1 \times \cdots \times \mathbb{S}^1. \tag{2.11.398}$$

For this reason, the level set is also called the *Liouville torus*. In particular, it has n nontrivial one-cycles $\gamma_1, \ldots, \gamma_n$.

We will now find a canonical transformation to a set of new coordinates

$$(q, p) \rightarrow (I, \phi). \tag{2.11.399}$$

These new coordinates are called *action-angle variables*. The action variables are defined by

$$I_j(\mathbf{f}) = \frac{1}{2\pi} \oint_{\gamma_j} p \cdot dq. \tag{2.11.400}$$

Note that, since ω vanishes on $M_{\mathbf{f}}$, these integrals only depend on the homology class of γ_i, by Stokes' theorem. They are functions of the integrals of motion $\mathbf{f}$ characterizing the Liouville torus. Let us now construct the generating function S of this canonical transformation. Let $\mathbf{x}_0$ be a point in $M_{\mathbf{f}}$, with coordinates q_0, and suppose that on a simply-connected neighborhood of $\mathbf{x}_0$ we can solve for p_i as:

$$p = p(q, I). \tag{2.11.401}$$

Let us consider the function

$$S(q, I) = \int_{q_0}^{q} p(q', I) \cdot dq', \tag{2.11.402}$$

where the integration path is contained in $M_{\mathbf{f}}$. Since ω vanishes on $M_{\mathbf{f}}$, this integral does not depend on the path chosen. Therefore, $S(q, I)$ is a well-defined function in a neigborhood of $\mathbf{x}_0$. In addition, we have that

$$p = \frac{\partial S}{\partial q}. \tag{2.11.403}$$

We now define the angle variables as

$$\phi = \frac{\partial S}{\partial I}. \tag{2.11.404}$$

These variables are indeed angles, i.e. they are multivalued, with period 2π. To see this, note that as we go around the one-cycle γ_i in the Liouville torus, the function S changes as

$$\Delta_i S = 2\pi I_i. \tag{2.11.405}$$

This induces a change in ϕ_j given by

$$\Delta_i \phi_j = \frac{\partial}{\partial I_j} \Delta_i S = 2\pi \delta_{ij}. \tag{2.11.406}$$

We conclude that ϕ_j parametrizes an angle around the nontrivial cycle γ_j.

We can now integrate the EOM. The Hamiltonian H only depends on the I, therefore Hamilton's equations read

$$\dot{I} = 0, \qquad \dot{\phi} = \frac{\partial H}{\partial I} = \omega, \tag{2.11.407}$$

and motion is given by linear evolution along the tori,

$$\phi(t) = \omega t + \phi(0). \tag{2.11.408}$$

The determination of the action and angle variables is particularly simple if the coordinates q are *separated variables*. The set of variables $\{q_i\}_{i=1,...,n}$ is separated if, on $M_{\mathbf{f}}$, the canonically conjugate momentum p_i depends only on q_i, and not on the other coordinates q_j, $j \neq i$. In this case, the motion splits into n separated one-dimensional motions, and the action variables are simply given by

$$I_j(\mathbf{f}) = \frac{1}{2\pi} \int_{\gamma_j} p_j \mathrm{d}q_j. \tag{2.11.409}$$

Let us now study the quantum theory corresponding to the integrable system. We will write the wavefunction in position space in the form,

$$\psi(\boldsymbol{q}, t) = A(\boldsymbol{q}, t) \exp\left\{\frac{\mathrm{i}}{\hbar}\sigma(\boldsymbol{q}, t)\right\}, \tag{2.11.410}$$

where $A(\boldsymbol{q}, t)$, $\sigma(\boldsymbol{q}, t)$ are real functions. We will assume that the quantum Hamiltonian $\mathsf{H}\,(\mathsf{q}, \mathsf{p})$ is an algebraic function of the p_i, $i = 1, \ldots, n$. If we plug the ansatz (2.11.410) into the time-dependent Schrödinger equation, we find

$$i\hbar\frac{\partial A}{\partial t} - A\frac{\partial \sigma}{\partial t} = \mathrm{e}^{-\mathrm{i}\sigma/\hbar}\mathsf{H}(\boldsymbol{q}, \mathsf{p})\mathrm{e}^{\mathrm{i}\sigma/\hbar}A, \tag{2.11.411}$$

where the first argument in the Hamiltonian is the classical position vector $\boldsymbol{q}$, and the second argument is the differential operator $\mathsf{p} = -\mathrm{i}\hbar\partial/\partial\boldsymbol{q}$. Since

$$\mathrm{e}^{-\mathrm{i}\sigma/\hbar}\mathsf{p}\,\mathrm{e}^{\mathrm{i}\sigma/\hbar} = \mathsf{p} + \frac{\partial\sigma}{\partial\boldsymbol{q}}, \tag{2.11.412}$$

we can write the equation (2.11.411) as

$$i\hbar\frac{\partial A}{\partial t} - A\frac{\partial\sigma}{\partial t} = \mathsf{H}\left(\boldsymbol{q}, \mathsf{p} + \frac{\partial\sigma}{\partial\boldsymbol{q}}\right)A. \tag{2.11.413}$$

We can analyze this equation systematically by performing a power series expansion in $\hbar$. We set

$$\sigma(\boldsymbol{q}, t) = \sigma_0(\boldsymbol{q}, t) + \mathcal{O}(\hbar), \qquad A(\boldsymbol{q}, t) = A_0(\boldsymbol{q}, t) + \mathcal{O}(\hbar), \tag{2.11.414}$$

and we note that the momentum operator contains one power of $\hbar$. We will restrict ourselves to the leading-order approximation for both σ and A. At leading order in $\hbar$, we find

$$\frac{\partial\sigma_0}{\partial t} + \mathsf{H}\left(\boldsymbol{q}, \frac{\partial\sigma_0}{\partial\boldsymbol{q}}\right) = 0. \tag{2.11.415}$$

This is simply the Hamilton–Jacobi equation

$$\frac{\partial\mathcal{S}(\boldsymbol{q}, t)}{\partial t} + \mathsf{H}\left(\boldsymbol{q}, \frac{\partial\mathcal{S}(\boldsymbol{q}, t)}{\partial\boldsymbol{q}}\right) = 0, \tag{2.11.416}$$

therefore we can identify $\sigma_0(\boldsymbol{q}, t) = \mathcal{S}(\boldsymbol{q}, t)$. As it is well-known, in a stationary motion with constant energy E, the Hamilton–Jacobi equation can be solved by

$$\sigma_0(\boldsymbol{q}, t) = S(\boldsymbol{q}, \boldsymbol{I}) - Et, \tag{2.11.417}$$

where $S(\boldsymbol{q}, \boldsymbol{I})$ is the function (2.11.402).

In order to analyze the next-to-leading term in the expansion of (2.11.413), we will use a trick attributable to Dirac. We multiply (2.11.413) by the function $A(\boldsymbol{q}, t)f(\boldsymbol{q})$, where $f(q)$ is an arbitrary real function with suitable decreasing properties at infinity, and we integrate over q. We will denote for simplicity

$$\langle g(q) \rangle = \int_{\mathbb{R}^n} g(q) \mathrm{d}^n q. \tag{2.11.418}$$

Then, we find the equation

$$\left\langle Af \left(i\hbar \frac{\partial A}{\partial t} - A \frac{\partial \sigma}{\partial t} \right) \right\rangle = \left\langle AfH \left(q, \mathsf{p} + \frac{\partial \sigma}{\partial q} \right) A \right\rangle. \tag{2.11.419}$$

The term in the r.h.s. can be written as the matrix element $\langle A|fH|A \rangle$. Subtracting from (2.11.419) its complex conjugate equation, and dividing by $i\hbar$, we obtain

$$2 \left\langle Af \frac{\partial A}{\partial t} \right\rangle = \frac{1}{i\hbar} \left\langle A \left[f, H \left(q, \mathsf{p} + \frac{\partial \sigma}{\partial q} \right) \right] A \right\rangle. \tag{2.11.420}$$

Let us now calculate the commutator appearing in the r.h.s. At leading order in $\hbar$, we have that

$$H \left(q, \mathsf{p} + \frac{\partial \sigma}{\partial q} \right) f = H \left(q, \frac{\partial \sigma}{\partial q} \right) f - i\hbar \sum_{k=1}^{n} \frac{\partial f}{\partial q_k} \left[\frac{\partial H(q,p)}{\partial p_k} \right]_{p = \frac{\partial S}{\partial q}} + \cdots. \tag{2.11.421}$$

Among the terms that we have not written down, there are contributions of order $\hbar$. These are obtained by the action of p on the q-dependence of the Hamiltonian, including the term $\partial \sigma / \partial q$. However, they are proportional to f, and they cancel in the commutator (the main reason for using Dirac's trick is that we are not obliged to calculate these terms in detail, and in fact they can be deduced from our final result). Note that, since we are interested in the leading-order behavior as $\hbar \to 0$, in the calculation above we have also replaced

$$\frac{\partial \sigma}{\partial q} \to \frac{\partial S}{\partial q}. \tag{2.11.422}$$

We conclude that

$$\frac{1}{i\hbar} \left[f, H \left(q, \mathsf{p} + \frac{\partial \sigma}{\partial q} \right) \right] = \sum_{k=1}^{n} \frac{\partial f}{\partial q_k} \left[\frac{\partial H(q,p)}{\partial p_k} \right]_{p = \frac{\partial S}{\partial q}} + \mathcal{O}(\hbar). \tag{2.11.423}$$

Equation (2.11.420) then reads,

$$\left\langle f \frac{\partial A^2}{\partial t} \right\rangle = \left\langle A^2 \sum_{k=1}^{n} \frac{\partial f}{\partial q_k} \left[\frac{\partial H(q,p)}{\partial p_k} \right]_{p = \frac{\partial S}{\partial q}} \right\rangle + \mathcal{O}(\hbar). \tag{2.11.424}$$

We now integrate by parts in the integral of the r.h.s. and replace A by its leading-order contribution in the $\hbar$ expansion, A_0. Since the resulting equality if valid for any f, we obtain

$$\frac{\partial A_0^2}{\partial t} = - \sum_{k=1}^{n} \frac{\partial}{\partial q_k} \left(A_0^2 \left[\frac{\partial H(q,p)}{\partial p_k} \right]_{p = \frac{\partial S}{\partial q}} \right). \tag{2.11.425}$$

Another way to write this equation is

$$\frac{\partial \log A_0}{\partial t} + \sum_{k=1}^{n} \frac{\partial H}{\partial p_k} \frac{\partial \log A_0}{\partial q_k} + \frac{1}{2} \sum_{k=1}^{n} \frac{\partial}{\partial q_k} \left(\frac{\partial H}{\partial p_k} \right) = 0. \tag{2.11.426}$$

In this equation, it is understood that the p_r's should be set to $\partial S/\partial q_r$ in $\partial H/\partial p_k$. We will now show that

$$A_0 = \left[\det \left(\frac{\partial^2 S}{\partial q_j \partial I_k} \right) \right]^{1/2}. \tag{2.11.427}$$

solves (2.11.426). Let us start by taking a derivative of the Hamilton–Jacobi equation (2.11.416) w.r.t. I_j. We obtain,

$$\sum_{k=1}^{n} \left(\frac{\partial H}{\partial p_k} \right)_{\boldsymbol{p} = \frac{\partial S}{\partial \boldsymbol{q}}} \frac{\partial^2 S}{\partial q_k \partial I_j} + \frac{\partial^2 S}{\partial I_j \partial t} = 0, \qquad j = 1, \dots, n. \tag{2.11.428}$$

By using the definition of the angle variables, we can rewrite this equation as

$$\sum_{k=1}^{n} \frac{\partial \phi_j}{\partial q_k} \left(\frac{\partial H}{\partial p_k} \right)_{\boldsymbol{p} = \frac{\partial S}{\partial \boldsymbol{q}}} = -\frac{\partial \phi_j}{\partial t} + \frac{\partial E}{\partial I_j}, \qquad j = 1, \dots, n. \tag{2.11.429}$$

Let us now introduce the matrix

$$\mathbf{M}_{jl} = \frac{\partial \phi_j}{\partial q_l}, \qquad j, l = 1, \dots, n, \tag{2.11.430}$$

as well as the matrix

$$\mathbf{H}_{kl} = \frac{\partial}{\partial q_l} \left(\frac{\partial H}{\partial p_k} \right)_{\boldsymbol{p} = \frac{\partial S}{\partial \boldsymbol{q}}}. \tag{2.11.431}$$

By taking a derivative of (2.11.429) w.r.t. q_l, we obtain the equation

$$\sum_{k=1}^{n} \frac{\partial \mathbf{M}}{\partial q_k} \frac{\partial H}{\partial p_k} + \mathbf{M} \mathbf{H} = -\frac{\partial \mathbf{M}}{\partial t}. \tag{2.11.432}$$

Like before, it is understood that the p_r's should be set to $\partial S/\partial q_r$ in $\partial H/\partial p_k$. Let us now multiply by $\mathbf{M}^{-1}$ and take the trace. We obtain,

$$\sum_{k=1}^{n} \mathrm{Tr} \left(\mathbf{M}^{-1} \frac{\partial \mathbf{M}}{\partial q_k} \right) \frac{\partial H}{\partial p_k} + \mathrm{Tr}\, \mathbf{H} = -\mathrm{Tr} \left(\mathbf{M}^{-1} \frac{\partial \mathbf{M}}{\partial t} \right). \tag{2.11.433}$$

We note that

$$\mathrm{Tr} \left(\mathbf{M}^{-1} \frac{\partial \mathbf{M}}{\partial q_k} \right) = \frac{\partial}{\partial q_k} \mathrm{Tr} \log \mathbf{M} = \frac{\partial}{\partial q_k} \log \det \mathbf{M}, \tag{2.11.434}$$

and a similar equation can be written down for the term in the r.h.s. We finally obtain,

$$\sum_{k=1}^{n} \frac{\partial H}{\partial p_k} \frac{\partial}{\partial q_k} \left(\log \det \mathbf{M} \right) + \sum_{k=1}^{n} \frac{\partial}{\partial q_k} \left(\frac{\partial H}{\partial p_k} \right) + \frac{\partial}{\partial t} \left(\log \det \mathbf{M} \right) = 0. \tag{2.11.435}$$

By comparing this equation with (2.11.426), we deduce that (2.11.426) is solved by

$$\log A_0 = \frac{1}{2} \log \det \mathbf{M}, \tag{2.11.436}$$

which is precisely (2.11.427). We conclude that the wavefunction

$$\psi(\boldsymbol{q}, t) \approx \psi_{\mathrm{WKB}}(\boldsymbol{q}) \mathrm{e}^{-\mathrm{i}Et/\hbar}, \tag{2.11.437}$$

where

$$\psi_{\mathrm{WKB}}(\boldsymbol{q}) = \left[\det\left(\frac{\partial^2 S}{\partial \boldsymbol{q} \partial \boldsymbol{I}}\right)\right]^{1/2} \exp\left(\frac{\mathrm{i}}{\hbar} S(\boldsymbol{q}, \boldsymbol{I})\right), \qquad (2.11.438)$$

solves the Schrödinger equation in the semiclassical limit $\hbar \to 0$. One can generate different solutions by picking the different branches of the multivalued function $S(\boldsymbol{q}, \boldsymbol{I})$.

It is easy to verify that, in one dimension, (2.11.438) agrees, up to normalization, with the basic WKB solutions (2.2.22). Indeed, in one dimension the action variable is given by

$$I = \frac{1}{2\pi} \oint p(q) \mathrm{d}q. \qquad (2.11.439)$$

We find,

$$\frac{\partial I}{\partial E} = \frac{m}{2\pi} \oint \frac{\mathrm{d}q}{\sqrt{p(q)}} = \frac{T}{2\pi}, \qquad (2.11.440)$$

where T is the period of motion. Therefore,

$$\frac{\partial^2 S}{\partial I \partial q} = \frac{\partial p}{\partial I} = \frac{2\pi m}{T} \frac{1}{p}, \qquad (2.11.441)$$

so that

$$\left(\frac{\partial^2 S}{\partial I \partial q}\right)^{1/2} \exp\left(\frac{\mathrm{i}}{\hbar} \int_{q_0}^{q} p(q') \mathrm{d}q'\right) = \sqrt{\frac{2\pi m}{T}} \frac{1}{\sqrt{p(q)}} \exp\left(\frac{\mathrm{i}}{\hbar} \int_{q_0}^{q} p(q') \mathrm{d}q'\right).$$
$$(2.11.442)$$

These are precisely the basic WKB solutions, up to an overall normalization.

We point out that a very similar analysis can be done to obtain the semiclassical expansion of the QM propagator $K(\boldsymbol{q}_f, \boldsymbol{q}_0; t_f, t_0)$. The starting point is the fact that the propagator satisfies the time-dependent Schrödinger equations (1.2.15) and (1.2.16). One finds,

$$K(\boldsymbol{q}_f, \boldsymbol{q}_0; t_f, t_0) \approx \frac{1}{(2\pi \mathrm{i}\hbar)^{n/2}} \left[-\det\left(\frac{\partial^2 S_c(\boldsymbol{q}_f, \boldsymbol{q}_0; t_f, t_0)}{\partial \boldsymbol{q}_f \partial \boldsymbol{q}_0}\right)\right]^{1/2}$$
$$\times \exp\left(\frac{\mathrm{i}}{\hbar} S_c(\boldsymbol{q}_f, \boldsymbol{q}_0; t_f, t_0)\right), \qquad (2.11.443)$$

where $S_c(\boldsymbol{q}_f, \boldsymbol{q}_0; t_f, t_0)$ is the action (1.2.58) evaluated at the classical trajectory between $\boldsymbol{q}_f$, $\boldsymbol{q}_0$. The overall normalization constant in (2.11.443) can be fixed by comparing the resulting formula to the explicit expression (1.2.29). The expression (2.11.443) is known as *Van Vleck formula* or *Van Vleck approximation* for the QM propagator. It reproduces the result (1.2.65) that we found for the quantum-mechanical propagators in one dimension. In order to justify that result completely, we have to show that higher corrections in the $\hbar$ expansion vanish for potentials that are at most quadratic. This will be done in the context of the path integral formulation in Chapter 4.

One interesting aspect of the higher-dimensional basic WKB solutions of the form (2.11.438) is that they maintain their structure under linear canonical transformations. In order to establish this, we have to first address the more general question of how wavefunctions in quantum mechanics transform under canonical transformations. This question was addressed by Dirac and it has a simple solution. Let us first consider the one-dimensional case. The generating functional of the canonical transformation is a function $F(q, Q)$ satisfying

$$p = \frac{\partial F}{\partial q}, \qquad P = -\frac{\partial F}{\partial Q}. \tag{2.11.444}$$

In general, the wavefunctions will be related by an integral operator

$$K(q, Q) = \langle q | Q \rangle, \tag{2.11.445}$$

as follows

$$\psi(q) = \langle q | \psi \rangle = \int_{\mathbb{R}} \langle q | Q \rangle \langle Q | \psi \rangle \mathrm{d}Q = \int_{\mathbb{R}} K(q, Q) \psi(Q) \mathrm{d}Q. \tag{2.11.446}$$

On the other hand, this kernel satisfies the equations

$$\langle q | \mathsf{p} | Q \rangle = -\mathrm{i}\hbar \frac{\partial}{\partial q} \langle q | Q \rangle = \frac{\partial F}{\partial q} \langle q | Q \rangle,$$
$$\langle q | \mathsf{P} | Q \rangle = \mathrm{i}\hbar \frac{\partial}{\partial Q} \langle q | Q \rangle = -\frac{\partial F}{\partial Q} \langle q | Q \rangle. \tag{2.11.447}$$

This is solved by

$$\langle q | Q \rangle = \mathcal{C} \mathrm{e}^{\mathrm{i}F(q,Q)/\hbar}, \tag{2.11.448}$$

where $\mathcal{C}$ is a normalization constant. It can be fixed (up to a phase) by the requirement that

$$\langle q | q' \rangle = \int_{\mathbb{R}} \langle q | Q \rangle \langle Q | q' \rangle \mathrm{d}Q = |\mathcal{C}|^2 \int_{\mathbb{R}} \mathrm{e}^{\mathrm{i}(F(q,Q)-F(q',Q))/\hbar} \mathrm{d}Q = \delta(q - q'). \tag{2.11.449}$$

We then conclude that

$$\psi(q) = \mathcal{C} \int_{\mathbb{R}} \mathrm{e}^{\mathrm{i}F(q,Q)/\hbar} \psi(Q) \mathrm{d}Q. \tag{2.11.450}$$

The generating functional of the inverse transformation is

$$F^{-1}(Q, q) = -F(q, Q), \tag{2.11.451}$$

so that

$$\psi(Q) = \mathcal{C} \int_{\mathbb{R}} \mathrm{e}^{-\mathrm{i}F(q,Q)/\hbar} \psi(q) \mathrm{d}q. \tag{2.11.452}$$

An important case occurs when the canonical transformation is linear,

$$\begin{pmatrix} Q \\ P \end{pmatrix} = \begin{pmatrix} a & b \\ c & d \end{pmatrix} \begin{pmatrix} q \\ p \end{pmatrix}, \qquad \begin{pmatrix} a & b \\ c & d \end{pmatrix} \in \mathrm{SL}(2, \mathbb{R}), \tag{2.11.453}$$

with inverse

$$\begin{pmatrix} q \\ p \end{pmatrix} = \begin{pmatrix} d & -b \\ -c & a \end{pmatrix} \begin{pmatrix} Q \\ P \end{pmatrix}. \tag{2.11.454}$$

If $b \neq 0$, the generating functional is given by

$$F(q, Q) = -\frac{1}{2b}\left(aq^2 + dQ^2 - 2qQ\right).$$ (2.11.455)

In this case, the normalization constant is easily fixed by (2.11.449), which gives

$$\mathcal{C} = \frac{1}{\sqrt{2\pi\hbar b}},$$ (2.11.456)

up to a phase.

The generalization to canonical transformations in $2n$ dimensions is straightforward. In particular, linear canonical transformations are given by linear symplectic transformations of the form

$$\begin{pmatrix} Q \\ P \end{pmatrix} = \begin{pmatrix} A & B \\ C & D \end{pmatrix}\begin{pmatrix} q \\ p \end{pmatrix}, \qquad \begin{pmatrix} A & B \\ C & D \end{pmatrix} \in \mathrm{Sp}(2n, \mathbb{R}),$$ (2.11.457)

where the $n \times n$ matrices A, B, C, D satisfy the constraints

$$BA^T = AB^T,$$

$$CD^T = DC^T,$$ (2.11.458)

$$DA^T - CB^T = \mathbf{1}.$$

Let us suppose that the matrix B is invertible. In this case, the generating functional for the canonical transformation is given by

$$F(q, Q) = -\frac{1}{2}\left(q^T B^{-1} A q + Q^T D B^{-1} Q - 2q^T B^{-1} Q\right),$$ (2.11.459)

and the integral kernel for the canonical transformation of the wavefunction is given by

$$K(q, Q) = \mathcal{C}e^{iF(q,Q)/\hbar},$$ (2.11.460)

where the normalization constant is now given by

$$\mathcal{C} = \frac{1}{(2\pi\hbar)^{n/2}(\det B)^{1/2}}.$$ (2.11.461)

Let us now consider the change of the WKB wavefunction under a linear canonical transformation. According to the discussion above, it is given by

$$\mathcal{C}\int_{\mathbb{R}^n} e^{-iF(q,Q)/\hbar}\psi_{\mathrm{WKB}}(q)dq,$$ (2.11.462)

where $\mathcal{C}$ is given by (2.11.461). When $\hbar \to 0$, the integral can be calculated in the saddle-point approximation. The saddle point q^* is given by

$$\frac{\partial}{\partial \mathbf{q}}\left(S(q, I) - F(q, Q)\right)\Big|_{q^*} = 0,$$ (2.11.463)

or, equivalently,

$$p(q^*, I) - \frac{\partial F(q^*, Q)}{\partial \mathbf{q}} = 0,$$ (2.11.464)

and it gives the functional relation between q^* and Q. We then find that (2.11.462) is given, in the saddle-point approximation, by

$$
\mathcal{C}(2\pi\hbar)^{n/2} \exp\left\{\frac{i}{\hbar}\left(S(q^*, I) - F(q^*, Q)\right)\right\}
$$

$$
\times \left[\det\left(\frac{\partial^2 S}{\partial q \partial I}\right)_{q^*}\right]^{1/2} \left[\det\left(\frac{\partial^2}{\partial q \partial q'}\left(S(q, I) - F(q, Q)\right)\right)_{q^*}\right]^{-1/2}. \tag{2.11.465}
$$

The leading exponential as $\hbar \to 0$ is easily evaluated if we take into account that

$$
dF = \mathbf{p} \cdot d\mathbf{q} - \mathbf{P} \cdot d\mathbf{Q}, \tag{2.11.466}
$$

therefore

$$
\exp\left\{\frac{i}{\hbar}\left(S(q^*, I) - F(q^*, Q)\right)\right\} = \exp\left\{\frac{i}{\hbar}\widetilde{S}(Q, I)\right\}, \tag{2.11.467}
$$

where

$$
\widetilde{S}(Q, I) = \int_{Q_0}^{Q} \mathbf{P}(Q', I) \cdot dQ', \tag{2.11.468}
$$

and $Q_0 = Q(q_0^*)$. Taking a derivative in (2.11.464), we deduce that

$$
\left.\frac{\partial \mathbf{p}}{\partial q'}\right|_{q^*} - \left.\frac{\partial^2 F}{\partial q \partial q'}\right|_{q^*} - \left.\frac{\partial^2 F}{\partial q \partial Q}\frac{\partial Q}{\partial q'}\right|_{q^*} = 0. \tag{2.11.469}
$$

It follows that

$$
\frac{\partial^2}{\partial q \partial q'}\left(S(q, I) - F(q, Q)\right)_{q^*} = B^{-1}\left.\frac{\partial Q}{\partial q'}\right|_{q^*}, \tag{2.11.470}
$$

where we have used (2.11.459). Therefore,

$$
\left[\det\left(\frac{\partial^2 S}{\partial q \partial I}\right)_{q^*}\right]^{1/2} \left[\det\left(\frac{\partial^2}{\partial q \partial q'}\left(S(q, I) - F(q, Q)\right)\right)_{q^*}\right]^{-1/2}
$$

$$
= (\det B)^{1/2}\left[\det\left(\frac{\partial^2 \widetilde{S}}{\partial Q \partial I}\right)\right]^{1/2}, \tag{2.11.471}
$$

since $F(q, Q)$ does not depend on I. By taking into account (2.11.461), we conclude that the integral (2.11.462) is given by

$$
\psi_{\text{WKB}}(Q) = \left[\det\left(\frac{\partial^2 \widetilde{S}}{\partial Q \partial I}\right)\right]^{1/2} \exp\left\{\frac{i}{\hbar}\widetilde{S}(Q, I)\right\}, \tag{2.11.472}
$$

which is the WKB wavefunction in the new coordinates Q, P.

By using the wavefunctions (2.11.438) we can derive an approximate quantization condition for integrable systems, which is also called the EBK quantization condition (for Einstein, Brillouin, and Keller). The basic observation is that the functions $S(q, I)$ in (2.11.438) are multivalued, as we saw in (2.11.405). This implies that, as we go around the cycle γ_j in the Liouville torus, the wavefunction changes as

$$
\exp\left(\frac{2\pi i}{\hbar}I_j\right), \qquad j = 1, \ldots, n. \tag{2.11.473}
$$

However, the exact wavefunction is single valued, therefore we must have

$$I_j(f) = \hbar m_j, \qquad m_j \in \mathbb{Z}, \qquad j = 1, \ldots, n. \tag{2.11.474}$$

These conditions constrain the values of the classically conserved quantities f, including the energy, and are therefore quantization conditions for the integrable system. They are sometimes called *EBK quantization conditions*.

In the case of a one-dimensional model, the EBK quantization condition reproduces the BS quantization condition (2.5.136) up to the shift $1/2$ in the quantum number. There is a refinement of (2.11.474) that gives the higher-dimensional version of this shift, and reads

$$I_j(f) = \hbar \left(m_j + \frac{\mu_j}{4} \right), \qquad m_j \in \mathbb{Z}, \qquad j = 1, \ldots, n. \tag{2.11.475}$$

The quantities μ_j are sometimes called Maslov indices. Heuristically, μ_i can be related to the number of times that the prefactor in (2.11.438) changes sign as we go around the one-cycle γ_j. Each of these changes of sign introduces a phase of $-\pi/2$ due to the square root. For example, in one dimension, if there are two turning points as we go around one cycle, as is usually the case, we have $\mu = 2$ and we recover the conventional BS formula.

Example 2.11.1 One of the simplest examples of a higher-dimensional integrable system is the motion of a particle in a three-dimensional, rotationally invariant potential $V(r)$. The problem can be easily solved in spherical coordinates. The conserved quantities are the Hamiltonian,

$$H = \frac{1}{2} \left(p_r^2 + \frac{p_\theta^2}{r^2} + \frac{p_\phi^2}{r^2 \sin^2 \theta} \right) + V(r), \tag{2.11.476}$$

the total angular momentum

$$J^2 = \sum_{i=1}^{3} J_i^2 = p_\theta^2 + \frac{p_\phi^2}{\sin^2 \theta}, \tag{2.11.477}$$

and the third component of the angular momentum,

$$J_3 = p_\phi. \tag{2.11.478}$$

In (2.11.476) we have set the mass to unity for simplicity.

Let us consider the surface in phase space M_f where these conserved quantities take fixed values. We can then solve for the momenta in terms of H, J^2, and J_3,

$$p_r = \sqrt{2 \left(H - V(r) \right) - \frac{J^2}{r^2}},$$

$$p_\theta = \sqrt{J^2 - \frac{J_3^2}{\sin^2 \theta}}, \tag{2.11.479}$$

$$p_\phi = J_3.$$

We note that these spherical coordinates provide a set of separated variables for this integrable system, since the momentum p_i only depends on its conjugate coordinate q_i.

The Hamilton–Jacobi function, given in (2.11.402), reads in this case,

$$S = \int^r \sqrt{2\left(H - V(r')\right) - \frac{J^2}{(r')^2}}\, dr' + \int^\theta \sqrt{J^2 - \frac{J_3^2}{\sin^2\theta'}}\, d\theta' + \int^\phi J_3 d\phi'.$$

(2.11.480)

The new angle variables are

$$\phi_H = \frac{\partial S}{\partial H}, \qquad \phi_{J^2} = \frac{\partial S}{\partial J^2}, \qquad \phi_{J_3} = \frac{\partial S}{\partial J_3}.$$

(2.11.481)

It follows from (2.11.407) that ψ_{J^2} and ψ_{J_3} are constant, while ψ_H evolves in time according to

$$\psi_H(t) = t + c,$$

(2.11.482)

where c is a constant. We deduce that

$$t - t_0 = \int_{r_0}^r \frac{dr}{\sqrt{2\left(H - V(r)\right) - \frac{J^2}{r^2}}},$$

(2.11.483)

which is the standard formula for the solution of the Kepler problem.

To apply the EBK quantization scheme, we consider the action variables I_r, I_θ, and I_ϕ. Their calculation is a standard problem in classical mechanics. The action I_ϕ is the simplest one,

$$I_\phi = \frac{1}{2\pi} \oint p_\phi d\phi = J_3.$$

(2.11.484)

For I_θ, we find

$$I_\theta = \frac{1}{\pi} \int_{\theta_1}^{\theta_2} \left(J^2 - \frac{J_3^2}{\sin^2\theta}\right)^{1/2} = J - |J_3|,$$

(2.11.485)

where $\theta_{1,2}$ are angular turning points that satisfy the equation

$$\sin^2\theta = \frac{J_3^2}{J^2}.$$

(2.11.486)

Finally, we have

$$I_r = \frac{1}{\pi} \int_{r_1}^{r_2} \sqrt{2(E - V(r)) - \frac{(I_\theta + |I_\phi|)^2}{r^2}}\, dr,$$

(2.11.487)

where $r_{1,2}$ are radial turning points. The EBK quantization conditions in this case are

$$I_r = n_r \hbar, \qquad I_\theta = n_\theta \hbar, \qquad I_\phi = n_\phi \hbar.$$

(2.11.488)

It is customary to introduce the principal quantum number n, the azimuthal quantum number n_ψ, and the magnetic quantum number m as

$$n = n_r + n_\theta + |n_\phi|, \qquad n_\psi = n_\theta + |n_\phi|, \qquad m = n_\phi.$$

(2.11.489)

In terms of these variables, we have $J_3 = m\hbar$, and the total angular momentum is quantized as

$$J = \hbar n_\psi.$$

(2.11.490)

We require $|n_\phi| \le n_\psi$ (since $J_3^2 \le J^2$). In the case of the Kepler problem, where

$$V(r) = -\frac{k}{r}, \tag{2.11.491}$$

which is the potential that governs e.g. the hydrogen atom, the action variable I_r can be computed in closed form:

$$I_r = -I_\theta - |I_\phi| - \frac{k}{2}\sqrt{\frac{2}{-E}}. \tag{2.11.492}$$

This gives the quantized energy spectrum in terms of the principal quantum number,

$$E_n = -\frac{k^2}{2\hbar^2}\frac{1}{n^2}, \qquad n \in \mathbb{Z}_{>0}. \tag{2.11.493}$$

We recover in this way the basic results of Bohr's theory of the hydrogen atom.   □

2.12 Semiclassical Quantization of the Periodic Toda Lattice

The EBK quantization scheme provides valuable information on the spectrum of integrable models, as we have seen in the case of central potentials in three dimensions. In this section we will consider a much more intricate integrable system called the *periodic Toda lattice*. Although the techniques to study this system are not elementary, it is a beautiful model with a rich structure that shows the power of semiclassical quantization methods.

The periodic Toda lattice consists of $n+1$ particles with positions q_i and momenta p_i. The EOM are given by

$$\dot{q}_i = p_i, \qquad \dot{p}_i = e^{q_{i-1}-q_i} - e^{q_i-q_{i+1}}, \qquad i = 1, \ldots, n+1. \tag{2.12.494}$$

These follow from the Hamiltonian

$$H = \frac{1}{2}\sum_{i=1}^{n+1} p_i^2 + \sum_{i=1}^{n+1} e^{q_i-q_{i+1}}, \tag{2.12.495}$$

where we have normalized $m = 1$ for all the particles. We set $(q_{n+1+j}, p_{n+1+j}) = (q_j, p_j)$, which enforces the periodicity condition. Clearly, the total momentum

$$P = \sum_{i=1}^{n+1} p_i \tag{2.12.496}$$

is a conserved quantity, together with the Hamiltonian itself.

Example 2.12.1 *The case $n = 1$.* The simplest case of the periodic Toda lattice occurs when $n = 1$. It is convenient to set to zero the total momentum.

$$P = p_1 + p_2 = 0. \tag{2.12.497}$$

Let us define the relative coordinate and momentum as

$$q_r = \frac{q_1 - q_2}{\sqrt{2}}, \qquad p_r = \frac{p_1 - p_2}{\sqrt{2}}. \tag{2.12.498}$$

The Hamiltonian reads, in terms of these variables,

$$H = \frac{p_r^2}{2} + 2\cosh(\sqrt{2}q_r). \tag{2.12.499}$$

It describes the motion of a nonrelativistic particle in a cosh potential. After quantization, the corresponding Schrödinger equation

$$-\frac{\hbar^2}{2}\psi''(q_r) + \left(2\cosh(\sqrt{2}q_r) - E\right)\psi(q_r) = 0 \tag{2.12.500}$$

is the so-called *modified Mathieu equation.*   □

The construction of the remaining conserved quantities can be done by using the technique of Lax matrices. There are various versions of this technique, but the most useful one for our purposes is due to Sklyanin. We first introduce Lax matrices, depending on a parameter μ, as follows:

$$L_i(\mu) = \begin{pmatrix} 0 & e^{q_i} \\ -e^{-q_i} & \mu - p_i \end{pmatrix}, \qquad i = 1, \ldots, n+1. \tag{2.12.501}$$

μ is sometimes called the *spectral parameter*. Note that the matrix $L_i(\mu)$ only involves the coordinates and momenta of the ith particle, and it satisfies

$$\det L_i(\mu) = 1, \qquad i = 1, \ldots, n. \tag{2.12.502}$$

Out of these matrices, we form the *monodromy matrix*:

$$T(\mu) = L_{n+1}(\mu) \cdots L_1(\mu) = \begin{pmatrix} A(\mu) & B(\mu) \\ C(\mu) & D(\mu) \end{pmatrix}. \tag{2.12.503}$$

We also have

$$\det T(\mu) = 1. \tag{2.12.504}$$

The entries of $T(\mu)$ are polynomials in μ whose coefficients are functions on phase space. It is easy to show that

$$A(\mu) = -e^{q_{n+1}-q_1}\mu^{n-1} + \cdots,$$

$$B(\mu) = e^{q_{n+1}}\left(\mu^n - \left(\sum_{i=1}^{n} p_i\right)\mu^{n-1} + \cdots\right),$$

$$C(\mu) = -e^{-q_1}\mu^n + e^{-q_1}\left(\sum_{i=2}^{n+1} p_i\right)\mu^{n-1} + \cdots,$$

$$D(\mu) = \mu^{n+1} - \left(\sum_{i=1}^{n+1} p_i\right)\mu^n + \left(\sum_{1 \leq i < j \leq n+1} p_i p_j - \sum_{i=1}^{n} e^{q_i - q_{i+1}}\right)\mu^{n-1} + \cdots.$$

$$\tag{2.12.505}$$

Let us define $t(\mu)$ by

$$2t(\mu) = \operatorname{Tr} T(\mu) = A(\mu) + D(\mu). \tag{2.12.506}$$

From its form, it is clear that it is a polynomial of degree $n + 1$ in μ, and we will write it as

$$2t(\mu) = \sum_{k=0}^{n+1} (-1)^k \mu^{n+1-k} H_k. \tag{2.12.507}$$

It follows from (2.12.505) that $H_0 = 1$ and that

$$H_1 = P, \qquad H_2 = \frac{P^2}{2} - H, \tag{2.12.508}$$

where H is the Hamiltonian.

We will now show that the $n + 1$ functions on phase space $H_1, \ldots, H_{n+1}$ are in involution. To do this, we need the technology developed in the modern study of integrable systems. Let $M(\mu)$ be a 2×2 matrix of the form

$$M(\mu) = \sum_{i,j=1}^{2} M(\mu)_{ij} E_{ij} = \begin{pmatrix} a(\mu) & b(\mu) \\ c(\mu) & d(\mu) \end{pmatrix}, \tag{2.12.509}$$

which acts on $\mathbb{C}^2$. Here, E_{ij} is the canonical basis for square matrices,

$$(E_{ij})_{kl} = \delta_{ik} \delta_{jl}. \tag{2.12.510}$$

We now introduce the following matrices, acting on $\mathbb{C}^2 \otimes \mathbb{C}^2$,

$$M^1(\mu) = M(\mu) \otimes I = \sum_{i,j=1}^{2} M(\mu)_{ij} \left(E_{ij} \otimes I \right) = \begin{pmatrix} a(\mu) & 0 & b(\mu) & 0 \\ 0 & a(\mu) & 0 & b(\mu) \\ c(\mu) & 0 & d(\mu) & 0 \\ 0 & c(\mu) & 0 & d(\mu) \end{pmatrix},$$

$$M^2(\mu) = I \otimes M(\mu) = \sum_{i,j=1}^{2} M(\mu)_{ij} \left(I \otimes E_{ij} \right) = \begin{pmatrix} a(\mu) & b(\mu) & 0 & 0 \\ c(\mu) & d(\mu) & 0 & 0 \\ 0 & 0 & a(\mu) & b(\mu) \\ 0 & 0 & c(\mu) & d(\mu) \end{pmatrix}. \tag{2.12.511}$$

Their product is given by

$$M^1(\mu)M^2(\mu') = \sum_{i,j,k,l=1}^{2} M_{ij}(\mu) M_{kl}(\mu') E_{ij} \otimes E_{kl}$$

$$= \begin{pmatrix} a(\mu)a\left(\mu'\right) & a(\mu)b\left(\mu'\right) & b(\mu)a\left(\mu'\right) & b(\mu)b\left(\mu'\right) \\ a(\mu)c\left(\mu'\right) & a(\mu)d\left(\mu'\right) & b(\mu)c\left(\mu'\right) & b(\mu)d\left(\mu'\right) \\ c(\mu)a\left(\mu'\right) & c(\mu)b\left(\mu'\right) & d(\mu)a\left(\mu'\right) & d(\mu)b\left(\mu'\right) \\ c(\mu)c\left(\mu'\right) & c(\mu)d\left(\mu'\right) & d(\mu)c\left(\mu'\right) & d(\mu)d\left(\mu'\right) \end{pmatrix}. \tag{2.12.512}$$

If the entries of $M(\mu)$ are functions on phase space, we define the matrix of Poisson brackets as,

$$\{M^1(\mu), M^2(\mu')\}$$

$$= \sum_{i,j,k,l=1}^{2} \{M_{ij}(\mu), M_{kl}(\mu')\} E_{ij} \otimes E_{kl}$$

$$= \begin{pmatrix} \{a(\mu), a(\mu')\} & \{a(\mu), b(\mu')\} & \{b(\mu), a(\mu')\} & \{b(\mu), b(\mu')\} \\ \{a(\mu), c(\mu')\} & \{a(\mu), d(\mu')\} & \{b(\mu), c(\mu')\} & \{b(\mu), d(\mu')\} \\ \{c(\mu), a(\mu')\} & \{c(\mu), b(\mu')\} & \{d(\mu), a(\mu')\} & \{d(\mu), b(\mu')\} \\ \{c(\mu), c(\mu')\} & \{c(\mu), d(\mu')\} & \{d(\mu), c(\mu')\} & \{d(\mu), d(\mu')\} \end{pmatrix}. \quad (2.12.513)$$

A direct calculation shows that

$$\{L_i^1(\mu), L_i^2(\mu')\} = \begin{pmatrix} 0 & 0 & 0 & 0 \\ 0 & 0 & 0 & -e^{q_i} \\ 0 & 0 & 0 & e^{q_i} \\ 0 & -e^{-q_i} & e^{-q_i} & 0 \end{pmatrix}. \quad (2.12.514)$$

Let us consider the matrix

$$C_{12} = \sum_{i,j=1}^{2} E_{ij} \otimes E_{ji} = \begin{pmatrix} 1 & 0 & 0 & 0 \\ 0 & 0 & 1 & 0 \\ 0 & 1 & 0 & 0 \\ 0 & 0 & 0 & 1 \end{pmatrix}. \quad (2.12.515)$$

A direct calculation gives,

$$\left[C_{12}, L_i^1(\mu) L_i^2(\mu')\right] = \begin{pmatrix} 0 & 0 & 0 & 0 \\ 0 & 0 & 0 & e^{q_i}(\mu - \mu') \\ 0 & 0 & 0 & -e^{q_i}(\mu - \mu') \\ 0 & e^{-q_i}(\mu - \mu') & -e^{-q_i}(\mu - \mu') & 0 \end{pmatrix}. \quad (2.12.516)$$

We conclude that

$$\{L_i^1(\mu), L_i^2(\mu')\} = \left[L_i^1(\mu) L_i^2(\mu'), r_{12}(\mu - \mu')\right], \quad (2.12.517)$$

where

$$r_{12}(\mu - \mu') = \frac{C_{12}}{\mu - \mu'}, \quad (2.12.518)$$

is called the *classical r-matrix*. The property (2.12.517) also holds for products of the matrices L_i. Indeed, we have

$$\{L_i^1(\mu)L_j^1(\mu), L_i^2(\mu')L_j^2(\mu')\}$$

$$= L_i^1(\mu)L_i^2(\mu')\{L_j^1(\mu), L_j^2(\mu')\} + \{L_i^1(\mu), L_i^2(\mu')\}L_j^1(\mu)L_j^2(\mu')$$

$$= L_i^1(\mu)L_i^2(\mu')\left[L_j^1(\mu)L_j^2(\mu'), r_{12}(\mu - \mu')\right] \tag{2.12.519}$$

$$+ \left[L_i^1(\mu)L_i^2(\mu'), r_{12}(\mu - \mu')\right]L_j^1(\mu)L_j^2(\mu')$$

$$= \left[L_i^1(\mu)L_i^2(\mu')L_j^1(\mu)L_j^2(\mu'), r_{12}(\mu - \mu')\right],$$

where we used that

$$A_1[A_2, R] + [A_1, R]A_2 = [A_1A_2, R], \tag{2.12.520}$$

for the standard matrix commutator. In the first line of (2.12.519) we used that the Poisson bracket of matrices with different indices is zero, since the matrices $L_i(\mu)$ only depend on the variables q_i, p_i. By induction, we conclude that (2.12.517) also holds for an arbitrary product of matrices L_i with different indices, in particular for the monodromy matrix $T(\mu)$, and we obtain

$$\{T^1(\mu), T^2(\mu')\} = \left[T^1(\mu)T^2(\mu'), r_{12}(\mu - \mu')\right]. \tag{2.12.521}$$

This implies the following properties:

$$\{A(\mu), A(\mu')\} = \{B(\mu), B(\mu')\} = \{C(\mu), C(\mu')\} = \{D(\mu), D(\mu')\} = 0, \tag{2.12.522}$$

as well as

$$\{A(\mu), B(\mu')\} = \frac{B(\mu)A(\mu') - A(\mu)B(\mu')}{\mu - \mu'},$$

$$\{A(\mu), D(\mu')\} = \frac{B(\mu)C(\mu') - C(\mu)B(\mu')}{\mu - \mu'}, \tag{2.12.523}$$

$$\{B(\mu), D(\mu')\} = \frac{B(\mu)D(\mu') - D(\mu)B(\mu')}{\mu - \mu'}.$$

It follows from the above results that

$$\{t(\mu), t(\mu')\} = 0, \tag{2.12.524}$$

for arbitrary μ, μ'. Therefore, the coefficients of $t(\mu)$ are in involution, i.e.

$$\{H_k, H_l\} = 0, \qquad k = 1, \ldots, n + 1. \tag{2.12.525}$$

In particular, it follows that

$$\{H, H_k\} = 0, \qquad k = 3, \ldots, n + 1, \tag{2.12.526}$$

so all these quantities are integrals of motion. It is not difficult to see that the H_k are independent (they involve different independent functions of the momenta, for example). Therefore, the periodic Toda chain is an integrable system in the sense of Liouville.

We now want to find action–angle variables. The first step is to find separated variables, i.e. we want to have a new set of canonically conjugate coordinates $(\tilde{q}_i, \tilde{p}_i)$ such that, on the surface in phase space where the Hamiltonians take constant values,

$\tilde{p}_i$ is only a function of $\tilde{q}_i$. The construction of separated variables goes as follows. As we saw in (2.12.505), $B(\mu)$ is a polynomial in μ of degree n, which can be written as

$$B(\mu) = e^{q_{n+1}} \prod_{k=1}^{n} \left(\mu - \mu_{\gamma_k} \right). \tag{2.12.527}$$

The zeros of this polynomial, μ_{γ_k}, $k = 1, \ldots, n$, are functions of the canonical coordinates q_i, p_i, $i = 1, \ldots, n + 1$. Let us also denote

$$\lambda_{\gamma_k} = A(\mu_{\gamma_k}). \tag{2.12.528}$$

Since

$$1 = \det T(\mu_{\gamma_k}) = A(\mu_{\gamma_k}) D(\mu_{\gamma_k}), \tag{2.12.529}$$

as well as

$$2t(\mu_{\gamma_k}) = A(\mu_{\gamma_k}) + D(\mu_{\gamma_k}), \tag{2.12.530}$$

it follows that the pairs $(\mu_{\gamma_k}, \lambda_{\gamma_k})$, $k = 1, \ldots, n$, belong to the so-called *spectral curve*, defined by

$$\Gamma(\mu, \lambda) = \lambda + \frac{1}{\lambda} - 2t(\mu) = 0. \tag{2.12.531}$$

We want to show now that

$$(\mu_{\gamma_k}, \log \lambda_{\gamma_k}), \qquad k = 1, \ldots, n, \tag{2.12.532}$$

form a set of canonical coordinates. In order to do this, we first need a technical result. Let $P(\mu)$ be an arbitrary polynomial in μ, i.e.

$$P(\mu) = \sum_{i \geq 0} a_i \mu^i. \tag{2.12.533}$$

Then, if F is an arbitrary function on phase space, we have

$$\begin{aligned}
\{F, P(\mu_{\gamma_k})\} &= \sum_{i \geq 0} \{F, a_i\} \mu_{\gamma_k}^i + \sum_{i \geq 0} a_i \{F, \mu_{\gamma_k}^i\} \\
&= \{F, P(\mu)\}_{\mu=\mu_{\gamma_k}} + \sum_{i \geq 0} a_i i \mu_{\gamma_k}^{i-1} \{F, \mu_{\gamma_k}\} \\
&= \{F, P(\mu)\}_{\mu=\mu_{\gamma_k}} + P'(\mu_{\gamma_k})\{F, \mu_{\gamma_k}\}.
\end{aligned} \tag{2.12.534}$$

We can now prove that

$$\{\mu_{\gamma_k}, \mu_{\gamma_{k'}}\} = 0. \tag{2.12.535}$$

To do this, we evaluate (2.12.534) for $F(\mu) = P(\mu) = B(\mu)$. We find,

$$\{B(\mu), B(\mu_{\gamma_k})\} = \{B(\mu), B(\mu')\}_{\mu'=\mu_{\gamma_k}} + B'(\mu_{\gamma_k})\{B(\mu), \mu_{\gamma_k}\} = 0. \tag{2.12.536}$$

Since $\{B(\mu), B(\mu')\} = 0$, as we found in (2.12.522), we deduce

$$\{B(\mu), \mu_{\gamma_k}\} = 0. \tag{2.12.537}$$

We again apply (2.12.534) with $F(\mu) = \mu_{\gamma_{k'}}$ and $P(\mu) = \mu$. This gives,

$$\{\mu_{\gamma_{k'}}, B(\mu_{\gamma_k})\} = \{\mu_{\gamma_{k'}}, B(\mu)\}_{\mu=\mu_{\gamma_k}} + B'(\mu_{\gamma_k})\{\mu_{\gamma_{k'}}, \mu_{\gamma_k}\} = 0. \qquad (2.12.538)$$

In view of (2.12.537), we conclude that (2.12.535) holds.

We want to evaluate now the bracket

$$\{\mu_{\gamma_k}, \lambda_{\gamma_k'}\}. \qquad (2.12.539)$$

We will first apply (2.12.534) to $F(\mu) = A(\mu)$, $P(\mu) = B(\mu)$. One finds,

$$\{A(\mu), B(\mu_{\gamma_k})\} = \{A(\mu), B(\mu')\}_{\mu'=\mu_{\gamma_k}} + B'(\mu_{\gamma_k})\{A(\mu), \mu_{\gamma_k}\} = 0. \qquad (2.12.540)$$

The first bracket can be calculated from (2.12.523),

$$\{A(\mu), B(\mu')\}_{\mu=\mu_{\gamma_k}} = \frac{B(\mu)A(\mu_{\gamma_k})}{\mu - \mu_{\gamma_k}}, \qquad (2.12.541)$$

and we deduce that

$$\{A(\mu), \mu_{\gamma_k}\} = -\frac{B(\mu)}{\mu - \mu_{\gamma_k}} \frac{A(\mu_{\gamma_k})}{B'(\mu_{\gamma_k})}. \qquad (2.12.542)$$

Let us now evaluate this Poisson bracket at $\mu = \mu_{\gamma_{k'}}$. Clearly, if $k' \neq k$, the r.h.s. of (2.12.542) vanishes, since $B(\mu_{\gamma_{k'}}) = 0$. If $k = k'$, we have

$$\lim_{\mu \to \mu_{\gamma_k}} \frac{B(\mu)}{\mu - \mu_{\gamma_k}} = B'(\mu_{\gamma_k}). \qquad (2.12.543)$$

We conclude that

$$\{A(\mu), \mu_{\gamma_k}\}_{\mu=\mu_{\gamma_{k'}}} = -\delta_{kk'} A(\mu_{\gamma_k}). \qquad (2.12.544)$$

To find the value of (2.12.539), we compute, by using (2.12.534),

$$\{\mu_{\gamma_k}, A(\mu_{\gamma_{k'}})\} = \{\mu_{\gamma_k}, A(\mu)\}_{\mu=\mu_{\gamma_{k'}}} + A'(\mu_{\gamma_k})\{\mu_{\gamma_k}, \mu_{\gamma_{k'}}\}. \qquad (2.12.545)$$

The last term in the r.h.s. vanishes, and the first term can be calculated using (2.12.544). The final result is

$$\{\mu_{\gamma_k}, \lambda_{\gamma_{k'}}\} = \delta_{kk'} \lambda_{\gamma_k}. \qquad (2.12.546)$$

Finally, we want to show that

$$\{\lambda_{\gamma_k}, \lambda_{\gamma_{k'}}\} = 0. \qquad (2.12.547)$$

To do this, we evaluate the Poisson bracket

$$\{A(\mu_{\gamma_k}), A(\mu_{\gamma_{k'}})\} = A'(\mu_{\gamma_{k'}})\{A(\mu), \mu_{\gamma_{k'}}\}_{\mu=\mu_{\gamma_k}} - A'(\mu_{\gamma_k})\{A(\mu), \mu_{\gamma_k}\}_{\mu=\mu_{\gamma_{k'}}}, \qquad (2.12.548)$$

where we took into account (2.12.535) and $\{A(\mu), A(\mu')\} = 0$. The r.h.s. of this equation vanishes due to (2.12.544), and we have then established (2.12.547).

The calculations above establish that (2.12.532) is indeed a set of canonical coordinates for the periodic Toda lattice. In addition, these variables are separated. The reason is that, as we mentioned before, each pair $(\mu_{\gamma_k}, \lambda_{\gamma_k})$ belongs to the spectral curve (2.12.531):

$$\Gamma\left(\mu_{\gamma_k}, \lambda_{\gamma_k}\right) = 0. \tag{2.12.549}$$

Therefore, when the conserved quantities take constant values, the canonical momentum $\log \lambda_{\gamma_k}$ is only a function of μ_{γ_k}. In addition, the dependence of λ_{γ_k} as a function of μ_{γ_k} is the same for all $k = 1, \ldots, n$, and it follows from the equation of the spectral curve. The action variables are simply given by

$$I_k = \frac{1}{2\pi} \oint_{\alpha_k} \log \lambda(\mu)\, d\mu, \qquad k = 1, \ldots, n, \tag{2.12.550}$$

where α_k are appropriate cycles describing the motion of the separated variables μ_{γ_k}. Since these variables live in the spectral curve, we can anticipate that the cycles α_k will be cycles of this curve, as it happens in one-dimensional motion. Indeed, this is the case. The spectral curve has $2n + 2$ branch points defined by

$$t(\mu) = \pm 1. \tag{2.12.551}$$

We will denote them by β_i, $i = 0, \ldots, 2n + 1$, and we will order them as

$$\beta_0 < \beta_1 < \cdots < \beta_{2n+1}. \tag{2.12.552}$$

Reality of λ requires that $t^2(\mu) - 1 > 0$. The n compact intervals in the real line where this happens are

$$[\beta_i, \beta_{i+1}], \quad i = 1, 3, \ldots, 2n - 1. \tag{2.12.553}$$

These intervals are sometimes called "forbidden zones" or "intervals of instability" for the spectral curve, and they provide the possible intervals of motion for the μ_{γ_k} coordinates. Therefore, the cycles α_k are simply the cycles surrounding the intervals of instability. This determines the action variables of the problem.

So far we have discussed the classical integrable system. Clearly, this system can be quantized by promoting the canonically conjugate variables (q_i, p_i), $i = 1, \ldots, n + 1$, to Heisenberg operators $(\mathsf{q}_i, \mathsf{p}_i)$, satisfying canonical commutation relations

$$[\mathsf{q}_i, \mathsf{p}_j] = i\hbar\delta_{ij}, \qquad i, j = 1, \ldots, n + 1. \tag{2.12.554}$$

The conserved quantities H_k, $k = 2, \ldots, n + 1$, become quantum operators H_k, and it is possible to show that they are mutually commuting operators. In other words, the quantization of the classical periodic Toda lattice leads to a *quantum integrable system*. One would like to obtain the spectrum of the complete set of commuting operators $\mathsf{H}_2, \ldots, \mathsf{H}_k$. Remarkably, one can find an exact solution to this problem, as first shown by M. Gutzwiller. This solution is relatively involved, but one can still find a reasonable approximation to the spectrum by using the EBK quantization conditions. Since the motion can be reduced to a one-dimensional motion along the intervals of instability, with two turning points, we can use the quantization conditions with Maslov indices $\mu_j = 2$. We then find:

$$I_k = \hbar \left(m_k + \frac{1}{2} \right), \qquad k = 1, \ldots, n, \tag{2.12.555}$$

where the action variables are given in (2.12.550). Let us now examinate in more detail the EBK quantization conditions for the Toda lattice with $n = 1$ and $n = 2$.

Example 2.12.2 *BS quantization condition for the modified Mathieu equation.* Let us examine in some detail the case $n = 1$. We have

$$2t(\mu) = \mu^2 - H, \tag{2.12.556}$$

where H is the Hamiltonian. Here there is only one interval of instability $[\mu_1, \mu_2]$, where $\mu_{1,2}$ solve

$$\mu^2 - H = -2. \tag{2.12.557}$$

This is illustrated in Figure 2.15. The action variable (2.12.550) can be written as

$$I(H) = \frac{1}{\pi} \int_{\mu_1}^{\mu_2} \cosh^{-1} \left(|t(\mu)| \right) d\mu. \tag{2.12.558}$$

This integral can be computed explicitly in terms of a hypergeometric function,

$$I(H) = \frac{H-2}{2} \, _2F_1 \left(\frac{1}{2}, \frac{1}{2}; 2; \frac{1}{2} \left(1 - \frac{H}{2} \right) \right). \tag{2.12.559}$$

The EBK quantization condition reads now

$$I(H) = \hbar \left(n + \frac{1}{2} \right), \qquad n \in \mathbb{Z}_{\geq 0}, \tag{2.12.560}$$

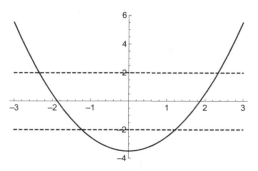

Figure 2.15   In the figure, we plot the curve $2t(\mu) = \mu^2 - H$ as a function of μ (for $H = 7/2$), and the lines ± 2 that intersect the curve at the branch points. There is a single interval of instability satisfying $\mu^2 - H \leq -2$.

Table 2.3. Comparison between the numerical values of the very first energy levels E_n for the periodic Toda lattice with $n = 1$ and the EKB approximation E_n^{BS}, which in this case is the BS approximation. We have set $\hbar = 1$.

m	E_n	E_n^{BS}
0	3.059175	3.030343
1	5.285126	5.259645
2	7.714580	7.691244

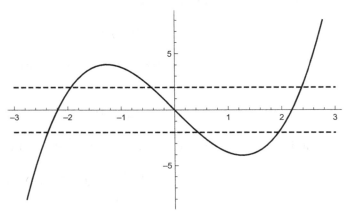

Figure 2.16   The intervals of instability for $n = 2$. In this figure we show the cubic polynomial $2t(\mu)$ in (2.12.561) for $H_1 = H_3 = 0$ and $H_2 = -4.8$.

and it agrees with the BS quantization condition for the modified Mathieu equation (2.12.500). In Table 2.3 we compare the numerical results for the very first energy levels, with the approximation obtained from (2.12.560).   □

Example 2.12.3 In the case $n = 2$, and after setting the total momentum to zero, the spectral curve reads

$$2t(\mu) = \mu^3 - H\mu + H_3, \qquad (2.12.561)$$

where H_3 is the conserved quantity

$$H_3 = p_1 p_2 p_3 - p_1 e^{q_2 - q_3} - p_2 e^{q_3 - q_1} - p_3 e^{q_1 - q_2}. \qquad (2.12.562)$$

In this case there are two intervals of instability, as shown in Figure 2.16. By using the EBK quantization conditions, we can obtain the approximate spectrum of the commuting Hamiltonians H and H_3. The allowed values of (H_3, H) for the very first levels, and with $\hbar = 1$, are shown in Figure 2.17 and tabulated in Table 2.4, where

Table 2.4. Spectrum of the periodic Toda lattice for $n = 2$ and $\hbar = 1$. The first column indicates the quantum numbers (m_1, m_2) that label the states. The second column is the approximate spectrum of (H_3, H) obtained with the EBK quantization condition. The last column is a numerical calculation of the spectrum.

(m_1, m_2)	$(H_3, H)^{\text{EBK}}$	(H_3, H)
$(0, 0)$	$(0, 4.77484016)$	$(0, 4.83741201)$
$(1, 0)$	$(2.41104487, 6.66856054)$	$(2.39338103, 6.72646625)$
$(1, 1)$	$(0, 8.58543359)$	$(0, 8.6385836)$

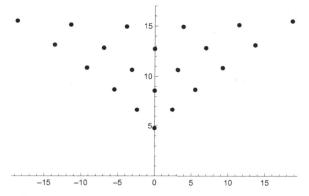

Figure 2.17   The EBK approximation to the spectrum of the commuting Hamiltonians of the periodic Toda lattice, when $n = 2$ and $\hbar = 1$. Each point represents a value (H_3, H).

the result of the EBK approximation is compared to a numerical calculation. We note that spectrum is invariant under $H_3 \rightarrow -H_3$, which corresponds to the exchange of the quantum numbers (m_1, m_2).   □

2.13 Bibliographical Notes

The WKB method is covered in many textbooks in quantum mechanics, although, as noted in Silverstone (1985), the connection formulae appearing in most of them are not correct. A honorable exception is the textbook by Galindo and Pascual (1990), which also follows the treatment of Silverstone (1985) in its presentation. In addition, the chapter on WKB in Galindo and Pascual (1990) is probably the best textbook reference I know on the subject.

The uniform WKB method was introduced in Langer (1949), Cherry (1950), Miller and Good (1953), and is also covered in Galindo and Pascual (1990). A detailed analysis of anharmonic oscillators with the uniform WKB method was performed in Álvarez and Casares (2000a, 2000b), building on previous work in Silverstone et al. (1985a) and Álvarez et al. (2002).

The all-orders Bohr–Sommerfeld quantization condition was first established in Dunham (1932), and explored in detail in Bender et al. (1977), Robnik and

Romanovski (2000), and Robnik and Salasnich (1997). In Robnik and Romanovski (2000) it is shown that higher order corrections to the Bohr–Sommerfeld quantization condition vanish for the harmonic oscillator (this is also shown in Voros (1983), although more indirectly). A rigorous mathematical proof of the oscillation theorem can be found in Berezin and Shubin (2012).

As mentioned in the introduction to the chapter, connection formulae in the WKB method have been a source of confusion and controversy; see for example Berry and Mount (1972) for a summary. The correct connection formula, which uses (lateral) Borel resummation, was first found by André Voros (1981, 1983), and independently by Silverstone (1985). A more mathematical derivation can be found in Kawai and Takei (2005). Simple applications of the Voros–Silverstone connection formulae are discussed in Shen and Silverstone (2004) and Shen et al. (2005).

The Voros–Silverstone connection formula is the foundation of the so-called exact (or complex) WKB method. This method was developed mostly by André Voros in the early 1980s. A first version of the method appears in the course Voros (1981), and it is fully developed in Voros (1983), where it is applied to the example of the pure quartic oscillator. The importance of working in the complex plane in order to fully capture exponentially small corrections was already noted by Balian et al. (1978, 1979).

Another important step in the development of the method was the exact quantization condition for the double-well potential, which was first conjectured by Zinn-Justin (1983) by using instanton methods. This quantization condition is explored in detail in Zinn-Justin and Jentschura (2004a, 2004b), where it is explained how to obtain the exact energies by Borel resummation of the perturbative expansion. A derivation of Zinn-Justin's quantization condition based on the uniform WKB method was first presented by Álvarez (2004), and extended to the periodic potential in Dunne and Ünsal (2014). The ideas of Voros were developed in Delabaere and Pham (1999), Delabaere et al. (1997), and Dillinger et al. (1993), where they were formulated mathematically in the context of the theory of resurgence by Jean Écalle. In these papers, Zinn-Justin's quantization condition for the double-well potential is derived by using the exact WKB method, and some other examples are analyzed in detail.

In this chapter we have focused on the Borel resummation of quantum periods to obtain exact quantization conditions. However one can also apply Borel resummation techniques to the all-orders WKB expansion of the wavefunction itself to obtain exact solutions of the Schrödinger equation; see e.g. Nikolaev (2020) for recent results and further references on this direction.

The P/NP relation (2.8.338) in the double-well potential was first noted by Álvarez (2004), following previous works in Hoe et al. (1982), Álvarez and Casares (2000a), Álvarez et al. (2002). It was further developed and exploited in Dunne and Ünsal (2014), Başar and Dunne (2015), Başar et al. (2017), and Codesido and Mariño (2018). The BenderWu code to compute perturbative expansions in one-dimensional quantum mechanics is due to Sulejmanpasic and Ünsal (2018).

The analysis of spectral functions for confining potentials in Section 2.7 follows the work of Voros (1980) and Voros (1983).

The connection between large-order behavior of the perturbative series and nonperturbative effects was pioneered by Lam (1968) and Bender and Wu (1973).

The application to the double-well potential is worked out by Brezin et al. (1977) and Zinn-Justin and Jentschura (2004a, 2004b).

The pure quartic potential was revisited in Grassi et al. (2015), where explicit computations of Borel resummations and energy levels are reported. A procedure to calculate the energy spectrum of the quartic potential, the double-well potential, and other potentials without explicit use of nonperturbative corrections can be found in Serone et al. (2017).

The exact WKB method has proved to be a very fruitful arena in mathematical physics. In the case of monic potentials (2.6.183), the quantum periods and the exact quantization conditions obtained using this method can be formulated in terms of integral equations appearing in integrable systems, leading to the so-called ODE/IM correspondence of Dorey and Tateo (1999); see Dorey et al. (2007) for a review. This correspondence can be extended to general polynomial potentials by using ideas from supersymmetric gauge theories; see e.g. Gaiotto et al. (2009, 2010), Gaiotto (2014), and Ito et al. (2019). In particular, Ito et al. (2019) use the (lateral) Borel resummation of the WKB expansion to test the exact quantization condition for the double-well potential.

General aspects of classical integrable systems are nicely summarized in Arnold (1989). An excellent, modern reference text is Babelon et al. (2003). Useful presentations of the WKB/EKB approximation for integrable systems can be found in De Almeida (1990) and Berry (1983). The derivation of (2.11.438) is based on results of Dirac (1981) and Van Vleck (1928), and it is well summarized in the book by Zinn-Justin (1996). The change of the wavefunction in quantum mechanics under canonical transformations was derived by Dirac (1933). The linear case is analyzed in detail in Moshinsky and Quesne (1971).

There is a large volume of literature on the periodic Toda lattice (both classical and quantum). Here we follow the presentation in Babelon et al. (2003), where one can also find references on the classical theory. Sklyanin's treatment is presented in Sklyanin (1985). Important papers on the quantum Toda lattice are due to Gutzwiller (1980, 1981) and Gaudin and Pasquier (1992), who propose, in addition, exact quantization conditions for the spectrum. The paper by Matsuyama (1992) compares the results of the EBK quantization condition with the numerical calculation of the spectrum. In more recent years it has been found that the exact spectrum of the periodic Toda lattice can be obtained in an efficient way by exploiting a surprising relationship to supersymmetric gauge theory found by Nekrasov and Shatashvili (2009). The equivalence between the Nekrasov–Shatashvili quantization conditions and the ones obtained by Gutzwiller (1980, 1981) and Gaudin and Pasquier (1992) has been established by Kozlowski and Teschner (2010).

3 The Phase Space Formulation of Quantum Mechanics

3.1 Introduction

The standard formulation of quantum mechanics, developed in the 1920s by Heisenberg, Born, Jordan, Schrödinger, Dirac, and others, culminated in the mathematical synthesis due to Von Neumann in the early 1930s. This synthesis is based on the theory of self-adjoint operators on a Hilbert space. Vectors in this Hilbert state represent pure states of the quantum system, while operators represent physical observables. A general, mixed state is represented by a density operator.

This mathematical language is clearly very different to that of classical physics, where states are probability distributions in phase space, and physical observables are functions on phase space. This mismatch leads to many conceptual and technical difficulties. In particular, we expect quantum mechanics to become equivalent to classical mechanics in the limit in which $\hbar$ is small as compared to the actions involved in the problem. However, it is not easy to see how functions and distributions on phase space can be obtained as limiting objects if one starts with operators in a Hilbert space. Clearly, a formulation of quantum mechanics in the mathematical language of classical mechanics would make this problem easier to solve, and would provide us with the rich geometric intuition associated with phase space.

Such a formulation was first sketched by Wigner in 1932 when he was trying to understand the semiclassical limit of quantum statistical mechanics. Wigner's approach was further developed by Groenewold and Moyal in 1940–1950, and various additions to the theory were made in subsequent years, leading to what we will call the *phase space formulation of quantum mechanics*. In this formulation, the basic object describing a state is Wigner's function, which is defined on phase space and has many of the properties of a probability distribution in classical statistical mechanics. However, we know that quantum mechanics is not a classical statistical theory, and indeed Wigner's function can be negative, in contrast to standard probability distributions. For this reason, it is often called a *quasi-probability distribution*.

In the phase space formulation of quantum mechanics, observables are again functions on phase space, but one has to modify the usual product of functions and to introduce an associative, noncommutative product called the *star product*. This product has the property that, as $\hbar \to 0$, one obtains the conventional product of functions used in classical mechanics. Therefore, the semiclassical limit of quantum mechanics is much easier to study in the phase space formulation.

3.2 Weyl–Wigner Quantization

In quantum mechanics, we typically start with a classical system that is then "quantized." This means, first of all, that we associate a Hilbert space $\mathcal{H}$ of states to the classical system. In addition, we should be able to associate every classical observable (i.e. to every real function on phase space) to a self-adjoint operator acting on $\mathcal{H}$. In the phase space formulation of quantum mechanics, we want to use functions in phase space as our observables. Therefore, it is natural to start our discussion by looking in more detail at the correspondence between classical functions and operators in quantum mechanics.

In this chapter, we will focus on the phase space for one-dimensional motion, and on its quantization. The extension to higher dimensions is straightforward.

Let us then consider the phase space $\mathbb{R}^2$, with coordinates (q, p), corresponding to a one-dimensional particle with position q and momentum p. The quantization of this theory leads to the Hilbert space $\mathcal{H} = L^2(\mathbb{R})$. The functions q and p on phase space become two operators q and p:

$$q \to \mathsf{q}, \qquad p \to \mathsf{p}, \tag{3.2.1}$$

acting on $L^2(\mathbb{R})$, and satisfying the canonical commutation relation

$$[\mathsf{q}, \mathsf{p}] = i\hbar. \tag{3.2.2}$$

If the Hamiltonian of the system is of the standard form,

$$H = \frac{p^2}{2m} + V(q), \tag{3.2.3}$$

its quantization leads to the Hamiltonian operator (1.3.83). However, the quantization prescription (3.2.1), when applied to general functions on phase space, is ambiguous, due to *ordering* issues. Let us consider, for example, the classical observable on $\mathbb{R}^2$,

$$q^2 p^2. \tag{3.2.4}$$

Clearly, the quantization of this function can not be $\mathsf{q}^2\mathsf{p}^2$, since this is not self-adjoint. We can fix this by considering the operator

$$\frac{1}{2}\left(\mathsf{q}^2\mathsf{p}^2 + \mathsf{p}^2\mathsf{q}^2\right), \tag{3.2.5}$$

and this is one possible quantization procedure (called "symmetrized pseudodifferential operator quantization"). However, another possibility is to consider the operator

$$\frac{1}{6}\left(\mathsf{q}^2\mathsf{p}^2 + \mathsf{p}^2\mathsf{q}^2 + \mathsf{p}\mathsf{q}\mathsf{p}\mathsf{q} + \mathsf{p}\mathsf{q}^2\mathsf{p} + \mathsf{q}\mathsf{p}^2\mathsf{q} + \mathsf{q}\mathsf{p}\mathsf{q}\mathsf{p}\right), \tag{3.2.6}$$

which is also self-adjoint. This procedure, in which we consider all possible orderings of a given monomial in q and p, is an example of *Weyl quantization*, and will be our choice of quantization prescription in this chapter.

Weyl quantization is a procedure for obtaining quantum operators starting with functions on the phase space $\mathbb{R}^2$ (the extension to $\mathbb{R}^{2n}$ is straightforward). More formally, it is a map:

$$W : \text{functions on phase space} \to \text{operators on } L^2(\mathbb{R}). \tag{3.2.7}$$

In order to define this map, we will first define it on polynomials, and then we will propose a more general framework. Given a monomial in q and p, we define its Weyl quantization or *Weyl transform* as

$$W(q^j p^k) = \frac{1}{(j+k)!} \sum_{\sigma \in S_{j+k}} \sigma(\mathsf{q}, \mathsf{q}, \ldots, \mathsf{q}, \mathsf{p}, \mathsf{p}, \ldots, \mathsf{p}), \qquad j, k \in \mathbb{Z}_{>0}. \qquad (3.2.8)$$

Here, S_n is the symmetric group of n elements, and for any set of n operators $A_1, \ldots, A_n$ and any $\sigma \in S_n$, we define

$$\sigma(A_1, \ldots, A_n) = A_{\sigma(1)} \cdots A_{\sigma(n)}. \qquad (3.2.9)$$

The map W is extended to arbitrary polynomials by linearity. It is easy to see that it satisfies

$$W\left((uq + vp)^j\right) = (u\mathsf{q} + v\mathsf{p})^j, \qquad u, v \in \mathbb{R}, \quad j \in \mathbb{Z}_{>0}. \qquad (3.2.10)$$

Example 3.2.1 Let

$$f(q) = \sum_{n=0}^{d} a_n q^n, \qquad (3.2.11)$$

be a polynomial. Let us calculate $W(f(q)p)$. By linearity, we have

$$W(f(q)p) = \sum_{n=0}^{d} a_n W(q^n p). \qquad (3.2.12)$$

On the other hand,

$$\begin{aligned}
W(q^n p) &= \frac{1}{n+1} \sum_{j=0}^{n} \mathsf{q}^{n-j} \mathsf{p} \mathsf{q}^j = \frac{1}{n+1} \sum_{j=0}^{n} \left(\mathsf{q}^n \mathsf{p} + j[\mathsf{p}, \mathsf{q}]\mathsf{q}^{n-1}\right) \\
&= \frac{1}{n+1}\left((n+1)\mathsf{q}^n \mathsf{p} + \frac{n(n+1)}{2}(-i\hbar)\mathsf{q}^{n-1}\right),
\end{aligned} \qquad (3.2.13)$$

and we conclude that

$$W(f(q)p) = \sum_{n=0}^{d} a_n \left(\mathsf{q}^n \mathsf{p} - \frac{i\hbar}{2} n \mathsf{q}^{n-1}\right) = f(\mathsf{q})\mathsf{p} - \frac{i\hbar}{2} f'(\mathsf{q}). \qquad (3.2.14)$$

$\square$

We would like to extend the map W to more general functions on $\mathbb{R}^2$. Let us introduce the following operators, which are sometimes called *Weyl operators*,

$$\mathsf{U}(u) = e^{iu\mathsf{q}}, \qquad \mathsf{V}(v) = e^{iv\mathsf{p}}, \qquad u, v \in \mathbb{R}. \qquad (3.2.15)$$

When acting on a function $\psi(q) \in L^2(\mathbb{R})$, one has

$$\left(\mathsf{U}(u)\psi\right)(q) = e^{iuq}\psi(q), \qquad \left(\mathsf{V}(v)\psi\right)(q) = \psi(q + \hbar v). \qquad (3.2.16)$$

We now recall a simplified version of the Baker–Campbell–Hausdorff formula: given two operators A, B, such that

$$[A, [A, B]] = [B, [A, B]] = 0, \tag{3.2.17}$$

one has

$$e^A e^B = e^{A+B+\frac{1}{2}[A,B]}. \tag{3.2.18}$$

Using this, we find that the operators (3.2.15) satisfy the *Weyl commutation relation*,

$$U(u)V(v) = e^{-i\hbar uv}V(v)U(u). \tag{3.2.19}$$

Let us now introduce the operator,

$$S(u, v) = e^{i\hbar uv/2}U(u)V(v). \tag{3.2.20}$$

By using again (3.2.18), we obtain

$$S(u, v) = e^{iuq+ivp}. \tag{3.2.21}$$

In order to motivate the extension of Weyl's map to more general functions, we note that, if we multiply (3.2.10) by $i^j/j!$, and we sum from $j = 0$ to $j = \infty$, we find

$$W\left(e^{iuq+ivp}\right) = S(u, v). \tag{3.2.22}$$

Let us now suppose that we have a function $f(q, p)$ that can be written as a Fourier transform

$$f(q, p) = \frac{1}{2\pi} \int_{\mathbb{R}^2} \check{f}(u, v)e^{i(uq+vp)}\mathrm{d}u\mathrm{d}v. \tag{3.2.23}$$

This happens, for example, if $f \in L^2(\mathbb{R}^2)$. We can now promote this function to an operator by using (3.2.22) as a guiding principle. It is then natural to define the *Weyl transform* of the function f, $W(f)$, as

$$W(f) = \frac{1}{2\pi} \int_{\mathbb{R}^2} \check{f}(u, v)S(u, v)\mathrm{d}u\mathrm{d}v. \tag{3.2.24}$$

This can be interpreted as a noncommutative version of the Fourier transform. The above formula is not very explicit, since it gives $W(f)$ as an integral of an operator. However, we can easily calculate its integral kernel, which we denote by

$$K_f(q, q') = \langle q|W(f)|q'\rangle. \tag{3.2.25}$$

We find,

$$\begin{aligned}
K_f(q, q') &= \frac{1}{2\pi} \int_{\mathbb{R}^2} \check{f}(u, v)e^{\frac{i\hbar}{2}uv+iuq}\langle q|e^{ivp}|q'\rangle\mathrm{d}u\mathrm{d}v \\
&= \frac{1}{2\pi} \int_{\mathbb{R}^2} \check{f}(u, v)e^{\frac{i\hbar}{2}uv+iuq}\langle q|q' - v\hbar\rangle\mathrm{d}u\mathrm{d}v \\
&= \frac{1}{2\pi} \int_{\mathbb{R}^2} \check{f}(u, v)e^{\frac{i\hbar}{2}uv+iuq}\delta(q - q' + v\hbar)\mathrm{d}u\mathrm{d}v \\
&= \frac{1}{2\pi\hbar} \int \check{f}\left(u, \frac{q'-q}{\hbar}\right)e^{\frac{1}{2}(q+q')u}\mathrm{d}u.
\end{aligned} \tag{3.2.26}$$

By using the explicit inverse,

$$\check{f}(u, v) = \frac{1}{2\pi} \int f(q, p) e^{-i(uq + vp)} dq dp, \qquad (3.2.27)$$

and plugging it in the final formula in (3.2.26), we find

$$K_f(q, q') = \frac{1}{2\pi\hbar} \int_{\mathbb{R}} f\left(\frac{q + q'}{2}, p\right) e^{\frac{i}{\hbar} p(q - q')} dp. \qquad (3.2.28)$$

The expression (3.2.28) produces the integral kernel of the operator $W(f)$ from a function $f(q, p)$ in phase space. We can also deduce that

$$\int_{\mathbb{R}^2} |K_f(q, q')|^2 dq dq' = \frac{1}{2\pi\hbar} \int_{\mathbb{R}^2} |\check{f}(u, v)|^2 du dv = \frac{1}{2\pi\hbar} \int_{\mathbb{R}^2} |f(q, p)|^2 dq dp, \qquad (3.2.29)$$

which can be obtained by applying Plancherel's theorem. Therefore, if $f \in L^2(\mathbb{R}^2)$, the kernel is square integrable, as a function on $\mathbb{R}^2$. Such a kernel defines a *Hilbert–Schmidt* (HS) operator. In particular, it is a well-defined operator on $L^2(\mathbb{R})$, in the sense that it maps functions in $L^2(\mathbb{R})$ to functions in $L^2(\mathbb{R})$. We will denote the space of HS operators on a Hilbert space $\mathcal{H}$ by $\mathcal{J}_2(\mathcal{H})$.

We conclude that the Weyl transform (3.2.24) can be regarded as a map

$$W: L^2(\mathbb{R}^2) \rightarrow \mathcal{J}_2\left(L^2(\mathbb{R})\right), \qquad (3.2.30)$$

from the set of square-integrable functions on phase space $\mathbb{R}^2$, to the space of HS operators on $L^2(\mathbb{R})$. We can construct an explicit inverse map

$$W^{-1}: \mathcal{J}_2(\mathcal{H}) \rightarrow L^2(\mathbb{R}^2), \qquad (3.2.31)$$

which is given by an inverse Fourier transform

$$W^{-1}(K) = K_W(q, p) = \int_{\mathbb{R}} K\left(q - \frac{q'}{2}, q + \frac{q'}{2}\right) e^{\frac{i}{\hbar} p q'} dq'. \qquad (3.2.32)$$

Therefore, given an operator K with kernel $K(q, q')$, we obtain a function in phase space $K_W(q, p)$. This function is called the *Wigner transform* of the operator K. We can also write

$$K_W(q, p) = \int_{\mathbb{R}} dq' \left\langle q - \frac{q'}{2} \left| K \right| q + \frac{q'}{2} \right\rangle e^{ipq'/\hbar}. \qquad (3.2.33)$$

Although the above formulae were obtained for square-integrable functions in phase space, they can be extended to arbitrary functions by using distributions. In particular, if $f(q, p)$ is a polynomial in q, p, we recover from (3.2.24), (3.2.28) the definition of Weyl transform (3.2.8). For example, if $f(q, p) = qp$, we find

$$K_f(q, q') = \frac{q + q'}{2} \int \frac{dp}{2\pi\hbar} p \, e^{\frac{i}{\hbar} p(q - q')} dp = \left\langle q \left| \frac{qp + pq}{2} \right| q' \right\rangle, \qquad (3.2.34)$$

as expected.

3.3 Star Product

We will now discuss some properties of the Weyl map and the Wigner transform.

The first property is that the *conjugation of phase space functions corresponds to Hermitian conjugation of operators*, i.e.

$$(W(f))^\dagger = W(f^*). \qquad (3.3.35)$$

This is easy to verify, since the kernel of the adjoint operator $(W(f))^\dagger$ is given by

$$K_f^\dagger(q, q') = K_f^*(q', q), \qquad (3.3.36)$$

therefore, from the expression (3.2.28), we find

$$K_f^\dagger(q, q') = \frac{1}{2\pi\hbar} \int_{\mathbb{R}} f^* \left(\frac{q + q'}{2}, p \right) e^{\frac{i}{\hbar} p(q - q')} dp = K_{f^*}(q, q'). \qquad (3.3.37)$$

Another important result concerns the *calculation of traces*. It follows immediately from (3.2.28) that the trace of the operator $W(f)$ (when it exists) is given by

$$\operatorname{Tr} W(f) = \int_{\mathbb{R}} K_f(q, q) dq = \frac{1}{2\pi\hbar} \int_{\mathbb{R}^2} f(q, p) dp dq. \qquad (3.3.38)$$

This can be also written as

$$\operatorname{Tr} \mathsf{K} = \frac{1}{2\pi\hbar} \int_{\mathbb{R}^2} K_\mathsf{W}(q, p) dp dq. \qquad (3.3.39)$$

The most important property of the Wigner transform is that the product of operators is mapped to the *star product* (or $\star$-product, for short) of functions. Let f_1, f_2 be two functions on phase space, $W(f_1)$, $W(f_2)$ be their Weyl transforms, and $W(f_1)W(f_2)$ their product. It turns out that the Wigner transform of this product operator, which is a function on phase space, can be written as a generalized, $\hbar$-dependent product of the functions f_1 and f_2 called the $\star$-product. We will then write,

$$f_1 \star f_2 = W^{-1}(W(f_1)W(f_2)). \qquad (3.3.40)$$

We will now show that the $\star$-product can be written explicitly as

$$(f_1 \star f_2)(q, p)$$
$$= \frac{1}{(2\pi)^2} \int_{\mathbb{R}^4} e^{-\frac{i\hbar}{2}(u_1 v_2 - u_2 v_1) + iq(u_1 + u_2) + ip(v_1 + v_2)} \check{f}_1(u_1, v_1) \check{f}_2(u_2, v_2) du_1 dv_1 du_2 dv_2. \qquad (3.3.41)$$

The proof of this is straightforward. We first note that

$$\left[W^{-1}(W(f_1)W(f_2)) \right](q, p) = \int_{\mathbb{R}} (W(f_1)W(f_2)) \left(q - \frac{1}{2}v, q + \frac{1}{2}v \right) e^{\frac{i}{\hbar} pv} dv. \qquad (3.3.42)$$

But

$$(W(f_1)W(f_2))(q_1, q_2) = \int_{\mathbb{R}} K_{f_1}(q_1, q'')K_{f_2}(q'', q_2)\mathrm{d}q''$$

$$= \frac{1}{(2\pi\hbar)^2} \int_{\mathbb{R}^3} \check{f}_1\left(u_1, \frac{q'' - q_1}{\hbar}\right) \check{f}_2\left(u_2, \frac{q_2 - q''}{\hbar}\right) \quad (3.3.43)$$

$$\times e^{\frac{i}{2}(q_1 + q'')u_1 + \frac{i}{2}(q'' + q_2)u_2} \mathrm{d}q'' \mathrm{d}u_1 \mathrm{d}u_2.$$

Therefore,

$$\left[W^{-1}(W(f_1)W(f_2))\right](q, p)$$

$$= \frac{1}{(2\pi\hbar)^2} \int_{\mathbb{R}^4} \check{f}_1\left(u_1, \frac{1}{\hbar}\left(q'' - q + \frac{v}{2}\right)\right) \check{f}_2\left(u_2, \frac{1}{\hbar}\left(q + \frac{v}{2} - q''\right)\right)$$

$$\times \exp\left\{\frac{i}{\hbar}pv + \frac{i}{2}\left(q - \frac{v}{2} + q''\right)u_1\right. \quad (3.3.44)$$

$$\left. + \frac{i}{2}\left(q'' + q + \frac{v}{2}\right)u_2\right\} \mathrm{d}q'' \mathrm{d}u_1 \mathrm{d}u_2 \mathrm{d}v.$$

We now change variables

$$v, q'' \to v_1, v_2, \quad (3.3.45)$$

with

$$\hbar v_1 = q'' - q + \frac{v}{2},$$

$$\hbar v_2 = q + \frac{v}{2} - q''. \quad (3.3.46)$$

This is inverted as

$$v = \hbar(v_1 + v_2),$$

$$q'' = q + \hbar\frac{v_1 - v_2}{2}, \quad (3.3.47)$$

and the determinant of the Jacobian is $\hbar^2$. We find, for the exponent,

$$\frac{i}{\hbar}pv + \frac{i}{2}\left(q - \frac{v}{2} + q''\right)u_1 + \frac{i}{2}\left(q'' + q + \frac{v}{2}\right)u_2$$

$$= -\frac{i\hbar}{2}(u_1 v_2 - u_2 v_1) + iq(u_1 + u_2) + ip(v_1 + v_2), \quad (3.3.48)$$

and we obtain the wished-for result.

Let us note that the $\star$-product inherits the properties of the product of operators: it is not commutative, but it is associative, i.e. one has

$$f_1 \star (f_2 \star f_3) = (f_1 \star f_2) \star f_3. \quad (3.3.49)$$

In addition, one has

$$\overline{f_1 \star f_2} = \overline{f_2} \star \overline{f_1}. \quad (3.3.50)$$

Example 3.3.1 *Star product of Gaussian functions.* A particularly important example of the $\star$-product occurs for Gaussian functions. Let us consider

$$f_1(q, p) = \exp\left(-a(q^2 + p^2)\right), \qquad f_2(q, p) = \exp\left(-b(q^2 + p^2)\right). \quad (3.3.51)$$

Their Fourier transforms are

$$\check{f}_1(u,v) = \frac{1}{2\pi} \int_{\mathbb{R}^2} e^{-a(q^2+p^2)} e^{-i(uq+vp)} dqdp = \frac{1}{2a} e^{-\frac{1}{4a}(u^2+v^2)}, \qquad (3.3.52)$$

and similarly for $\check{f}_2(u,v)$ (with $a \leftrightarrow b$). The integral in (3.3.41) is given by four inverse Fourier transforms of Gaussians, and after some elementary calculations, one obtains,

$$(f_1 \star f_2)(q,p) = \frac{1}{1+ab\hbar^2} \exp\left[-\frac{a+b}{1+ab\hbar^2}\left(q^2+p^2\right)\right]. \qquad (3.3.53)$$

Note that, when $\hbar \to 0$,

$$(f_1 \star f_2)(q,p) \approx \exp\left(-(a+b)\left(q^2+p^2\right)\right) = (f_1 \cdot f_2)(q,p), \qquad (3.3.54)$$

and becomes the ordinary product. The product law (3.3.53) is called the hyperbolic tangent $\star$-composition law of Gaussians, since if we write

$$a = \frac{1}{\hbar}\tanh(\theta_1), \qquad b = \frac{1}{\hbar}\tanh(\theta_2), \qquad (3.3.55)$$

we have that

$$\frac{a+b}{1+ab\hbar^2} = \frac{1}{\hbar}\tanh(\theta_1+\theta_2). \qquad (3.3.56)$$

$\square$

The formula (3.3.41) for the star product involves the Fourier transform of the functions $f_{1,2}(q,p)$, which only exists as a function under certain conditions (for example, if $f_{1,2} \in L^2(\mathbb{R}^2)$, as we have assumed). However, this formula can be extended to more general functions. One way to do this is to use the following form of the star product. Let us write (3.3.41) as

$$(f_1 \star f_2)(q,p)$$
$$= \frac{1}{(2\pi)^2} \int_{\mathbb{R}^4} \check{f}_1(u_1,v_1) e^{i(qu_1+pv_1)} e^{-\frac{i\hbar}{2}(u_1v_2-u_2v_1)} e^{i(qu_2+pv_2)} \check{f}_2(u_2,v_2) du_1 dv_1 du_2 dv_2. \qquad (3.3.57)$$

The exponent

$$e^{i(qu_1+pv_1)} e^{-\frac{i\hbar}{2}(u_1v_2-u_2v_1)} e^{i(qu_2+pv_2)}, \qquad (3.3.58)$$

can be written as

$$e^{i(qu_1+pv_1)} \exp\left[\frac{i\hbar}{2}\overset{\leftrightarrow}{\Lambda}\right] e^{i(qu_2+pv_2)}, \qquad (3.3.59)$$

where the operator $\overset{\leftrightarrow}{\Lambda}$ is given by

$$\overset{\leftrightarrow}{\Lambda} = \overset{\leftarrow}{\partial}_q \overset{\rightarrow}{\partial}_p - \overset{\leftarrow}{\partial}_p \overset{\rightarrow}{\partial}_q, \qquad (3.3.60)$$

and the arrows indicate the direction in which the derivatives act. We find that the r.h.s. of (3.3.41) is given by

$$\left(\int e^{i(qu_1+pv_1)} \check{f}_1(u_1,v_1) \frac{du_1 dv_1}{2\pi}\right) \exp\left[\frac{i\hbar}{2}\overset{\leftrightarrow}{\Lambda}\right] \left(\int e^{i(qu_2+pv_2)} \check{f}_2(u_2,v_2) \frac{du_2 dv_2}{2\pi}\right), \qquad (3.3.61)$$

i.e.

$$(f_1 \star f_2)(q, p) = f_1(q, p) \exp\left[\frac{i\hbar}{2} \overset{\leftrightarrow}{\Lambda}\right] f_2(q, p). \tag{3.3.62}$$

This expression is extremely useful to obtain asymptotic, formal $\hbar$ expansions of the $\star$-product. At next-to-leading order in an expansion in $\hbar$, we find

$$(f_1 \star f_2)(q, p) = f_1(q, p) f_2(q, p) + \frac{i\hbar}{2} \{f_1, f_2\} + \cdots, \tag{3.3.63}$$

where $\{\cdot, \cdot\}$ is the Poisson bracket (2.11.396). An alternative form of the product involves the so-called *Bopp shifts*:

$$f_1(q, p) \star f_2(q, p) = f_1\left(q, p - \frac{i\hbar}{2}\overset{\rightarrow}{\partial}_q\right) f_2\left(q, p + \frac{i\hbar}{2}\overset{\leftarrow}{\partial}_q\right)$$

$$= f_1\left(q + \frac{i\hbar}{2}\overset{\rightarrow}{\partial}_p, p - \frac{i\hbar}{2}\overset{\rightarrow}{\partial}_q\right) f_2(q, p) \tag{3.3.64}$$

$$= f_1(q, p) f_2\left(q - \frac{i\hbar}{2}\overset{\leftarrow}{\partial}_p, p + \frac{i\hbar}{2}\overset{\leftarrow}{\partial}_q\right).$$

An obvious consequence of this discussion is that the algebra of operators in quantum mechanics (w.r.t. their standard product) becomes, through the Wigner–Weyl transform, the $\star$-algebra of functions on phase space. In particular, if A, B are two operators acting on $L^2(\mathbb{R})$, the Wigner transform of their product is the $\star$-product of Wigner transforms.

$$(AB)_W = A_W \star B_W, \tag{3.3.65}$$

or equivalently,

$$W^{-1}(AB) = W^{-1}(A) \star W^{-1}(B). \tag{3.3.66}$$

Example 3.3.2 Let us give some simple examples of the $\star$-product, by using (3.3.62). One has,

$$q \star p = qp + \frac{i\hbar}{2}, \qquad p \star q = qp - \frac{i\hbar}{2}. \tag{3.3.67}$$

More generally, if $f(q)$ is a polynomial in q, we have

$$f(q) \star p = f(q)p + \frac{i\hbar}{2} f'(q), \tag{3.3.68}$$

which can be obtained from (3.2.14) by applying W^{-1}.   □

An easy consequence of (3.3.41) is that

$$\int dqdp \, f_1 \star f_2 = \int dqdp \, f_2 \star f_1 = \int dqdp \, f_1 f_2. \tag{3.3.69}$$

Indeed, when we integrate (3.3.41), the integrations over p and q lead to two delta functions that set $u_1 = -u_2$, $v_1 = -v_2$, and we finally find

$$\int \check{f}_1(u, v) \check{f}_2(-u, -v) dudv = \int dqdp \, f_1 f_2, \tag{3.3.70}$$

where the $\hbar$ dependence drops out.

If the star product is the analogue of the product of operators for Wigner transforms, the analogue of the commutator is the *Moyal bracket*, defined by

$$[f_1, f_2]_\star = f_1 \star f_2 - f_2 \star f_1. \tag{3.3.71}$$

By using the explicit expression (3.3.62) for the $\star$-product, we can write

$$[f_1, f_2]_\star = 2i f_1 \sin\left(\frac{\hbar}{2} \overleftrightarrow{\Lambda}\right) f_2. \tag{3.3.72}$$

In the limit $\hbar \to 0$, one finds

$$[f_1, f_2]_\star = i\hbar\{f_1, f_2\} + \mathcal{O}(\hbar^3), \tag{3.3.73}$$

where $\{\,,\}$ is the classical Poisson bracket. This makes Dirac's correspondence between quantum commutators and Poisson brackets more transparent. Note that, if f_1, f_2 are at most quadratic in q, p, one has

$$[f_1, f_2]_\star = i\hbar\{f_1, f_2\}, \tag{3.3.74}$$

with no further corrections.

3.4 Quantum Mechanics in Phase Space

Weyl–Wigner quantization suggests that, instead of formulating quantum mechanics in terms of wave functions in $L^2(\mathbb{R})$, we should be able to formulate it in terms of functions on phase space, provided that we use the star product instead of the usual product. This leads to the *phase space* or *Wigner formulation* of quantum mechanics.

The starting point of the phase space formulation is the description of states. In the conventional formulation of quantum mechanics, a state is described by a density operator. In particular, a pure state $|\psi\rangle$ leads to the density operator

$$\rho_\psi = |\psi\rangle\langle\psi|. \tag{3.4.75}$$

We can obtain a function in phase space associated to the state $|\psi\rangle$ by simply considering the Wigner transform of the corresponding density operator (with an extra $2\pi\hbar$ factor). The function

$$f_\psi(q, p) = \frac{1}{2\pi\hbar} W^{-1}(\rho_\psi) = \frac{1}{2\pi\hbar} \int_\mathbb{R} \psi^*\left(q + \frac{\xi}{2}\right) \psi\left(q - \frac{\xi}{2}\right) e^{\frac{i}{\hbar}p\xi} d\xi, \tag{3.4.76}$$

is called the *Wigner function* or *Wigner distribution* associated to the wavefunction ψ. The overall factor has been chosen in such a way that, if $|\psi\rangle$ is normalized, $f_\psi(q, p)$ satisfies

$$\int_{\mathbb{R}^2} f_\psi(q, p) dq dp = 1. \tag{3.4.77}$$

In addition, the Wigner function $f_\psi(q, p)$ is square normalizable as a function on $\mathbb{R}^2$. To see this, we note that, if the wavefunction ψ has unit norm, the corresponding density operator satisfies $\rho_\psi^2 = \rho_\psi$, therefore the Wigner function f_ψ associated to ψ is an idempotent element of the $\star$-algebra, up to a normalization factor:

$$f_\psi \star f_\psi = \frac{1}{2\pi\hbar} f_\psi. \tag{3.4.78}$$

By using (3.3.69), we find

$$\int_{\mathbb{R}^2} f^2(q,p)\mathrm{d}q\mathrm{d}p = \frac{1}{2\pi\hbar}. \tag{3.4.79}$$

Since $f_\psi(q,p)$ is real, we conclude that it belongs to $L^2(\mathbb{R}^2)$. Another way to see this is to use the map (3.2.30), and to use the fact that the operator ρ_ψ is manifestly HS.

The Wigner function provides all the necessary information on the quantum probability distributions associated to the state $|\psi\rangle$. Indeed, from its definition it is easy to check that

$$\int_{\mathbb{R}} f_\psi(q,p)\mathrm{d}p = |\psi(q)|^2, \tag{3.4.80}$$

which is the probability distribution for q. Similarly, one finds that

$$\int_{\mathbb{R}} f_\psi(q,p)\mathrm{d}q = |\widehat{\psi}(p)|^2, \tag{3.4.81}$$

where

$$\widehat{\psi}(p) = \frac{1}{\sqrt{2\pi\hbar}} \int_{-\infty}^{\infty} \mathrm{e}^{-ipq/\hbar}\psi(q)\mathrm{d}q. \tag{3.4.82}$$

Averages of operators can be also easily computed in this formulation. Indeed, we have

$$\langle\psi|O|\psi\rangle = \mathrm{Tr}\left(\rho_\psi O\right) = \langle O_W\rangle_{f_\psi}, \tag{3.4.83}$$

where the average of a function $g(q,p)$, in the state defined by the Wigner function, f_ψ, is defined as

$$\langle g\rangle_{f_\psi} = \int_{\mathbb{R}^2} f_\psi(q,p)g(q,p)\mathrm{d}q\mathrm{d}p. \tag{3.4.84}$$

In deriving this equation we have used (3.3.39), (3.3.65), and (3.3.69). We note the average (3.4.84) is precisely what one would compute in classical statistical mechanics for a probability distribution $f_\psi(q,p)$. However, this analogy is only partial, since Wigner functions, in contrast to classical probability distributions on phase space, can be *negative*. This is the price to pay for describing quantum mechanics in this classical language.

Another property of Wigner functions that does not have a classical counterpart is that they are bounded:

$$|f_\psi(q,p)| \leq \frac{1}{\pi\hbar}. \tag{3.4.85}$$

To see how this comes about, let us consider the wavefunctions

$$\psi_1(q') = \frac{1}{\sqrt{2}}\mathrm{e}^{-ipq'/\hbar}\psi\left(q + \frac{q'}{2}\right), \qquad \psi_2(q') = \frac{1}{\sqrt{2}}\psi\left(q - \frac{q'}{2}\right). \tag{3.4.86}$$

They have unit norm, and we can write Wigner's function as the product of these two normalized wavefunctions.

$$f_\psi(q,p) = \frac{1}{\pi\hbar} \int_\mathbb{R} \psi_1^*(q')\psi_2(q')\mathrm{d}q'. \tag{3.4.87}$$

We can now use Schwarz's inequality to deduce that

$$\left|f_\psi(q,p)\right| \le \frac{1}{\pi\hbar}\|\psi_1\|\|\psi_2\| = \frac{1}{\pi\hbar}. \tag{3.4.88}$$

It is easy to see that if ψ is even or odd, the bound is saturated when $q = p = 0$.

Example 3.4.1 *Wigner function of a coherent state.* A coherent state is described by the normalized wavefunction

$$\psi(q) = \frac{1}{(2\pi\Delta^2)^{\frac{1}{4}}} \exp\left\{\frac{\mathrm{i}}{\hbar}\langle p\rangle q - \frac{1}{4\Delta^2}(q - \langle q\rangle)^2\right\}, \tag{3.4.89}$$

where $\Delta > 0$ is a parameter characterizing the state. As is well known, coherent states minimize the product of uncertainties appearing in Heisenberg's principle. For the wavefunction above, we have

$$\Delta q = \Delta, \qquad \Delta p = \frac{\hbar}{2\Delta}. \tag{3.4.90}$$

The corresponding Wigner function is given by:

$$f_\psi(q,p) = \frac{1}{(2\pi)^{3/2}\Delta} \mathrm{e}^{-\frac{1}{2\Delta^2}(q-\langle q\rangle)^2} \int_\mathbb{R} \mathrm{e}^{\mathrm{i}(p-\langle p\rangle)y - \frac{\hbar^2}{8\Delta^2}y^2}\mathrm{d}y, \tag{3.4.91}$$

where we have set $y = \hbar q'$ in (3.4.76). The Fourier transform of the Gaussian can be calculated immediately, and leads to:

$$f_\psi(x,p) = \frac{1}{\pi\hbar} \exp\left\{-\frac{1}{2\Delta^2}(q - \langle q\rangle)^2 - \frac{2\Delta^2}{\hbar^2}(p - \langle p\rangle)^2\right\}. \tag{3.4.92}$$

This is a Gaussian function in phase space, localized around the point $q = \langle q\rangle$, $p = \langle p\rangle$, and with dispersions Δ and $\hbar/(2\Delta)$. Note that this Wigner function, obtained from a coherent state, is everywhere positive. Surprisingly, the converse is true, as proved by Hudson in 1974: a Wigner function that is everywhere positive must be of the form (3.4.92). Since coherent states can be regarded as semiclassical states in Hilbert space, the negativity of the Wigner function has been proposed as an index of nonclassicality of the corresponding state.   □

As the Wigner function lives on phase space, it is interesting to ask what its behavior is under canonical transformations. We will perform the analysis in the one-dimensional case (the analysis in higher dimensions is a straightforward generalization). By using the explicit expression (2.11.450) and the definition of Wigner function, we find

$$f_\psi(q,p) = \frac{|\mathcal{C}|^2}{2\pi\hbar} \int_{\mathbb{R}^3} \mathrm{e}^{\mathrm{i}(F(q-\xi/2,Q_2)-F^*(q+\xi/2,Q_1)+p\xi)/\hbar}\psi^*(Q_1)\psi(Q_2)\mathrm{d}Q_1\mathrm{d}Q_2\mathrm{d}\xi. \tag{3.4.93}$$

We change variables to

$$Q = \frac{Q_1 + Q_2}{2}, \qquad Z = Q_2 - Q_1, \tag{3.4.94}$$

and we use the definition of Wigner function $\widetilde{f}_\psi(Q, P)$ in the transformed coordinates Q, P to write

$$\psi^*(Q - Z/2)\psi(Q + Z/2) = \int_{\mathbb{R}} \widetilde{f}_\psi(Q, P)e^{iPZ/\hbar}. \qquad (3.4.95)$$

We conclude that

$$f_\psi(q, p) = \int_{\mathbb{R}^2} \mathcal{T}(q, p; Q, P)\widetilde{f}_\psi(Q, P)dQdP, \qquad (3.4.96)$$

where the integral kernel $\mathcal{T}(q, p; Q, P)$ is given by

$$\mathcal{T}(q, p; Q, P) = \frac{|\mathcal{C}|^2}{2\pi\hbar} \int_{\mathbb{R}^2} e^{i(F(q-\xi/2,Q+Z/2)-F^*(q+\xi/2,Q-Z/2)+p\xi+PZ)/\hbar}d\xi dZ. \qquad (3.4.97)$$

Let us determine the integral kernel in the case of linear canonical transformations. Since

$$F(q - \xi/2, Q + Z/2) - F^*(q + \xi/2, Q - Z/2)$$
$$= -\frac{1}{2b}\left(-2aq\xi + 2dQZ + 2\xi Q - 2qZ\right), \qquad (3.4.98)$$

the integration in (3.4.97) is immediate and one finds

$$\mathcal{T}(q, p; Q, P) = \delta(Q - aq - bp)\delta(P - cq - dp), \qquad (3.4.99)$$

where we used that $ad - bc = 1$ and the explicit expression (2.11.456) for $\mathcal{C}$. Therefore, under linear canonical transformations, the Wigner functions are related by the change of variables underlying the transformation,

$$f_\psi(q, p) = \widetilde{f}_\psi(aq + bp, cq + dp). \qquad (3.4.100)$$

This contrasts to the transformation rule of the wavefunction, which involves a nontrivial integral transform.

So far we have stated the kinematical aspects of the phase space formulation of quantum mechanics: states are described by Wigner functions, and observables are described by functions in phase space, which are multiplied by the star product. We now introduce a dynamical principle, i.e. the analogue of the time-dependent Schrödinger's equation. This dynamical principle is the *Moyal equation of motion*,

$$i\hbar\frac{\partial f}{\partial t} = [H, f]_\star, \qquad (3.4.101)$$

which describes the evolution in time of the Wigner function. As $\hbar \to 0$, this equation becomes Liouville's equation for the time evolution of a probability density. Moyal's EOM can be derived from the time evolution of the density matrix in the conventional formulation,

$$i\hbar\frac{\partial \rho_\psi}{\partial t} = [H, \rho_\psi], \qquad (3.4.102)$$

and then taking Wigner transforms on both sides. Moyal's EOM can be solved by using the *Moyal propagator* $U_\star(q, p; t)$, which is the Wigner transform of the evolution operator:

$$U_\star(q, p; t) = U(t, 0)_W. \qquad (3.4.103)$$

The Moyal propagator can be defined by the following property: if we are given a Wigner function at $t = 0$, $f(q, p; 0)$, its time evolution is given by

$$f(q, p; t) = U_\star(q, p; t) \star f(q, p; 0) \star U_\star^{-1}(q, p; t), \qquad (3.4.104)$$

where the inverse is defined w.r.t. the star product, i.e.

$$U_\star(q, p; t) \star U_\star^{-1}(q, p; t) = 1. \qquad (3.4.105)$$

It follows from its definition that the Moyal propagator satisfies the composition property

$$U_\star(q, p; t_1) \star U_\star(q, p; t_2) = U_\star(q, p; t_1 + t_2). \qquad (3.4.106)$$

By using (3.4.104) and Moyal's EOM, one deduces that $U_\star(q, p; t)$ satisfies the Wigner transform of the evolution equation (1.2.5), namely

$$i\hbar \frac{\partial U_\star}{\partial t} = H \star U_\star, \qquad (3.4.107)$$

with initial condition

$$U_\star(q, p; 0) = 1. \qquad (3.4.108)$$

In the case of time-independent Hamiltonians, the Moyal propagator can be obtained as

$$U_\star(q, p; t) = e_\star^{-itH/\hbar}, \qquad (3.4.109)$$

where the $\star$-*exponential* of a function in phase space is defined as

$$e_*^g = \sum_{n=0}^{\infty} \frac{1}{n!} g \star \cdots \star g. \qquad (3.4.110)$$

It also follows that

$$U_\star^{-1}(q, p; t) = e_\star^{itH/\hbar}. \qquad (3.4.111)$$

Example 3.4.2 Let us suppose that H is of the standard nonrelativistic form [as in (1.3.83)], and $V(q)$ is at most quadratic in q. Then, the Moyal bracket in (3.4.101) reduces to its classical limit, the Poisson bracket

$$[H, f]_\star = i\hbar\{H, f\} = i \left(\frac{\partial H}{\partial q} \frac{\partial f}{\partial p} - \frac{\partial H}{\partial p} \frac{\partial f}{\partial q} \right) = i\hbar \left(\frac{\partial V}{\partial q} \frac{\partial f}{\partial p} - \frac{p}{m} \frac{\partial f}{\partial q} \right). \quad (3.4.112)$$

It follows that the EOM for Wigner's function is

$$\frac{\partial f}{\partial t} = \frac{\partial V}{\partial q} \frac{\partial f}{\partial p} - \frac{p}{m} \frac{\partial f}{\partial q}. \qquad (3.4.113)$$

This can be solved with the method of characteristics: we consider the functions of t, $q(t)$, $p(t)$, characterized by the standard classical Hamiltonian evolution

$$\dot{q} = \frac{p(t)}{m}, \qquad \dot{p} = -\frac{\partial V}{\partial q}, \qquad (3.4.114)$$

and with initial conditions

$$q(0) = q, \qquad p(0) = p. \qquad (3.4.115)$$

If is easy to check that

$$f(q, p, t) = f(q(-t), p(-t), 0),\qquad(3.4.116)$$

satisfies Moyal's EOM. This just means that, in these cases (including the free particle and the harmonic oscillator), Wigner's function evolves by following classical trajectories.   □

Very often we are interested in Wigner functions associated to an orthonormal, discrete basis $|\psi_n\rangle$ of the Hilbert space $\mathcal{H} = L^2(\mathbb{R})$. These functions satisfy the completeness condition

$$\sum_n |\psi_n\rangle\langle\psi_n| = \mathbf{1}.\qquad(3.4.117)$$

Let us denote by $f_n(q, p)$ the Wigner function associated to the state $|\psi_n\rangle$. By considering the Wigner transform of the resolution of the identity (3.4.117), we obtain the completeness relation for Wigner functions,

$$\sum_n f_n(q, p) = \frac{1}{2\pi\hbar}.\qquad(3.4.118)$$

A typical basis for the Hilbert space in ordinary quantum mechanics is provided by the stationary states, or eigenstates, of the Hamiltonian. In the phase space formulation, a stationary Wigner function f_n satisfies the so-called $\star$-genvalue equation,

$$H \star f_n = f_n \star H = E_n f.\qquad(3.4.119)$$

It follows from Moyal's EOM that f_n is then time-independent,

$$\frac{\partial f_n}{\partial t} = 0.\qquad(3.4.120)$$

The $\star$-genvalue equation can be also derived from the eigenvalue equation in the standard formulation,

$$H|\psi_n\rangle = E_n|\psi_n\rangle,\qquad(3.4.121)$$

since in this case

$$H\rho_{\psi_n} = \rho_{\psi_n}H = E_n\rho_{\psi_n},\qquad(3.4.122)$$

and the Wigner transform of this equation gives (3.4.119). Let us note that $\star$-genfunctions corresponding to different energies are $\star$-*orthogonal* if

$$H \star f_n = f_n \star H = E_n f_n, \qquad H \star f_m = f_m \star H = E_m f_m,\qquad(3.4.123)$$

with $E_n \neq E_m$, then

$$f_n \star f_m = 0.\qquad(3.4.124)$$

Indeed, by associativity of the $\star$-product,

$$f_n \star H \star f_m = (f_n \star H) \star f_m = E_n f_n \star f_m = f_n \star (H \star f_m) = E_m f_n \star f_m,\qquad(3.4.125)$$

and $\star$-orthogonality follows.

We note that, as a consequence of (3.4.119), the Moyal propagator satisfies

$$U_\star(q, p; t) \star f_n(q, p) = e^{-itE_n/\hbar} f_n(q, p). \tag{3.4.126}$$

This, combined with (3.4.118), leads to the following spectral decomposition of the Moyal propagator,

$$U_\star(q, p; t) = U_\star(q, p; t) \star 2\pi\hbar \sum_n f_n(q, p) = 2\pi\hbar \sum_n e^{-itE_n/\hbar} f_n(q, p), \tag{3.4.127}$$

which is of course the counterpart in phase space of (1.2.13).

In order to provide a complete description of the theory, it is important to consider *off-diagonal Wigner functions*, or Moyal's matrix elements. These are defined as follows. Let $|\psi_n\rangle$ be, as before, an orthonormal basis of the Hilbert space, and let us consider the operator

$$\rho_{mn} = |\psi_n\rangle\langle\psi_m|. \tag{3.4.128}$$

Then, the off-diagonal Wigner function $f_{mn}(q, p)$ is defined by the Wigner transform,

$$f_{mn}(q, p) = \frac{1}{2\pi\hbar} \left(\rho_{mn}\right)_{\mathrm{W}} = \frac{1}{2\pi\hbar} \int_{\mathbb{R}} \psi_n\left(q + \frac{\xi}{2}\right) \psi_m^*\left(q - \frac{\xi}{2}\right) e^{-\frac{i}{\hbar}p\xi} d\xi. \tag{3.4.129}$$

Note that this function is no longer real, and it satisfies the conjugation property

$$f_{mn}^*(q, p) = f_{nm}(q, p). \tag{3.4.130}$$

The off-diagonal Wigner function makes it possible to calculate general expectation values of operators, since

$$\langle\psi_m|O|\psi_n\rangle = \mathrm{Tr}\left(O\rho_{mn}\right) = \int_{\mathbb{R}^2} O_W(q, p) f_{mn}(q, p) dq dp. \tag{3.4.131}$$

The two obvious properties

$$\mathrm{Tr}\rho_{mn} = \delta_{mn}, \qquad \rho_{mn}\rho_{kl} = \delta_{ml}\rho_{kn}, \tag{3.4.132}$$

lead to

$$\int_{\mathbb{R}^2} f_{mn}(q, p) dq dp = \delta_{mn}, \qquad f_{mn} \star f_{kl} = \frac{1}{2\pi\hbar}\delta_{ml} f_{kn}. \tag{3.4.133}$$

In particular, by integrating the last relation, we conclude that

$$\int_{\mathbb{R}^2} f_{mn}(q, p) f_{lk}^*(q, p) dq dp = \frac{1}{2\pi\hbar}\delta_{ml}\delta_{kn}. \tag{3.4.134}$$

The off-diagonal Wigner functions satisfy a completeness relation in phase space, given by

$$\sum_{m,n} f_{mn}(q, p) f_{mn}(q', p') = \frac{1}{2\pi\hbar}\delta(q - q')\delta(p - p'). \tag{3.4.135}$$

This relation can be derived from the following completeness relation in the tensor product Hilbert space $\mathcal{H} \otimes \mathcal{H}$:

$$\sum_{n,m} |\psi_n\rangle \otimes |\psi_m\rangle\langle\psi_n| \otimes \langle\psi_m| = \mathbf{1}. \tag{3.4.136}$$

The off-diagonal Wigner functions associated to eigenstates of the energy satisfy generalized $\star$-genvalue equations of the form

$$H \star f_{mn} = E_n f_{mn}, \qquad f_{mn} \star H = E_m f_{mn}. \tag{3.4.137}$$

In the above, we have supposed that we have a discrete spectrum, but it is easy to generalize these results to the continuum spectrum: we replace the sums over states by integrals over the energy, and the Kronecker deltas by Dirac deltas. For example, the completeness condition (3.4.118) becomes

$$\int_E f_E(q, p) \mathrm{d}E = \frac{1}{2\pi\hbar}. \tag{3.4.138}$$

Example 3.4.3 *The linear potential.* Let us consider a one-dimensional particle with Hamiltonian

$$H(q, p) = \frac{p^2}{2} + q. \tag{3.4.139}$$

Let us solve the $\star$-genvalue equation. By using Bopp shifts (3.3.64), we obtain,

$$(H(q, p) - E) \star f(q, p) = \left[\left(q + \frac{i\hbar}{2} \partial_p \right) + \frac{1}{2} \left(p - \frac{i\hbar}{2} \partial_q \right)^2 - E \right] f(q, p) = 0. \tag{3.4.140}$$

The imaginary part of this equation says that

$$\left(\partial_p - p \partial_q \right) f(q, p) = 0, \tag{3.4.141}$$

i.e.

$$f(q, p) = f(H). \tag{3.4.142}$$

Set $u = H(q, p)$. Then, the real part reads

$$\left(u - \frac{\hbar^2}{8} \partial_u^2 - E \right) f(u) = 0. \tag{3.4.143}$$

This is solved by the Airy function, namely

$$f(u) = c_1 \mathrm{Ai} \left[\frac{2}{\hbar^{2/3}} (u - E) \right] + c_2 \mathrm{Bi} \left[\frac{2}{\hbar^{2/3}} (u - E) \right]. \tag{3.4.144}$$

The physical requirement that the Wigner function decays when $q \to \infty$ implies that $c_2 = 0$, so

$$f(q, p) = \frac{1}{\pi \hbar^{5/3}} \mathrm{Ai} \left[\frac{2}{\hbar^{2/3}} (H(q, p) - E) \right]. \tag{3.4.145}$$

The normalization constant can be fixed by using the condition (3.4.138) and the fact that

$$\int_{\mathbb{R}} \mathrm{Ai}(x) \mathrm{d}x = 1. \tag{3.4.146}$$

In the case of the Hamiltonian (3.4.139), it is also possible to compute the Moyal propagator in closed form. We can use formula (3.4.109) to calculate it. Let us assume that $f(H) = f(H(q, p))$ is a function on phase space that depends on p and q only through its dependence on (3.4.139). Then, by using Bopp shifts, it is easy to show that

$$H \star f(H) = Hf(H) - \frac{\hbar^2}{8} f''(H).$$ (3.4.147)

It follows from the definition of the $\star$-exponential that the Moyal propagator is, in this case, a function of H only. To find it, let us consider the following ansatz

$$U_\star(H; t) = \exp\left[-a(t)H - b(t)\right].$$ (3.4.148)

Then, the equation (3.4.107) reads,

$$a'(t)H + b'(t) = \frac{i}{\hbar} H - \frac{i\hbar}{8} a^2(t),$$ (3.4.149)

i.e.

$$a'(t) = \frac{i}{\hbar}, \qquad b'(t) = -\frac{i\hbar}{8} a^2(t).$$ (3.4.150)

The initial condition

$$U_\star(H; 0) = 1,$$ (3.4.151)

implies that

$$a(0) = b(0) = 0.$$ (3.4.152)

Therefore, we have

$$a(t) = \frac{it}{\hbar}, \qquad b(t) = \frac{it^3}{24\hbar}.$$ (3.4.153)

We conclude that

$$U_\star(H; t) = \exp\left[-\frac{it}{\hbar} H - \frac{it^3}{24\hbar}\right].$$ (3.4.154)

$\square$

Example 3.4.4 *The uncertainty principle in the phase space formulation.* In his correspondence with Moyal, Dirac was suspicious of the phase space formulation of quantum mechanics. He believed that the uncertainty principle made it impossible to talk about functions that were well defined for q and p simultaneously. Let us see in detail how the phase space formulation accommodates Heisenberg's uncertainty principle, following an argument made by Curtright and Zachos.

As a first step, let $g(q, p)$ be a function on phase space with complex values. Let us show that

$$\langle \overline{g} \star g \rangle > 0,$$ (3.4.155)

where the average is evaluated as in (3.4.84) with an arbitrary Wigner function f. This goes as follows:

$$
\frac{1}{2\pi\hbar} \langle \overline{g} \star g \rangle = \frac{1}{2\pi\hbar} \int_{\mathbb{R}^2} f \cdot (\overline{g} \star g)\, dpdq = \int_{\mathbb{R}^2} (\overline{f} \star f) \cdot (\overline{g} \star g)\, dpdq
$$

$$
= \int_{\mathbb{R}^2} (\overline{f} \star f) \star (\overline{g} \star g)\, dpdq = \int_{\mathbb{R}^2} \overline{f} \star (f \star \overline{g}) \star g\, dpdq
$$

$$
= \int_{\mathbb{R}^2} \overline{f} \cdot (f \star \overline{g}) \star g\, dpdq = \int_{\mathbb{R}^2} (f \star \overline{g}) \star (g \star \overline{f})\, dpdq
$$

$$
= \int_{\mathbb{R}^2} (f \star \overline{g}) \star \overline{(f \star \overline{g})}\, dpdq = \int_{\mathbb{R}^2} (f \star \overline{g}) \cdot \overline{(f \star \overline{g})}\, dpdq \geq 0.
$$

$$(3.4.156)$$

We have exploited the fact that Wigner functions are real. In the first line we have used the property (3.4.78), and in the last line we have used (3.3.50). We have also used the property (3.3.69) repeatedly.

Let us now choose $g(q, p) = \alpha + \beta q + \gamma p$, with $\alpha, \beta, \gamma \in \mathbb{R}$. Then

$$
\langle \overline{g} \star g \rangle = \left\langle (\alpha + \beta q + \gamma p) \star (\alpha + \beta q + \gamma p) \right\rangle
$$

$$
= \begin{pmatrix} \alpha & \beta & \gamma \end{pmatrix} \left\langle \begin{pmatrix} 1 & q & p \\ q & q^2 & q \star p \\ p & p \star q & p^2 \end{pmatrix} \right\rangle \begin{pmatrix} \alpha \\ \beta \\ \gamma \end{pmatrix} \geq 0.
$$

$$(3.4.157)$$

Since this is valid for any α, β, γ, the determinant of the matrix must be positive, i.e.

$$
\begin{vmatrix} 1 & \langle q \rangle & \langle p \rangle \\ \langle q \rangle & \langle q^2 \rangle & \langle qp \rangle + \frac{i\hbar}{2} \\ \langle p \rangle & \langle qp \rangle - \frac{i\hbar}{2} & \langle p^2 \rangle \end{vmatrix} \geq 0.
$$

$$(3.4.158)$$

We expand to find

$$
0 \leq \langle q^2 \rangle \langle p^2 \rangle + \langle p \rangle \langle q \rangle \left\langle pq + \frac{i\hbar}{2} \right\rangle + \langle p \rangle \langle q \rangle \left\langle pq - \frac{i\hbar}{2} \right\rangle
$$

$$
- \langle p \rangle^2 \langle q^2 \rangle - \langle p^2 \rangle \langle q \rangle^2 - \left\langle pq - \frac{i\hbar}{2} \right\rangle \left\langle pq + \frac{i\hbar}{2} \right\rangle,
$$

$$(3.4.159)$$

and rearranging we obtain

$$
(\Delta q)^2 (\Delta p)^2 \geq (\langle pq \rangle - \langle p \rangle \langle q \rangle)^2 + \frac{\hbar^2}{4} \geq \frac{\hbar^2}{4}.
$$

$$(3.4.160)$$

This is precisely Heisenberg's uncertainty principle. $\square$

3.5 The Harmonic Oscillator in Phase Space

As usual, the harmonic oscillator provides an ideal testing ground where the phase space formulation can be made explicit. To simplify our notation, in this section we will set $m = \omega = 1$, so that the classical Hamiltonian reads,

$$H(q, p) = \frac{1}{2}(q^2 + p^2).\tag{3.5.161}$$

As is well known, the eigenstates of the quantum Hamiltonian are given, in position space, by the wavefunctions

$$\psi_n(q) = \left(\frac{1}{\pi\hbar}\right)^{\frac{1}{4}} \frac{e^{-q^2/2\hbar}}{\sqrt{2^n n!}} H_n\left(\frac{q}{\hbar^{1/2}}\right),\tag{3.5.162}$$

where $H_n(q)$ are Hermite polynomials. The off-diagonal Wigner function is given by

$$
\begin{aligned}
f_{mn}(q, p) = \frac{1}{2\pi} \left(\frac{1}{\pi\hbar}\right)^{\frac{1}{2}} \frac{e^{-q^2/\hbar}}{\sqrt{2^n n! \, 2^m m!}} \\
\times \int_{\mathbb{R}} dy \, e^{ipy} e^{-\frac{\hbar}{4}y^2} H_n\left(\frac{q}{\hbar^{1/2}} - \frac{\hbar^{1/2}y}{2}\right) H_m\left(\frac{q}{\hbar^{1/2}} + \frac{\hbar^{1/2}y}{2}\right),
\end{aligned}\tag{3.5.163}
$$

where we have set $\xi = \hbar y$ in (3.4.129). After the change of variables

$$y - \frac{2ip}{\hbar} = \frac{2z}{\hbar^{1/2}},\tag{3.5.164}$$

we find,

$$f_{mn}(q, p) = \frac{(-1)^n}{\pi^{3/2}\hbar} \frac{e^{-(q^2+p^2)/\hbar}}{\sqrt{2^n n! \, 2^m m!}} \int_{\mathbb{R}} dz \, e^{-z^2} H_m\left(z + \frac{q + ip}{\hbar^{1/2}}\right) H_n\left(z - \frac{q - ip}{\hbar^{1/2}}\right),\tag{3.5.165}$$

where we have taken into account the symmetry of the Hermite polynomials under a parity transformation,

$$H_n(-q) = (-1)^n H_n(q).\tag{3.5.166}$$

We now use the following property of the Hermite polynomials,

$$\int_{\mathbb{R}} dx \, e^{-x^2} H_n(x + u) H_m(x + v) = 2^m \pi^{1/2} n! \, v^{m-n} L_n^{m-n}(-2uv),\tag{3.5.167}$$

where $L_n^k(x)$ are generalized Laguerre polynomials, to find

$$f_{mn}(q, p) = \frac{(-1)^n}{\pi\hbar} \sqrt{\frac{n!}{m!}} e^{i(m-n)\theta} \left(\frac{4H}{\hbar}\right)^{\frac{m-n}{2}} e^{-\frac{2H(q,p)}{\hbar}} L_n^{m-n}\left(\frac{4H(q,p)}{\hbar}\right),\tag{3.5.168}$$

where

$$\theta = \tan^{-1}\left(\frac{p}{q}\right).\tag{3.5.169}$$

In the diagonal case $m = n$, we obtain the Wigner functions associated with the eigenstates of the quantum harmonic oscillator,

$$f_n(q, p) = \frac{(-1)^n}{\pi\hbar} e^{-\frac{2H(q,p)}{\hbar}} L_n\left(\frac{4H(q,p)}{\hbar}\right).\tag{3.5.170}$$

In Figure 3.1 we show the Wigner functions corresponding to the ground state, f_0, and to the third excited state f_3. Notice that f_3 is not positive, and since it corresponds to an odd eigenfunction it reaches the value $-1/(\pi\hbar)$ at the origin, so it cannot be interpreted as a classical probability distribution.

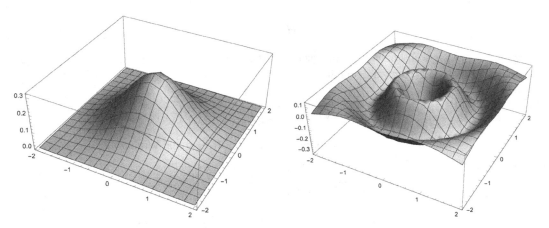

Figure 3.1 The Wigner functions (3.5.187) for (left) the ground state $n = 0$ and (right) the third state $n = 3$ of the harmonic oscillator, as a function of (q,p). We are choosing units in which $\omega = m = \hbar = 1$.

The above derivation of the Wigner functions relies on previous, known results for the eigenfunctions in the conventional formulation. However, we should be able to obtain these functions directly in the phase space formulation by solving the $\star$-genvalue equations (3.4.119). By using (3.3.64), we find

$$H \star f = \left[\frac{1}{2}\left(p - \frac{i\hbar}{2}\partial_q\right)^2 + \frac{1}{2}\left(q + \frac{i\hbar}{2}\partial_p\right)^2\right] f(q,p) = E f(q,p), \qquad (3.5.171)$$

which can be written as

$$\left[H - \frac{\hbar^2}{8}\left(\partial_p^2 + \partial_q^2\right) + \frac{i\hbar}{2}\left(q\partial_p - p\partial_q\right) - E\right] f(q,p) = 0. \qquad (3.5.172)$$

Since the function $f(q,p)$ is real, the imaginary part of the equation should vanish:

$$\left(q\partial_p - p\partial_q\right) f(q,p) = 0. \qquad (3.5.173)$$

This means that $f(q,p)$ is a function of the "radial" coordinate only, $f(q,p) = f(y)$, where

$$y = \frac{4H}{\hbar} = \frac{2(p^2 + q^2)}{\hbar}. \qquad (3.5.174)$$

A simple calculation shows that the real part of the $\star$-genvalue equation becomes,

$$\left[\frac{y}{4} - y\partial_y^2 - \partial_y - e\right] f(y) = 0, \qquad (3.5.175)$$

where $e = E/\hbar$. Let us now set,

$$f(y) = e^{-y/2}\phi(y). \qquad (3.5.176)$$

Then, the ODE for $f(y)$ becomes,

$$\left(y\partial_y^2 + (1 - y)\partial_y + e - \frac{1}{2}\right)\phi(y) = 0. \qquad (3.5.177)$$

When compared to the differential equation for the confluent hypergeometric equation,

$$\left(y\partial_y^2 + (b - y)\partial_y - a\right)\phi(y) = 0, \tag{3.5.178}$$

and we conclude that $\phi(y)$ solves this equation with

$$b = 1, \qquad a = 1/2 - e. \tag{3.5.179}$$

There are two linearly independent solutions of (3.5.178), usually denoted by $M(a, b, y)$ and $U(a, b, y)$. Therefore, the general solution can be written as

$$\phi(y) = c_1 M(a, 1, y) + c_2 U(a, 1, y). \tag{3.5.180}$$

We have to impose appropriate boundary conditions for $f(y)$. First of all, we want $f(y)$ to decay as $y \to \infty$, so that $f \in L^2(\mathbb{R}^2)$ and it satisfies the normalization condition (3.4.77). Since

$$M(a, 1, y) \sim \frac{1}{\Gamma(a)} e^y y^{a-1}, \qquad y \to \infty, \tag{3.5.181}$$

we must have $c_1 = 0$ unless $a \in \mathbb{Z}_{\leq 0}$. Let us now look at the behavior near $y = 0$. The function $U(a, 1, y)$ behaves as,

$$U(a, 1, y) = -\frac{1}{\Gamma(a)}\left(\log(y) + \psi(a) + 2\gamma_E\right) + \mathcal{O}\left(y \log y\right), \tag{3.5.182}$$

and it diverges at $y = 0$ unless

$$\frac{1}{\Gamma(a)} = 0. \tag{3.5.183}$$

Physically, $f(y)$ should not diverge at $y = 0$. In fact, such a divergence would contradict the property (3.4.85) of Wigner functions. We then conclude that physically acceptable solutions are of the form

$$\phi(y) \propto U(-n, 1, y), \qquad n = 0, 1, 2, \ldots. \tag{3.5.184}$$

This leads to the usual quantization of the energy levels,

$$e = n + 1/2. \tag{3.5.185}$$

We also note that $U(-n, 1, y)$ is, up to a normalization, a Laguerre polynomial:

$$U(-n, 1, y) = (-1)^n n! \, L_n(y), \qquad n = 0, 1, 2, \ldots \tag{3.5.186}$$

Therefore, the $\star$-genfunctions are of the form

$$f_n(q, p) = c_n (-1)^n e^{-2H/\hbar} L_n\left(\frac{4H}{\hbar}\right), \tag{3.5.187}$$

where c_n is an appropriate normalization constant. This can be fixed by imposing (3.4.77). We find,

$$\int_{\mathbb{R}^2} f_n(q, p) \mathrm{d}q\mathrm{d}p = \frac{1}{2}(-1)^n c_n \pi\hbar \int_0^\infty e^{-y/2} L_n(y)\mathrm{d}y = c_n \pi\hbar, \tag{3.5.188}$$

so that $c_n = (\pi\hbar)^{-1}$ and we recover (3.5.170).

The above calculation can be generalized to obtain the off-diagonal Wigner functions. The generalized $\star$-genvalue equations (3.4.137) give two partial differential equations for $f_{mn}(q, p)$,

$$\left\{ H - \frac{\hbar^2}{8} \left(\partial_q^2 + \partial_p^2 \right) \pm \frac{i\hbar}{2} \left(q\partial_p - p\partial_q \right) \right\} f_{mn}(q, p) = E_{n,m} f_{mn}(p, q), \quad (3.5.189)$$

where we have used the Bopp shifts (3.3.64). It is useful to use polar coordinates on phase space and write f_{mn} as a function of θ and y, where

$$\theta = \tan^{-1}\left(\frac{p}{q} \right), \quad (3.5.190)$$

and y is the variable introduced in (3.5.174). The dependence w.r.t. θ is easily obtained by subtracting the two equations in (3.5.189):

$$i\hbar \frac{\partial f_{mn}}{\partial \theta} = E_n - E_m, \quad (3.5.191)$$

therefore

$$f_{mn}(q, p) = e^{i(m-n)\theta} f_{mn}(y). \quad (3.5.192)$$

It is easy to see that $f_{mn}(y)$ now satisfies the ODE

$$\left[\frac{y}{4} - y\partial_y^2 - \partial_y + \frac{(m-n)^2}{4y} - e_{mn} \right] f_{mn}(y) = 0, \quad (3.5.193)$$

where

$$e_{mn} = \frac{E_m + E_n}{2\hbar}. \quad (3.5.194)$$

The ODE above is solved by

$$f_{mn}(y) = e^{-y/2} y^{\frac{m-n}{2}} L_n^{m-n}(y), \quad (3.5.195)$$

and we recover in this way the result (3.5.168), up to a normalization.

Another quantity that can be calculated exactly for the harmonic oscillator is the Moyal propagator $U_\star$. By a calculation similar to the one leading to (3.5.172), we find,

$$H \star f(H) = H f(H) - \frac{\hbar^2}{4} \left(f'(H) + H f''(H) \right). \quad (3.5.196)$$

As in the example of the linear potential, it is clear that $U_\star$ only depends on H, and it satisfies the equation

$$i\hbar \frac{\partial U_\star}{\partial t} = H U_\star - \frac{\hbar^2}{4} \left(U_\star'(H) + H U_\star''(H) \right). \quad (3.5.197)$$

To solve this equation, we use again the ansatz (3.4.148), and we deduce the following first-order ODEs for $a(t)$, $b(t)$,

$$i\hbar \dot{a}(t) = -1 + \frac{\hbar^2}{4} a^2(t), \qquad i\hbar \dot{b}(t) = -\frac{\hbar^2}{4} a(t), \quad (3.5.198)$$

with the initial conditions

$$a(0) = b(0) = 0. \quad (3.5.199)$$

They can be integrated immediately to give,

$$a(t) = \frac{2i}{\hbar} \tan\left(\frac{t}{2} + c_1\right), \qquad b(t) = \log\left[\cos\left(\frac{t}{2} + c_1\right)\right] + c_2. \tag{3.5.200}$$

The initial conditions fix $c_1 = c_2 = 0$. We conclude that

$$U_\star(H; t) = \frac{1}{\cos\left(\frac{t}{2}\right)} \exp\left[-\frac{2i}{\hbar} \tan\left(\frac{t}{2}\right) H\right]. \tag{3.5.201}$$

Note that the composition property (3.4.106) is in this case equivalent to the hyperbolic tangent composition of Gaussians (3.3.53). After changing the variable to

$$\frac{it}{\hbar} = \beta, \tag{3.5.202}$$

we obtain the Wigner transform of the unnormalized density matrix,

$$\left(e^{-\beta H}\right)_\star = \text{sech}\left(\frac{\beta\hbar}{2}\right) \exp\left(-\frac{2}{\hbar} \tanh\left(\frac{\beta\hbar}{2}\right) H\right). \tag{3.5.203}$$

3.6 The WKB Expansion in Phase Space

One advantage of the phase space formulation of quantum mechanics is that we use from the very beginning mathematical objects that are identical to those appearing in classical physics. This makes the semiclassical limit of quantum mechanics much more transparent. For example, quantum observables are functions on phase space, but they have to be multiplied by Moyal's product. Since this product becomes the ordinary product as $\hbar \to 0$, there is a smooth interpolation between the classical and the quantum theory. In addition, the star product can be expressed as an expansion in powers of $\hbar$, as we saw in (3.3.62). Therefore, the phase space formulation is particularly suited to study $\hbar$-corrections to many different quantities, such as wavefunctions, quantization conditions, and partition functions in statistical mechanics (in fact, the Wigner function was introduced by Wigner in order to calculate quantum corrections to the classical, canonical partition function!).

It turns out that many of these semiclassical expansions can be expressed in terms of functions $\mathcal{G}_r$ that are defined as follows. Let H be an operator (for example, the Hamiltonian). Given any function $f(H)$, we can expand it around $H_W(q, p)$, which is a c-number, as

$$f(H) = \sum_{r \geq 0} \frac{1}{r!} f^{(r)}(H_W) \left(H - H_W(q, p)\right)^r. \tag{3.6.204}$$

The Wigner transform of this equation gives

$$f(H)_W = \sum_{r \geq 0} \frac{1}{r!} f^{(r)}(H_W) \mathcal{G}_r, \tag{3.6.205}$$

where

$$\mathcal{G}_r = \left[\left(H - H_W(q, p)\right)^r\right]_W, \tag{3.6.206}$$

and the Wigner transform is evaluated at the *same* point (q, p). The quantities $\mathcal{G}_r$ can be computed by using the Moyal product, and they have a formal series expansion in powers of $\hbar$. One trivially has $\mathcal{G}_0 = 1$, $\mathcal{G}_1 = 0$. The first nontrivial function is $\mathcal{G}_2$. We have,

$$\mathcal{G}_2 = H_W \star H_W - H_W^2 = H_W \left\{ \cos\left(\frac{\hbar \overleftrightarrow{\Lambda}}{2}\right) - 1 \right\} H_W$$

$$= -\frac{\hbar^2}{4} \left[\frac{\partial^2 H_W}{\partial q^2} \frac{\partial^2 H_W}{\partial p^2} - \left(\frac{\partial^2 H_W}{\partial q \partial p}\right)^2 \right] + \mathcal{O}(\hbar^4).$$

(3.6.207)

For $\mathcal{G}_3$, one finds:

$$\mathcal{G}_3 = H_W \star H_W \star H_W - 3(H_W \star H_W)H_W + 2H_W^3$$

$$= -\frac{\hbar^2}{4} \left[\left(\frac{\partial H_W}{\partial q}\right)^2 \frac{\partial^2 H_W}{\partial p^2} + \left(\frac{\partial H_W}{\partial p}\right)^2 \frac{\partial^2 H_W}{\partial q^2} - 2 \frac{\partial H_W}{\partial q} \frac{\partial H_W}{\partial p} \frac{\partial^2 H_W}{\partial q \partial p} \right] + \mathcal{O}(\hbar^4).$$

(3.6.208)

Some important properties of the functions $\mathcal{G}_r$ can be obtained from the Wigner transform of the resolvent. If we apply (3.6.205) to the resolvent operator $G(z)$, we obtain

$$G_W(z) \equiv [G(z)]_W = \sum_{r \geq 0} \frac{\mathcal{G}_r}{(z - H_W)^{r+1}}.$$

(3.6.209)

On the other hand, the resolvent satisfies

$$(z - H)G(z) = G(z)(z - H) = \mathbf{1},$$

(3.6.210)

therefore

$$(z - H_W) \star G_W(z) = G_W(z) \star (z - H_W) = 1,$$

(3.6.211)

and taking the sum of the first two quantities, we obtain the equation

$$(z - H_W) \cos\left(\frac{\hbar \overleftrightarrow{\Lambda}}{2}\right) G_W(z) = 1.$$

(3.6.212)

Since the operator implementing the star product is even in $\hbar$, it is easy to see that $G_W(z)$ is also even in $\hbar$ (we assume here that H_W is independent of $\hbar$, or at least that it does not contain odd powers of $\hbar$). In addition, (3.6.212) leads to an equation of the form,

$$(z - H_W)G_W(z) + \sum_{n \geq 1} \hbar^{2n} \mathcal{H}_n G_W(z) = 1,$$

(3.6.213)

where $\mathcal{H}_n$ is a differential operator in ∂_q, ∂_p of order $2n$. Let us now write the $\hbar$ expansion of $G_W(z)$ in the form,

$$G_W(z) = \sum_{n \geq 0} G_n(z)\hbar^{2n}.$$

(3.6.214)

Together with (3.6.213), we deduce that

$$G_0(z) = \frac{1}{z - H_{\mathrm{W}}},$$

$$G_n(z) = -\frac{1}{z - H_{\mathrm{W}}} \sum_{s=1}^{n} \mathcal{H}_s G_{n-s}, \qquad n \geq 1. \tag{3.6.215}$$

For example,

$$G_1(z) = -\frac{1}{z - H_{\mathrm{W}}} \mathcal{H}_1 \left(\frac{1}{z - H_{\mathrm{W}}} \right). \tag{3.6.216}$$

Since $\mathcal{H}_1$ is a differential operator of order 2, we conclude that $G_1(z)$ has the form,

$$G_1(z) = \frac{\mathcal{G}_2^{(1)}}{(z - H_{\mathrm{W}})^3} + \frac{\mathcal{G}_3^{(1)}}{(z - H_{\mathrm{W}})^4}. \tag{3.6.217}$$

It is easy to prove by induction that $G_n(z)$ will be a sum of the form

$$G_n(z) = \sum_{r=2}^{3n} \frac{\mathcal{G}_r^{(n)}}{(z - H_{\mathrm{W}})^{r+1}}. \tag{3.6.218}$$

Indeed, let us assume that the maximal power of $(z - H_{\mathrm{W}})^{-1}$ in $G_{n-s}(z)$ is $3(n-s)+1$, for $s = 1, \ldots, n-1$. According to the second equation in (3.6.215), in order to obtain $G_n(z)$ we should act with $\mathcal{H}_s$ on $G_{n-s}(z)$ and multiply the result by $(z - H_{\mathrm{W}})^{-1}$. In this way we obtain a term where the power of $(z - H_{\mathrm{W}})^{-1}$ is $3(n-s) + 1 + 2s + 1 = 3n - s + 2$ (since an operator of order $2s$ increases the power in the denominator by $2s$). The maximum value obtained in this way is $3n + 1$ and corresponds to $s = 1$. The minimum value of the power of $(z - H_{\mathrm{W}})^{-1}$ is always 3, coming from the term involving $G_0(z)$ in the second equation of (3.6.215). We conclude that

$$G_{\mathrm{W}}(z) = \frac{1}{z - H_{\mathrm{W}}} + \sum_{n=1}^{\infty} \sum_{r=2}^{3n} \frac{\hbar^{2n} \mathcal{G}_r^{(n)}}{(z - H_{\mathrm{W}})^{r+1}}. \tag{3.6.219}$$

If we now compare this expansion to (3.6.209), we finally obtain the expansion

$$\mathcal{G}_r = \sum_{n \geq r/3}^{\infty} \mathcal{G}_r^{(n)} \hbar^{2n}. \tag{3.6.220}$$

This means that the expansion in r is in fact a *semiclassical* expansion, since at every order in $\hbar^2$ only a finite number of values of r contribute in (3.6.205). Note that the functions $\mathcal{G}_r$ are universal: they depend on the operator H, but not on the function $f(\mathrm{H})$.

The formalism we have developed makes it possible to derive various semiclassical expansions. One can use it, for example, to obtain the semiclassical expansion of the thermal partition function, which is sometimes called the *Wigner–Kirkwood expansion* (this expansion was in fact the original motivation to introduce the phase space formulation of quantum mechanics). The Wigner transform of the unnormalized density operator (1.2.17) defines a function in phase space that can be computed, as in the case of the evolution operator, by using the $\star$-exponential,

$$\left(e^{-\beta \mathrm{H}} \right)_{\mathrm{W}} = e_\star^{-\beta H_W}. \tag{3.6.221}$$

On the other hand, if we apply the equality (3.6.205), we obtain

$$\left(e^{-\beta H}\right)_{\mathrm{W}} = \left(\sum_{r=0}^{\infty} \frac{(-\beta)^r}{r!} \mathcal{G}_r\right) e^{-\beta H_{\mathrm{W}}}. \tag{3.6.222}$$

In particular, we can calculate the canonical partition function (1.2.23) by using (3.3.39). One obtains,

$$Z(\beta) = \sum_{r=0}^{\infty} \frac{(-\beta)^r}{r!} \int_{\mathbb{R}^2} \frac{dqdp}{2\pi\hbar} \mathcal{G}_r(q,p) e^{-\beta H_{\mathrm{W}}(q,p)}. \tag{3.6.223}$$

As emphasized above, this is a semiclassical expansion. The leading order as $\hbar \to 0$ is precisely

$$\int_{\mathbb{R}^2} \frac{dqdp}{2\pi\hbar} e^{-\beta H_{\mathrm{W}}(q,p)}, \tag{3.6.224}$$

which is nothing but the *classical*, thermal partition function for a theory with Hamiltonian $H_{\mathrm{W}}(q,p)$. The next-to-leading correction in $\hbar^2$ to $Z(\beta)$ is given by

$$\int_{\mathbb{R}^2} \frac{dqdp}{2\pi\hbar} e^{-\beta H_{\mathrm{W}}(q,p)} \left(\frac{\beta^2}{2} \mathcal{G}_2^{(1)}(q,p) - \frac{\beta^3}{6} \mathcal{G}_3^{(1)}(q,p)\right). \tag{3.6.225}$$

Further corrections can be computed systematically. Note in addition, that the Wigner transform of the un-normalized density matrix gives an explicit generating function for the functions $\mathcal{G}_r$.

Example 3.6.1 In the case of the harmonic oscillator, we have that

$$\sum_{r=0}^{\infty} \frac{(-\beta)^r}{r!} \mathcal{G}_r = \operatorname{sech}\left(\frac{\beta\hbar}{2}\right) \exp\left[-\left(\frac{2}{\hbar} \tanh\left(\frac{\beta\hbar}{2}\right) - \beta\right) H_{\mathrm{W}}\right]. \tag{3.6.226}$$

It is easy to derive from this expression that $\mathcal{G}_r$ is a polynomial in H_{W} of degree

$$\sigma_r = \frac{r}{3}, \qquad \frac{r-1}{3} - 1, \qquad \frac{r-2}{3}, \tag{3.6.227}$$

depending on whether r is 0,1 or 2 modulo 3, respectively. Also, the maximum power of $\hbar$ in $\mathcal{G}_r$ is $\hbar^r$ if r is even, and $\hbar^{r-1}$ if r is odd. We have, for the very first $\mathcal{G}_r$,

$$\mathcal{G}_2 = -\frac{\hbar^2}{4}, \quad \mathcal{G}_3 = -\frac{\hbar^2}{2} H_{\mathrm{W}}, \quad \mathcal{G}_4 = \frac{5\hbar^4}{16}, \quad \mathcal{G}_5 = \frac{9\hbar^4}{2} H_{\mathrm{W}}. \tag{3.6.228}$$

$\square$

Example 3.6.2 Let us consider the following operator

$$O = e^q + e^p + e^{-q-p}. \tag{3.6.229}$$

It is possible to show that its inverse $\rho = O^{-1}$ exists and is of trace class. Let us calculate $\operatorname{Tr}\rho$ up to next-to-leading order in $\hbar$. To do this, we use (3.6.205) for the function $f(x) = x^{-1}$. We find,

$$\rho_{\mathrm{W}} = \frac{1}{O_{\mathrm{W}}} + \frac{\mathcal{G}_2}{O_{\mathrm{W}}^3} - \frac{\mathcal{G}_3}{O_{\mathrm{W}}^4} + \cdots, \tag{3.6.230}$$

where

$$O_{\mathrm{W}} = e^q + e^p + e^{-q-p}, \tag{3.6.231}$$

is the Wigner transform of O. The calculation of $\mathcal{G}_{2,3}$ at leading order is straightforward by using (3.6.207), (3.6.208):

$$\mathcal{G}_2 = -\frac{\hbar^2}{4}\left(e^{q+p} + e^{-q} + e^{-p}\right) + \mathcal{O}(\hbar^4),$$

$$\mathcal{G}_3 = -\frac{\hbar^2}{4}e^{-2(p+q)}\left(-6e^{2(p+q)} + e^{3p+q} + e^{p+3q} + e^{4p+3q} + e^{3p+4q} + e^p + e^q\right) + \mathcal{O}(\hbar^4).$$

$$(3.6.232)$$

We then conclude that, up order $\hbar^2$, the Wigner transform of ρ is

$$\rho_W = \frac{1}{O_W} - \frac{9\hbar^2}{40 O_W^4} + \mathcal{O}(\hbar^4). \qquad (3.6.233)$$

Another, shorter, way of deriving this result is to start with the identity

$$\rho_W \star O_W = O_W \star \rho_W = 1, \qquad (3.6.234)$$

compute the product with (3.3.62), and solve in power series of $\hbar^2$. We finally obtain,

$$\mathrm{Tr}\,\rho = \frac{1}{2\pi\hbar}\int_{\mathbb{R}^2}\rho_W(q,p)\mathrm{d}p\mathrm{d}q = \frac{1}{2\pi\hbar}\int_{\mathbb{R}^2}\frac{\mathrm{d}q\mathrm{d}p}{O_W(q,p)} - \frac{9\hbar}{8\pi}\int_{\mathbb{R}^2}\frac{\mathrm{d}q\mathrm{d}p}{O_W^4(q,p)} + \mathcal{O}(\hbar^4).$$

$$(3.6.235)$$

These integrals can be computed in terms of Gamma functions, and we find

$$\mathrm{Tr}\,\rho = \frac{1}{2\pi\hbar}\frac{\Gamma(\frac{1}{3})^3}{3}\left\{1 - \frac{\hbar^2}{72} + \mathcal{O}(\hbar^4)\right\}. \qquad (3.6.236)$$

It turns out that one can obtain an exact expression for this trace; see the bibliographical notes in Section 3.9 for references.   □

We will now address a second application of the formalism developed above, which gives an expression for the one-dimensional WKB wavefunction in terms of the functions $\mathcal{G}_r$. Let us consider a wavefunction written in the WKB form,

$$\psi(q) = A(q)e^{iS(q)/\hbar}, \qquad (3.6.237)$$

where $A(q)$ is a real function. We assume in addition that this wavefunction solves the stationary Schrödinger equation

$$\mathsf{H}(q,\mathsf{p})\,\psi(q) = E\psi(q), \qquad (3.6.238)$$

but is not necessarily normalizable. Let us calculate the Wigner function associated to $\psi(q)$. From its definition (3.4.76) we obtain,

$$f_{\mathrm{WKB}}(q,p) = \frac{1}{2\pi\hbar}\int_{\mathbb{R}}A\left(q - \frac{x}{2}\right)A\left(q + \frac{x}{2}\right)$$

$$\times \exp\left[\frac{i}{\hbar}\left(S\left(q - \frac{x}{2}\right) - S\left(q + \frac{x}{2}\right) + px\right)\right]\mathrm{d}x. \qquad (3.6.239)$$

Let $p_E(q)$ be one of the branches of the function defined by the classical momentum, i.e. by the implicit equation

$$H_W(q,p) = E. \qquad (3.6.240)$$

We now define

$$\Sigma(q, x) = S\left(q - \frac{x}{2}\right) - S\left(q + \frac{x}{2}\right) + p_E(q)x. \tag{3.6.241}$$

Then, the Wigner WKB function can be written as

$$
\begin{aligned}
f_{\text{WKB}}(q, p) &= A\left(q + \frac{i\hbar}{2}\partial_p\right) A\left(q - \frac{i\hbar}{2}\partial_p\right) \\
&\quad \times \exp\left\{\frac{i}{\hbar}\Sigma\left(q, -i\hbar\partial_p\right)\right\} \int_{\mathbb{R}} e^{i(p - p_E(q))x/\hbar} \frac{dx}{2\pi\hbar} \\
&= A\left(q + \frac{i\hbar}{2}\partial_p\right) A\left(q - \frac{i\hbar}{2}\partial_p\right) \exp\left\{\frac{i}{\hbar}\Sigma\left(q, -i\hbar\partial_p\right)\right\} \delta(p - p_E(q)).
\end{aligned}
\tag{3.6.242}
$$

This expression leads to a formal power series in $\hbar$, where each term is a distribution in phase space. It has the structure,

$$f_{\text{WKB}}(q, p) = \sum_{s \geq 0} \alpha_s(q)\delta^{(s)}(p - p_E(q)), \tag{3.6.243}$$

where

$$\alpha_0(q) = A^2(q). \tag{3.6.244}$$

We note that the coefficients $\alpha_s(q)$ in the expansion above can be obtained as

$$\alpha_s(q) = \frac{(-1)^s}{s!} \int_{\mathbb{R}} f_{\text{WKB}}(q, p)(p - p_E(q))^s dp, \qquad s \geq 0. \tag{3.6.245}$$

This equation can be obtained by plugging (3.6.243) in the integral and using that

$$\int_{\mathbb{R}} \delta^{(s)}(x)f(x)\, dx = s!\,(-1)^s f^{(s)}(0). \tag{3.6.246}$$

Let us now assume that $f_{\text{WKB}}(q, p)$ corresponds to a stationary state of energy E. Clearly, the function on phase space

$$\delta(E - H)_{\text{W}}, \tag{3.6.247}$$

solves the $\star$-genvalue equation (3.4.119) with energy E. On the other hand, it follows from (3.6.205) that

$$f_{\text{WKB}}(q, p) = \delta(E - H)_{\text{W}} = \sum_{r \geq 0}^{\infty} \frac{(-1)^r}{r!} \mathcal{G}_r \delta^{(r)}(E - H_{\text{W}}), \tag{3.6.248}$$

where the derivative of the delta function is w.r.t. the energy E. Let us now compare the expressions (3.6.243) and (3.6.248). Our choice of $p_E(q)$ allows us to re-express the delta functions localizing on $H_{\text{W}}(q, p) = E$, in terms of delta functions localizing on $p = p_E(q)$. In addition, it follows from (3.6.245) that

$$
\begin{aligned}
A^2(q) &= \sum_{r \geq 0} \frac{(-1)^r}{r!} \frac{d^r}{dE^r} \int_{\mathbb{R}} \mathcal{G}_r(q, p)\delta(E - H_{\text{W}})dp \\
&= \sum_{r \geq 0} \frac{(-1)^r}{r!} \frac{d^r}{dE^r} \left(\left|\frac{\partial p_E}{\partial E}\right| \mathcal{G}_r(q, p_E(q))\right),
\end{aligned}
\tag{3.6.249}
$$

where we have used that

$$\delta(E - H_{\mathrm{W}}) = \left|\frac{\partial H_{\mathrm{W}}}{\partial p}\right|^{-1}\delta(p - p_E(q)). \tag{3.6.250}$$

By comparing (3.6.237) with (2.2.12), we note that $A^2(q) = 1/P(q)$, where $P(q)$ is the function (2.2.7) appearing in the all-orders WKB method. Therefore, (3.6.249) produces an expression for $1/P(q)$ in terms of the functions $\mathcal{G}_r(q, p)$ obtained from Weyl–Wigner quantization. In fact, it generalizes the WKB expansion of the wavefunction to Hamiltonians that are not necessarily of the standard form (1.3.83).

Example 3.6.3 At next-to-leading order in $\hbar^2$, we find

$$\frac{1}{P(q)} = \left|\frac{\partial p_E}{\partial E}\right| + \left\{\frac{1}{2}\frac{\partial^2}{\partial E^2}\left(\left|\frac{\partial p_E}{\partial E}\right|\mathcal{G}_2^{(1)}\right) - \frac{1}{6}\frac{\partial^3}{\partial E^3}\left(\left|\frac{\partial p_E}{\partial E}\right|\mathcal{G}_3^{(1)}\right)\right\}\hbar^2 + \mathcal{O}(\hbar^4). \tag{3.6.251}$$

For a standard Hamiltonian of the form (1.3.83), it is easy to verify that this expression reproduces the results obtained in (2.2.13).   □

As a final application of this formalism, we will now derive the all-orders WKB quantization condition in the context of the phase space formulation of quantum mechanics. The resulting approach has the advantage that it can be applied to a large class of quantum Hamiltonians, not necessarily of the standard form (1.3.83). Let us consider the operator

$$\mathsf{N}(E) = \theta(E - \mathsf{H}), \tag{3.6.252}$$

where $\theta(x)$ is the Heaviside step function. The trace of this operator will be denoted by $n(E)$. By evaluating it in a basis of eigenfunctions of H, we find

$$n(E) = \mathrm{Tr}\,\mathsf{N}(E) = \sum_{n=0}^{\infty}\theta(E - E_n). \tag{3.6.253}$$

Therefore, $n(E)$ counts the number of eigenstates whose energy is less than E. We can now apply the Wigner transform formula (3.6.205) to perform the semiclassical expansion of the operator (3.6.252). One finds,

$$\mathsf{N}(E)_{\mathrm{W}} = \theta(E - H_{\mathrm{W}}) + \sum_{r=2}^{\infty}\frac{(-1)^r}{r!}\mathcal{G}_r\delta^{(r-1)}(E - H_{\mathrm{W}}). \tag{3.6.254}$$

The trace (3.6.253) then has the semiclassical expansion

$$\begin{aligned} n(E) &= \int_{H_{\mathrm{W}}(q,p)\leq E}\frac{\mathrm{d}q\mathrm{d}p}{2\pi\hbar} + \sum_{r=2}^{\infty}\frac{(-1)^r}{r!}\int_{\mathbb{R}^2}\frac{\mathrm{d}q\mathrm{d}p}{2\pi\hbar}\mathcal{G}_r\delta^{(r-1)}(E - H_{\mathrm{W}}) \\ &= \int_{H_{\mathrm{W}}(q,p)\leq E}\frac{\mathrm{d}q\mathrm{d}p}{2\pi\hbar} + \sum_{r=2}^{\infty}\frac{(-1)^r}{r!}\frac{\mathrm{d}^r}{\mathrm{d}E^r}\int_{\mathbb{R}^2}\frac{\mathrm{d}q\mathrm{d}p}{2\pi\hbar}\mathcal{G}_r\theta(E - H_{\mathrm{W}}). \end{aligned} \tag{3.6.255}$$

Note that the first term is the volume of the region in phase space enclosed by the classical orbit $H_{\mathrm{W}}(q, p) = E$. This is precisely the quantity (2.5.142) appearing in the BS quantization condition. The all-orders quantization condition is

$$n(E) = n + \frac{1}{2}, \qquad n = 0, 1, 2, \ldots. \tag{3.6.256}$$

As a formal power series expansion in $\hbar$, the function $n(E)$ agrees with the l.h.s. of (2.5.165), divided by $2\pi\hbar$.

Example 3.6.4 *The harmonic oscillator, revisited.* By using the results in this section, it is easy to show that, for the harmonic oscillator, all the higher order corrections in $n(E)$ vanish. These corrections involve the terms

$$\frac{\mathrm{d}^{r-1}}{\mathrm{d}E^{r-1}} \int \frac{\mathrm{d}q\mathrm{d}p}{2\pi\hbar} \mathcal{G}_r \delta(E - H_{\mathrm{W}}), \qquad r \geq 2, \tag{3.6.257}$$

We know from Example 3.6.1 that, in the case of the harmonic oscillator, the $\mathcal{G}_r$ are polynomials in H_{W} of degree σ_r, where σ_r is given in (3.6.227). Therefore, (3.6.257) is of the form

$$\frac{\mathrm{d}^{r-1}}{\mathrm{d}E^{r-1}} \int \frac{\mathrm{d}q\mathrm{d}p}{2\pi\hbar} \left(\sum_{k=0}^{\sigma_r} c_k H_{\mathrm{W}}^k \right) \delta(E - H_{\mathrm{W}}) = \frac{\mathrm{d}^{r-1}}{\mathrm{d}E^{r-1}} \left\{ \sum_{k=0}^{\sigma_r} c_k E^k \int \frac{\mathrm{d}q\mathrm{d}p}{2\pi\hbar} \delta(E - H_{\mathrm{W}}) \right\}$$

$$= \frac{\pi}{\omega} \frac{\mathrm{d}^{r-1}}{\mathrm{d}E^{r-1}} \left\{ \sum_{k=0}^{\sigma_r} c_k E^k \right\},$$
$$\tag{3.6.258}$$

with $r \geq 2$. Clearly, the degree of the polynomial in E is at least one unit less than the order of the derivative operator acting on it, and all the corrections vanish. □

3.7 The Semiclassical Limit of the Wigner Function

In Section 3.6 we have derived a WKB expansion for the Wigner function as a power series expansion in $\hbar$, in (3.6.243) and (3.6.248). This expansion is in terms of distributions supported on the classical orbit $p = p_E(q)$ in phase space. Therefore, in the classical limit, the Wigner distribution is concentrated on the classically allowed region of phase space. However, this form is not very useful in determining the shape of the Wigner function when $\hbar$ is small but nonzero. In particular, the expression (3.6.243) does not reflect the rapidly oscillating behavior of the Wigner function inside the region bounded by the classical orbit. For this reason, it is useful to revisit the problem of calculating the Wigner function with a slightly different approach that takes into account finite $\hbar$ effects in a different way.

Let us consider the basic WKB wavefunction obtained in (2.11.438) in one dimension:

$$\psi_{\mathrm{WKB}}(q) = \frac{1}{\sqrt{2\pi}} \left(\frac{\partial p}{\partial I} \right)^{1/2} \exp\left(\frac{\mathrm{i}}{\hbar} S(q, I) \right), \tag{3.7.259}$$

where

$$S(q, I) = \int^q p(x, I)\mathrm{d}x. \tag{3.7.260}$$

Here, I is the value of the action variable characterizing the solution, and $p(q, I)$ is the momentum as a function of q and I [we recall that the action variables in a general integrable system are given in (2.11.400)]. We also note that the action variable can be written as a function of q and p, $I(q, p)$. The overall constant in (3.7.259) is chosen so as to have properly normalized functions, as we will check later on. If we use the definition (3.4.76) of the Wigner distribution, we obtain

$$f_{\text{WKB}}(q, p) = \frac{1}{\pi h} \int_{\mathbb{R}} \frac{\exp\left\{\frac{i}{\hbar}\left[\int_{q-X}^{q+X} p(x, I)dx - 2pX\right]\right\}}{\left[I_p(q + X)I_p(q - X)\right]^{1/2}} \, dX, \qquad (3.7.261)$$

where we have changed variables so that $X = -\xi/2$, and we have denoted

$$I_p(q \pm X) = \frac{\partial I\left(q \pm X, p(q \pm X, I)\right)}{\partial p}, \qquad I_q(q \pm X) = \frac{\partial I\left(q \pm X, p(q \pm X, I)\right)}{\partial q}. \qquad (3.7.262)$$

The expression (3.6.243) is obtained by expanding the integral around $X = 0$. We have,

$$\int_{q-X}^{q+X} p(x, I)dx - 2pX \approx 2(p(q, I) - p)X. \qquad (3.7.263)$$

Therefore, the integral reads, at leading order in this expansion,

$$f_{\text{WKB}}(q, p) \approx \frac{1}{2\pi} \frac{\partial p}{\partial I} \delta\left(p(q, I) - p\right)) = \frac{1}{2\pi} \delta\left(I(q, p) - I\right), \qquad (3.7.264)$$

which is the first term in (3.6.243). This expression is appropriately normalized, since

$$\frac{1}{2\pi} \int_{\mathbb{R}^2} \delta\left(I(q, p) - I\right) \, dqdp = \frac{1}{2\pi} \int d\theta = 1, \qquad (3.7.265)$$

where we have changed variables from (q, p) to the action-angle variables (I, θ).

However, as pointed out by Berry, at small $\hbar$ the integral should be better evaluated by a standard saddle-point calculation. The saddle is defined by the equation

$$\frac{\partial}{\partial X} \Sigma(X; q, p) = 0, \qquad (3.7.266)$$

where

$$\Sigma(X; q, p) = \int_{q-X}^{q+X} p(x, I)dx - 2pX, \qquad (3.7.267)$$

which leads to the equation

$$p(q + X, I) + p(q - X, I) = 2p. \qquad (3.7.268)$$

Since $\Sigma(-X; q, p) = -\Sigma(X; q, p)$, the equation (3.7.268) has a symmetry $X \to -X$, therefore it has two different solutions that we will denote by $\pm X_0$. It has in addition a nice geometrical interpretation. Let us consider the curve in phase space defined by the classical momentum $p(q, I)$, and let (q, p) be a point inside this curve. Then, the values $\pm X_0$ are such that (q, p) is the midpoint of a *chord* between the two points,

$$(q \pm X_0, p\left(q \pm X_0, I\right)), \qquad (3.7.269)$$

as shown in Figure 3.2. We will denote the points (3.7.269) as 2, 1, respectively [we numbered them in the order they are encountered in the motion of the classical system around the curve $p(q, I)$]. The value of $\Sigma(X; q, p)$ at $X = X_0$, i.e. at the

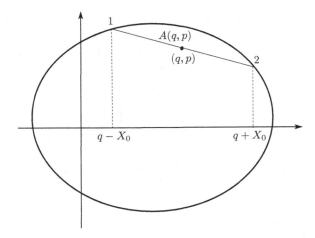

Figure 3.2 The saddle point of the integral (3.7.261) defines a chord in phase space.

point 2, is the area $A(q,p)$ between the curve $p(q,I)$ and the chord joining 1 and 2. We then obtain,

$$\Sigma\,(\pm X_0; q, p) = \pm A(q, p), \tag{3.7.270}$$

and more explicitly,

$$A(q,p) = \int_{q-X_0}^{q+X_0} p(x, I)\mathrm{d}x - 2pX_0. \tag{3.7.271}$$

We also note that

$$\frac{\partial^2}{\partial X^2}\Sigma(X; q, p)\Big|_{\pm X_0} = \pm\left(\frac{\partial p}{\partial q}(q + X_0, I) - \frac{\partial p}{\partial q}(q - X_0, I)\right). \tag{3.7.272}$$

On the other hand, since $I(p(q,I),q)$ is the constant value of the action on the classical orbit, we have that

$$\frac{\mathrm{d}I}{\mathrm{d}q} = \frac{\partial I}{\partial p}\frac{\partial p(q,I)}{\partial q} + \frac{\partial I}{\partial q} = 0. \tag{3.7.273}$$

When performing the Gaussian integral, we have to combine (3.7.272) with the denominator appearing in (3.7.261). By taking (3.7.273) into account, we obtain,

$$-\left(\frac{\partial p}{\partial q}(q + X_0, I) - \frac{\partial p}{\partial q}(q - X_0, I)\right)I_p(q + X_0)I_p(q - X_0)$$
$$= I_q(q + X_0)I_p(q - X_0) - I_q(q - X_0)I_p(q + X_0). \tag{3.7.274}$$

We will denote the r.h.s. of this equation by

$$I_q(2)I_p(1) - I_q(1)I_p(2). \tag{3.7.275}$$

It is now easy to calculate the integral in the saddle-point approximation. The contribution of the saddles at $X = \pm X_0$ gives, at leading order,

$$\frac{1}{\pi\sqrt{\pm i\hbar}}\frac{e^{\pm iA(q,p)/\hbar}}{\left(I_q(2)I_p(1) - I_q(1)I_p(2)\right)^{1/2}}, \tag{3.7.276}$$

and they add to

$$f_{\text{WKB}}(q,p) \approx \frac{2\cos\left[A(q,p)/\hbar - \frac{\pi}{4}\right]}{\pi\sqrt{h}\left(I_q(2)I_p(1) - I_q(1)I_p(2)\right)^{1/2}}. \tag{3.7.277}$$

This expression is valid inside the curve $p = p(q,I)$ in phase space. It tells us that there is a nontrivial, rapidly oscillating structure in that region. As we anticipated above, the delta function distribution (3.7.264) in the strict $\hbar \to 0$ limit hides a rich structure at small but non-zero $\hbar$.

In the above derivation we have not taken into account some important subtleties. Let us consider more the saddle-point equation (3.7.268) in greater depth. It is clear that, given a value of q, there is a minimal value of p, $p_{\min}$, which occurs when $q + X_0$ is the turning point of the motion q_+. This value is given by

$$p(2q - q_+, I) = 2p_{\min}. \tag{3.7.278}$$

When $0 < p < p_{\min}$, the function (3.7.267) does not have a stationary point, and therefore it would seem that $f_{\text{WKB}}(q,p) \approx 0$ in the saddle-point approximation. However, in performing the calculation above, we have described the bound state with a wavefunction of the form (3.7.259), involving a single exponential. The correct starting point should rather be the WKB wavefunction (2.5.137), which is a sum of two exponentials. The Wigner transform of this wavefunction is the sum of four contributions, and in calculating (3.7.277) we have only taken one of them into account. It turns out that the sum of all these contributions always has a saddle point described by the chord construction, and one always obtains the result (3.7.277). In this result, the function $A(q,p)$ is not necessarily given by (3.7.271), but it is always equal to the area inside the circular segment determined by the saddle point.

The result (3.7.277) has one important drawback: it diverges near the classical trajectory in phase space described by the function $p(q,I)$. Indeed, when the point (q,p) approaches this curve, the endpoints 1 and 2 of the chord come together, and the denominator of (3.7.277) goes to zero. The reason for this divergence is simply that, as we approach the curve, the two saddle points of the integral (3.7.261) coalesce, and the standard, quadratic saddle point approximation is no longer valid. We need a generalized saddle-point method, which is similar in spirit to the uniform WKB approximation near turning points. The general idea of this method is the following. Let us suppose that we have to calculate an integral of the form

$$\int_C g(z)e^{f(z,\alpha)/\hbar}\mathrm{d}z, \tag{3.7.279}$$

where C is an appropriate integration contour and α is a control parameter. We will assume that $f(z,\alpha)$ has two critical points $z_{1,2}$ that coalesce when $\alpha = 0$, say. To evaluate the integral, we change variables from z to u, where the new variable u is defined by the equation,

$$f(z,\alpha) = \frac{u^3}{3} - \zeta(\alpha)u + \rho(\alpha). \tag{3.7.280}$$

The coefficients $\zeta(\alpha)$, $\rho(\alpha)$ are determined as follows. The saddle points satisfy

$$f'(z,\alpha)\frac{\mathrm{d}z}{\mathrm{d}u} = u^2 - \zeta(\alpha), \tag{3.7.281}$$

therefore, in terms of the u variable, they are given by

$$u_{1,2} = \pm\zeta^{1/2}(\alpha), \tag{3.7.282}$$

which correspond to $z_{1,2}$. Since they come together when $\alpha = 0$, we must have $\zeta(0) = 0$. In addition, we have that

$$f(z_{1,2}, \alpha) = \mp\frac{2}{3}\zeta^{3/2}(\alpha) + \rho(\alpha), \tag{3.7.283}$$

and we deduce the two equations,

$$\frac{4}{3}\zeta^{3/2}(\alpha) = f(z_2, \alpha) - f(z_1, \alpha),$$
$$\rho(\alpha) = \frac{1}{2}\left(f(z_1, \alpha) + f(z_2, \alpha)\right). \tag{3.7.284}$$

From these equations we can deduce the values of $\zeta(\alpha)$, $\rho(\alpha)$. We note that, by taking an additional derivative w.r.t. u in (3.7.281), we obtain

$$f''(z_{1,2}, \alpha)\left(\frac{dz}{du}\right)^2_{z_{1,2}} = \pm 2\zeta^{1/2}(\alpha), \tag{3.7.285}$$

which will be useful later on. In order to proceed with the integration, we have to re-express the function $g(z)$ as a function of u,

$$g(z)\frac{dz}{du} = \sum_{m\geq 0} p_m(\alpha)(u^2 - \zeta(\alpha))^m + \sum_{m\geq 0} q_m(\alpha)u(u^2 - \zeta(\alpha))^m, \tag{3.7.286}$$

where we have assumed regularity of $g(z)$ at the saddle points. Once this has been done, one can write the original integral (3.7.279) in terms of Airy functions.

The calculation of the saddle-point integral (3.7.261) is a good illustration of this generalized saddle-point method. The function $f(z, \alpha)$ is in this case the function $\Sigma(X; q, p)$, where the point in phase space (q, p) plays the role of parameter α. We write

$$\Sigma(X; q, p) = \frac{u^3}{3} - \zeta(q, p)u + \rho(q, p). \tag{3.7.287}$$

Since

$$\Sigma(\pm X_0; q, p) = \pm A(q, p), \tag{3.7.288}$$

we deduce immediately that $\rho(q, p) = 0$ and that

$$\zeta(q, p) = \left(\frac{3}{2}A(q, p)\right)^{2/3}. \tag{3.7.289}$$

In order to calculate the leading order approximation to the integral, we just calculate the first terms in the expansion of $g(z)$. We note that, in general,

$$g(z_{1,2})\frac{dz}{du}\bigg|_{z_{1,2}} = p_0(\alpha) \pm q_0(\alpha)\zeta^{1/2}(\alpha). \tag{3.7.290}$$

In our case,

$$g(X; q, p) = \left[\frac{\partial I}{\partial p}(q + X)\frac{\partial I}{\partial p}(q - X)\right]^{-1/2}. \tag{3.7.291}$$

By using (3.7.272) and (3.7.285), we find

$$g\left(\pm X_0; q, p\right) \left.\frac{dX}{du}\right|_{\pm X_0} = \sqrt{2}\,\frac{\left(\frac{3}{2}A(q,p)\right)^{1/6}}{\left(I_q(2)I_p(1) - I_q(1)I_p(2)\right)^{1/2}}. \tag{3.7.292}$$

This, in particular, means that $q_0(\alpha) = 0$. At leading order, the Wigner function is then given by

$$f_{\text{WKB}}(q,p) \approx \frac{\sqrt{2}}{\pi h}\,\frac{\left(\frac{3}{2}A(q,p)\right)^{1/6}}{\left(I_q(2)I_p(1) - I_q(1)I_p(2)\right)^{1/2}}\int_{\mathbb{R}} \exp\left\{\frac{i}{\hbar}\left(\frac{u^3}{3} - \zeta(\alpha)u\right)\right\}du. \tag{3.7.293}$$

By using the representation of the Airy function given by

$$\text{Ai}(z) = \frac{1}{2\pi}\int_{\mathbb{R}} e^{izt + it^3/3}\,dt, \tag{3.7.294}$$

we finally obtain the following semiclassical approximation for the Wigner function, due to Berry,

$$f_{\text{WKB}}(q,p) \approx \frac{\sqrt{2}}{\pi \hbar^{2/3}}\,\frac{\left(\frac{3}{2}A(q,p)\right)^{1/6}}{\left(I_q(2)I_p(1) - I_q(1)I_p(2)\right)^{1/2}}\,\text{Ai}\left[-\left(\frac{3A(q,p)}{2\hbar}\right)^{2/3}\right]. \tag{3.7.295}$$

This expression has various interesting properties. First of all, when the point (q,p) is inside the classical orbit, $A(q,p)$ is large as compared to $\hbar$, and one can approximate this result by using the leading term in the asymptotics of the Airy function for negative argument (see the second equation in (B.32)). In this way, one recovers our previous expression (3.7.277). What happens in the exterior of the classical orbit? In this case, there is no real chord joining points 1 and 2, but (3.7.268) has complex solutions. Since the r.h.s. of (3.7.268) is always real, and $p(q, I)$ is a real function, it follows that X_0, is either real or purely imaginary. The same holds for $A(q,p)$, in view of (3.7.271). Therefore, in the exterior of the classical orbit, both X_0 and $A(q,p)$ must be purely imaginary. The correct analytic continuation is such that $A(q,p)$ has a phase $3\pi/2$. Therefore, outside the region limited by the classical orbit, the argument of the Airy function in (3.7.295) is positive and large, and the semiclassical expression for the Wigner function decays outside the region enclosed by the classical orbit. It should be noted, however, that (3.7.295) has, in general, additional singularities away from the classical orbit. These involve coalescing of multiple saddle points, in general, therefore they require a more careful treatment of the saddle-point approximation.

It can be verified that, as $\hbar \to 0$, one recovers the delta function expression (3.7.264). In order to do this, it is useful to obtain an intermediate expression near the classical orbit. This is the limit in which X_0 is small. One can easily find approximate expressions for X_0 and $A(q,p)$ in this limit. From (3.7.268), we find

$$X_0^2 \approx 2\,\frac{p - p(q,I)}{\partial^2 p(q,I)/\partial q^2}, \tag{3.7.296}$$

and from (3.7.271), we obtain

$$A(q, p) \approx \frac{4\sqrt{2}}{3} \sqrt{\frac{(p(q, I) - p)^3}{-\partial^2 p(q, I)/\partial q^2}}.$$ (3.7.297)

We also write, near the classical orbit,

$$\frac{\partial I}{\partial p} (p - p(q, I)) \approx I(q, p) - I.$$ (3.7.298)

By using these equations, one finds the so-called transitional approximation,

$$f_{\text{WKB}}(q, p) \approx \frac{1}{\pi} \left(\frac{1}{\hbar^2 B(q, p)} \right)^{1/3} \text{Ai} \left[2 \frac{I(q, p) - I}{\hbar^{2/3} B^{1/3}(q, p)} \right],$$ (3.7.299)

where

$$B(q, p) = -\frac{\partial^2 p(q, I)}{\partial q^2} \left(\frac{\partial I(q, p)}{\partial p} \right)^3.$$ (3.7.300)

We can write the function $B(q, p)$ in a slightly more symmetric form, as follows. If we take a further derivative of (3.7.273), we obtain

$$I_{pp} \left(\frac{\partial p(q, I)}{\partial q} \right)^2 + 2I_{pq} \frac{\partial p(q, I)}{\partial q} + I_p \frac{\partial^2 p(q, I)}{\partial q^2} + I_{qq} = 0.$$ (3.7.301)

We can use this equation, together with (3.7.273), to re-express derivatives of the $p(q, I)$ in terms of derivatives of $I(q, p)$ w.r.t. q and p, and we find

$$B(q, p) = I_p^2 I_{qq} + I_q^2 I_{pp} - 2I_p I_q I_{pq}.$$ (3.7.302)

The transitional approximation describes the behavior of the Wigner distribution near the classical orbit. Since the Airy function satisfies

$$\lim_{\epsilon \to 0} \frac{1}{\epsilon} \text{Ai} \left(\frac{x}{\epsilon} \right) = \delta(x),$$ (3.7.303)

the limit of (3.7.299) as $\hbar \to 0$ gives (3.7.264). Note that the exact solution for the Wigner function in a linear potential, (3.4.145), agrees precisely with (3.7.299). This is similar to the fact that, when solving the Schrödinger equation, uniform WKB approximations based on Airy functions are exact for the linear potential.

Example 3.7.1 As an example of Berry's semiclassical approximation for the Wigner function, let us consider the case of the harmonic oscillator. The Hamiltonian is given by (3.5.161), where we set $\omega = m = 1$ for simplicity. The action variable is given by

$$I = E,$$ (3.7.304)

where E is the conserved energy, therefore as a function of q and p, we have

$$I(q, p) = H(q, p).$$ (3.7.305)

It follows that

$$p(q, E) = \sqrt{2E - q^2}.$$ (3.7.306)

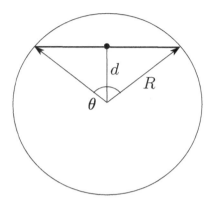

Figure 3.3 The geometry of chords in a circle of radius R. The chord spans an angle θ and its midpoint is at a distance d from the center of the circle.

In order to calculate (3.7.295), we need an explicit expression for the function $A(q, p)$. This can be obtained by recalling the elementary geometry of chords in a circle of radius R, as shown in Figure 3.3. The chord is determined by an angle θ. The midpoint of the chord is at a distance d from the origin of the circle, which is related to θ by the elementary trigonometric relation

$$\theta = 2 \arccos\left(\frac{d}{R}\right). \tag{3.7.307}$$

The area above the chord is given by

$$A = \frac{R^2}{2}(\theta - \sin\theta). \tag{3.7.308}$$

In our case, we have that

$$R^2 = 2E, \qquad d^2 = p^2 + q^2, \tag{3.7.309}$$

and we conclude that

$$A(q, p) = 2E\left\{\arccos\left(\sqrt{\frac{H(q, p)}{E}}\right) - \sqrt{\frac{H(q, p)}{E}}\sqrt{1 - \frac{H(q, p)}{E}}\right\}. \tag{3.7.310}$$

The denominator in (3.7.277) can be also computed geometrically. We note that

$$\left|I_q(2)I_p(1) - I_p(2)I_q(1)\right| = |q(2)p(1) - q(1)p(2)|. \tag{3.7.311}$$

This is the modulus of the wedge product of the vectors shown in Figure 3.3. It is given by

$$R^2 \sin\theta = 4E\sqrt{\frac{H(q, p)}{E}}\sqrt{1 - \frac{H(q, p)}{E}}. \tag{3.7.312}$$

The above compact expressions for $A(q, p)$ and the denominator (3.7.312) have been obtained in the regime where $H(q, p) < E$, but there is an obvious analytic continuation for $H(q, p) > E$. In particular,

$$A(q, p) = 2E e^{3\pi i/2}\left\{\cosh^{-1}\left(\sqrt{\frac{H(q, p)}{E}}\right) - \sqrt{\frac{H(q, p)}{E}}\sqrt{\frac{H(q, p)}{E} - 1}\right\}. \tag{3.7.313}$$

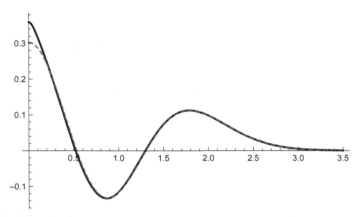

Figure 3.4 A plot of (full line) Berry's semiclassical Wigner function (3.7.295) as compared to the (dashed line) exact function (3.5.170), along the q axis, for $n = 2$ and $\hbar = \omega = 1$. In order to avoid the singularity at $q = p = 0$, we make the plot for a fixed value $p = 1/10$.

For the denominator, we simply consider the modulus of (3.7.312). When we plug the expressions (3.7.310), (3.7.313), and (3.7.312) into (3.7.295), we obtain an excellent approximation for the Wigner functions of the bound states in a harmonic oscillator. In Figure 3.4, we plot Berry's semiclassical Wigner function (3.7.295) as compared to the exact function (3.5.170), along the q axis, for $n = 2$ and $\hbar = \omega = 1$. The expression (3.7.295) is singular at $q = p = 0$, and to avoid this singularity we make the plot for a fixed value of $p = 1/10$. As we see, (3.7.295) is very close to the exact result, except for small values of q, and the detailed oscillating structure of the exact solution is fully captured by the Airy function in (3.7.295). Indeed, the result (3.7.295) is equivalent in this case to the leading term of the asymptotic expansion for Laguerre polynomials obtained by Erdélyi. Near the classical orbit $H(q, p) = E$, the "transitional approximation" is given by

$$f_{\text{WKB}}(q, p) \approx \frac{1}{\pi \xi} \text{Ai} \left[\frac{2}{\xi} (H(q, p) - E) \right], \qquad \xi = \left(2\hbar^2 E \right)^{1/3}. \qquad (3.7.314)$$

$\square$

In the analysis above, we have obtained the WKB form of the Wigner function by considering the WKB form of the standard wavefunction, and then performing the integral transform as in (3.4.76). It should be possible to derive many aspects of Berry's result by working directly in phase space and developing a small $\hbar$ expansion for the Wigner function. This can be done as follows. Let us consider the following ansatz for the Wigner function:

$$f(q, p) = \exp \left(\frac{i}{\hbar} \sum_{n=0}^{\infty} \sigma_n(q, p) \hbar^n \right). \qquad (3.7.315)$$

Let us assume that this function satisfies the $\star$-genvalue equations (3.4.119) with energy E. By using Bopp shifts, we find that the leading order function $\sigma_0(q, p)$ in the WKB ansatz satisfies the two PDEs:

$$H \left(q \mp \frac{1}{2} \frac{\partial \sigma_0}{\partial p}, p \pm \frac{1}{2} \frac{\partial \sigma_0}{\partial q} \right) = E. \qquad (3.7.316)$$

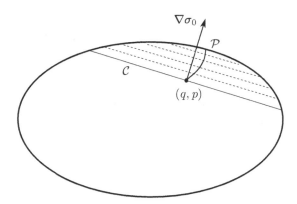

Figure 3.5 The geometric setting for the calculation of the integral (3.7.318).

If (q, p) is a point inside the classical region bounded by $H(q, p) = E$, the above conditions say that $x = (q, p)$ is the midpoint of a chord $\mathcal{C}$, i.e. that the points

$$x_{\mp} = x \pm \frac{1}{2}\xi, \qquad \xi = \left(\frac{\partial \sigma_0}{\partial p}, -\frac{\partial \sigma_0}{\partial q}\right), \tag{3.7.317}$$

are on the classical torus $H(q, p) = E$. The chord, and the vector ξ, are orthogonal to the gradient vector of the surface $\sigma_0(q, p) = $ constant. To identify the function $\sigma_0(q, p)$, let us consider all the chords parallel to $\mathcal{C}$, and filling the region between $\mathcal{C}$ and the classical torus (see Figure 3.5). Let $\mathcal{P}$ be the path obtained by joining the midpoints of the chords, and going from (q, p) to the classical torus. Then, if we assign the value $\sigma_0 = 0$ to points in the classical torus, we have that

$$\sigma_0(q, p) = -\int_{\mathcal{P}} \nabla \sigma_0 \cdot d\mathbf{r}. \tag{3.7.318}$$

Since the modulus of $\nabla \sigma_0$ is the length of the chord, and the product of the two vectors gives the projection of $d\mathbf{r}$ on the orthogonal direction to the chords, the above integral computes (up to a sign) the area between the chord $\mathcal{C}$ and the classical torus. In this way, we recovery Berry's results directly from the $\star$-genvalue equations. One can extend this approach and find PDEs for the higher order corrections σ_n, which can, in principle, be solved recursively.

3.8 The Husimi Distribution

There are other formulations of quantum mechanics in terms of distributions in phase space. The most popular one, after Wigner's distribution, is the Husimi distribution, which has the added advantage of being positive definite. The Husimi distribution requires the choice of a reference harmonic oscillator with mass m and frequency ω. This choice defines creation and annihilation operators,

$$\mathsf{a} = \frac{1}{\sqrt{2\hbar}}\left(\sqrt{m\omega}\,\mathsf{q} + \frac{i}{\sqrt{m\omega}}\mathsf{p}\right), \qquad \mathsf{a}^{\dagger} = \frac{1}{\sqrt{2\hbar}}\left(\sqrt{m\omega}\,\mathsf{q} - \frac{i}{\sqrt{m\omega}}\mathsf{p}\right). \tag{3.8.319}$$

The values of m, ω will parametrize the resulting distribution. This can be thought of as choosing complex coordinates in phase space,

$$z = \frac{1}{\sqrt{2}}\left(\sqrt{m\omega}q - \frac{i}{\sqrt{m\omega}}p\right), \qquad z^* = \frac{1}{\sqrt{2}}\left(\sqrt{m\omega}q + \frac{i}{\sqrt{m\omega}}p\right). \qquad (3.8.320)$$

Once this reference harmonic oscillator has been chosen, we can define the corresponding coherent states as follows. First of all, the vacuum is defined by

$$a|0\rangle = 0. \qquad (3.8.321)$$

Let us now consider the operator

$$\Omega_{p,q} = \exp\left(\frac{i}{\hbar}(p\mathsf{q} - q\mathsf{p})\right). \qquad (3.8.322)$$

The coherent state is defined by

$$|\Omega_{p,q}\rangle = \Omega_{p,q}|0\rangle. \qquad (3.8.323)$$

It will be also denoted by $|z\rangle$. In the position representation, it is given by

$$\beta_{p,q}(x) = \langle x|\Omega_{p,q}\rangle = e^{-\frac{ipq}{2\hbar}}e^{ipx/\hbar}\langle x|e^{-iq\mathsf{p}/\hbar}|0\rangle = e^{-\frac{ipq}{2\hbar}}e^{ipx/\hbar}\langle x - q|0\rangle$$
$$= \left(\frac{m\omega}{\pi\hbar}\right)^{1/4}e^{-\frac{ipq}{2\hbar}}e^{ipx/\hbar - \frac{m\omega}{2\hbar}(x-q)^2}, \qquad (3.8.324)$$

where we have used (3.2.18) to write

$$\Omega_{p,q} = e^{-\frac{ipq}{2\hbar}}e^{ipq/\hbar}e^{-iq\mathsf{p}/\hbar}. \qquad (3.8.325)$$

The *Husimi distribution* associated to a density operator ρ is defined by

$$\rho_{\mathrm{H}}(q,p) = \frac{1}{2\pi\hbar}\langle\Omega_{p,q}|\rho|\Omega_{p,q}\rangle. \qquad (3.8.326)$$

An important property of this distribution is that it can be obtained from the Wigner distribution by a two-dimensional convolution with a Gaussian function, i.e. we have

$$\rho_{\mathrm{H}}(q,p) = \frac{1}{\pi\hbar}\int_{\mathbb{R}^2}\frac{dq'\,dp'}{2\pi\hbar}\exp\left(-\frac{m\omega}{\hbar}(q-q')^2 - \frac{1}{\hbar m\omega}(p-p')^2\right)\rho_{\mathrm{W}}(q',p'). \qquad (3.8.327)$$

To verify this, we first calculate (3.8.326) as

$$\rho_{\mathrm{H}}(q,p) = \frac{1}{2\pi\hbar}\int_{\mathbb{R}^2}dx\,dy\,\beta_{p,q}^*(x)\beta_{p,q}(y)\rho(x,y)$$
$$= \frac{1}{2\pi\hbar}\left(\frac{m\omega}{\pi\hbar}\right)^{1/2}\int_{\mathbb{R}^2}dx\,dy\,e^{-ip(x-y)/\hbar}e^{-\frac{m\omega}{2\hbar}((x-q)^2+(y-q)^2)}\rho(x,y). \qquad (3.8.328)$$

On the other hand, from (3.8.327) we obtain

$$\rho_{\mathrm{H}}(q,p) = \frac{1}{2\pi^2\hbar^2}\int_{\mathbb{R}^3}dp'\,dq'\,dz\,e^{-\frac{m\omega}{\hbar}(q-q')^2 - \frac{1}{\hbar m\omega}(p-p')^2 - \frac{izp'}{\hbar}}\rho\left(q' + \frac{z}{2}, q' - \frac{z}{2}\right). \qquad (3.8.329)$$

If we integrate over p', and identify $x = q' + z/2$, $y = q' - z/2$, we can immediately see that (3.8.328) is equal to (3.8.329). An obvious consequence of the definition (3.8.326) is that the Husimi distribution for a pure state with density operator $\rho = |\psi\rangle\langle\psi|$, which we will denote by $H_\psi(q, p)$, is positive definite:

$$H_\psi(q, p) = \frac{1}{2\pi\hbar} \left| \langle \Omega_{p,q} | \psi \rangle \right|^2 = \frac{1}{2\pi\hbar} \left| \int_{\mathbb{R}} \beta^*_{p,q}(x)\psi(x)\mathrm{d}x \right|^2. \tag{3.8.330}$$

This is an appealing property of the Husimi distribution, and makes it more akin to a classical probability distribution.

Example 3.8.1 *Husimi distribution for the harmonic oscillator.* Let us consider a harmonic oscillator of mass m and frequency ω. We recall that the normalized eigenstates of the Hamiltonian can be constructed as

$$|n\rangle = \frac{\left(a^\dagger\right)^n}{\sqrt{n!}} |0\rangle. \tag{3.8.331}$$

Let us consider the Husimi distribution with the same parameters m and ω. We first note that the state (3.8.323) can be written as

$$|\Omega_{p,q}\rangle = \mathrm{e}^{-|z|^2/2\hbar} \mathrm{e}^{-a^\dagger z^*/\sqrt{\hbar}} |0\rangle = \mathrm{e}^{-|z|^2/2\hbar} \sum_{n \geq 0} \frac{(z^*)^n}{\hbar^{n/2}\sqrt{n!}} |n\rangle. \tag{3.8.332}$$

Therefore,

$$\langle \Omega_{p,q} | n \rangle = \mathrm{e}^{-|z|^2/2\hbar} \frac{z^n}{\hbar^{n/2}\sqrt{n!}}, \tag{3.8.333}$$

and the Husimi distribution associated to the pure state $|\psi\rangle = |n\rangle$ is given by

$$2\pi\hbar H_n(q, p) = \frac{1}{n!} \mathrm{e}^{-\frac{H}{\hbar\omega}} \left(\frac{H}{\hbar\omega} \right)^n, \tag{3.8.334}$$

where $H(q, p)$ is the classical Hamiltonian of the harmonic oscillator. $\quad\square$

One important aspect of the Husimi distribution for pure states is that it is given by the square of a simpler function, called the *Bargmann transform*. We define the Bargmann transform of the state $|\psi\rangle$ as

$$\psi_B(z) = \mathrm{e}^{\frac{|z|^2}{2\hbar}} \langle \Omega_{p,q} | \psi \rangle. \tag{3.8.335}$$

Explicitly, we have

$$\psi_B(z) = \left(\frac{m\omega}{\pi\hbar} \right)^{1/4} \int_{\mathbb{R}} \exp\left\{ -\frac{1}{\hbar} \left(\frac{m\omega}{2} x^2 + \frac{z^2}{2} - \sqrt{2m\omega}xz \right) \right\} \psi(x) \, \mathrm{d}x. \tag{3.8.336}$$

This expression makes it manifest that the Bargmann transform depends only on z, and not on z^*. Therefore, it provides a holomorphic representation of the wavefunction. The expression (3.8.330) can now be written as,

$$H_\psi(q, p) = \frac{1}{2\pi\hbar} \mathrm{e}^{-|z|^2/\hbar} \left| \psi_B(z) \right|^2. \tag{3.8.337}$$

The Bargmann transform can be understood in the framework of canonical transformations. Indeed, one can regard the change of variables from q, p to the complex variables z, z^* as a linear canonical transformation:

$$\begin{pmatrix} z \\ -iz^* \end{pmatrix} = \sqrt{\frac{m\omega}{2}} \begin{pmatrix} 1 & -i\,(m\omega)^{-1} \\ -i & (m\omega)^{-1} \end{pmatrix} \begin{pmatrix} q \\ p \end{pmatrix}. \tag{3.8.338}$$

The corresponding generating functional is given by

$$F(z, q) = -\frac{1}{2i} \left(m\omega q^2 + z^2 - 2\sqrt{2m\omega}\,qz \right), \tag{3.8.339}$$

which satisfies

$$-iz^* = \frac{\partial F}{\partial z}, \qquad p = -\frac{\partial F}{\partial q}. \tag{3.8.340}$$

The Bargmann transform is therefore the canonical transformation of the wavefunction (2.11.450) induced by the linear transformation (3.8.338). We can now use the general formula (2.11.472) to obtain the WKB approximation to the Bargmann transform. The new canonical momentum is $P = -iz^*$, and we obtain

$$\psi_B^{\text{WKB}}(z) = \left(\frac{\partial z^*}{\partial I} \right)^{1/2} \exp\left(\frac{1}{\hbar} \int^z z_E^*(z')\mathrm{d}z' \right), \tag{3.8.341}$$

where the function $z_E^*(z)$ is defined implicitly by the classical energy curve

$$H(z, z^*) = E. \tag{3.8.342}$$

Example 3.8.2 The classical Hamiltonian for the harmonic oscillator reads

$$H(z, z^*) = zz^*, \tag{3.8.343}$$

where we have set $\omega = 1$ for simplicity. The exact expression for the Bargmann function associated to the eigenstate $|n\rangle$ can be read from (3.8.333),

$$\psi_B^{(n)}(z) = \frac{z^n}{\hbar^{n/2}\sqrt{n!}}. \tag{3.8.344}$$

On the other hand, the WKB approximation to the Bargmann function with energy E reads

$$\psi_B^{\text{WKB}}(z) = z^{-1/2} \exp\left(\frac{1}{\hbar} \int^z \frac{E}{z'}\mathrm{d}z' \right) = z^{-1/2 + E/\hbar}. \tag{3.8.345}$$

Since $E/\hbar = n + 1/2$, we see that the WKB approximation is in this case exact, and it captures the full dependence on the holomorphic variable z. □

We know that, in one-dimensional systems, the Wigner distribution, in the semiclassical limit, is concentrated on the classical torus, and the behavior near the torus is described by an Airy function. Is there an analogue of this for the Husimi distribution? To address this issue, let us consider the expression (3.8.337) in terms of the Bargmann transform, as well as the WKB approximation for the latter, (3.8.341). We write,

$$H_\psi(z, z^*) \approx \frac{\mathcal{N}}{|\partial_z H|} e^{A_0(z,z^*)/\hbar}, \tag{3.8.346}$$

where $\mathcal{N}$ is a normalization factor and

$$A_0(z, z^*) = S_0(z) + S_0(z^*) - |z|^2, \qquad S_0(z) = \int^z z_E^*(z')\mathrm{d}z'. \tag{3.8.347}$$

The maximum of the function $A_0(z, z^*)$ satisfies

$$\frac{\partial A_0}{\partial z} = \frac{\mathrm{d}S_0}{\mathrm{d}z} - z^* = 0, \qquad \frac{\partial A_0}{\partial z^*} = \frac{\mathrm{d}S_0^*}{\mathrm{d}z^*} - z = 0, \tag{3.8.348}$$

which gives

$$z^* = z_E^*(z), \tag{3.8.349}$$

i.e. the maximum occurs at the classical energy curve. Let us now expand the Husimi distribution around a point z_c, z_c^* in this curve. We have

$$A_0(z_c + \xi, z_c^* + \bar{\xi}) = A_0(z_c, \bar{z}_c) + \Delta_2 A_0 + \cdots, \tag{3.8.350}$$

where

$$\Delta_2 A_0 = \frac{1}{2}\left(\frac{\partial^2 A_0}{\partial z^2}\xi^2 + 2\xi\bar{\xi}\frac{\partial^2 A_0}{\partial z \partial z^*} + \bar{\xi}^2\frac{\partial^2 A_0}{\partial (z^*)^2}\right) + \cdots \tag{3.8.351}$$

The second derivatives are evaluated at the point z_c, z_c^* in the classical energy curve, and they are given by

$$\frac{\partial^2 A_0}{\partial z^2} = \frac{\mathrm{d}z_E^*}{\mathrm{d}z}, \qquad \frac{\partial^2 A_0}{\partial z \partial z^*} = -1, \qquad \frac{\partial^2 A_0}{\partial (z^*)^2} = \frac{\mathrm{d}z_E}{\mathrm{d}z^*}. \tag{3.8.352}$$

On the other hand, we have that

$$\frac{\mathrm{d}z_E^*}{\mathrm{d}z} = -\frac{\partial_z H}{\partial_{z^*} H}, \qquad \frac{\mathrm{d}z_E}{\mathrm{d}z^*} = -\frac{\partial_{z^*} H}{\partial_z H}, \tag{3.8.353}$$

where all the derivatives are evaluated on the classical trajectory. We conclude that, up to second order in $\omega, \bar{\omega}$,

$$\Delta_2 A_0 = -\frac{1}{2}\frac{(H(z, z^*) - E)^2}{|\partial_z H|^2}, \tag{3.8.354}$$

and

$$\mathrm{H}_\psi(z, z^*) \approx \frac{\mathcal{N}}{|\partial_z H|}\exp\left(-\frac{1}{2\hbar}\frac{(H(z, z^*) - E)^2}{|\partial_z H|^2}\right). \tag{3.8.355}$$

We conclude that Husimi's distribution is concentrated along the classical orbit, and it decays away from it. This is in contrast with the Wigner function, which decays outside the classical torus but oscillates rapidly inside it.

3.9 Bibliographical Notes

The phase space formulation of quantum mechanics has its origins in the seminal work of Wigner (1932). A classic review of the subject can be found in Hillery et al. (1984), while the treatise by Curtright et al. (2014) is a modern (yet concise) textbook presentation. Important papers in the development of the subject are collected in

Zachos et al. (2005). The introduction to this collection is also a useful reference. The subject has been recently revived in a beautiful series of papers by Curtright et al. (1998, 2001) and Curtright and Zachos (2001). Our exposition in Sections 3.2 and 3.3 is also influenced by the book by Takhtajan (2008). Another useful review is by Tatarskii (1983).

A pedagogical survey of the quantization problem can be found in Hall (2013). Hudson's theorem, giving necessary and sufficient conditions for the positivity of Wigner's function, can be found in Hudson (1974). The negativity of Wigner's function as an indicator of nonclassicality has been proposed by Kenfack and Życzkowski (2004). The phase space formulation of Heisenberg's uncertainty principle in Example 3.4.4 is based on Curtright and Zachos (2001). An interesting historical account of the relationship between Moyal and Dirac can be found in Moyal (2006). The calculation of the off-diagonal Wigner functions for the harmonic oscillator follows the derivations in Bartlett and Moyal (1949) and Fairlie (1964). The derivation of the Moyal propagators for the linear potential and the harmonic oscillator is based on Bayen et al. (1978). The transformation of the Wigner function under canonical transformations is presented in Curtright et al. (1998).

The semiclassical expansion developed in Section 3.6 was obtained by Voros (1977) and Grammaticos and Voros (1979). A useful discussion and derivation of the all-orders WKB quantization condition in the phase space formalism can be found in Colin de Verdière (2005). The operator studied in Example 3.6.2 is analyzed in detail in Kashaev and Mariño (2016). The semiclassical calculation of the Wigner function in Section 3.7 is due to Berry (1977). As we mentioned in the text, the actual calculation of the Wigner function associated to a WKB bound state is more subtle than that is presented in Berry (1977). A more detailed presentation in the case of the harmonic oscillator can be found in Giannopoulou and Makrakis (2017). The extension of Berry's calculation to the off-diagonal case was addressed in de Almeida (1984) and also in Giannopoulou and Makrakis (2017). The structure of semiclassical Wigner functions has been explored by De Almeida (1990) and by de Almeida and Hannay (1982), among others. The asymptotic expansion of Laguerre polynomials mentioned in Section 3.7 can be found in, for example, section 2 of Temme (1990). The WKB approach to the $\star$-genvalue equations has been developed by Veble et al. (2002).

The Husimi distribution is reviewed in, for example, Lee (1995). The connection to the Bargmann representation and its WKB approximation is studied in Voros (1989). The semiclassical limit of the Husimi distribution in phase space is worked out in Kurchan et al. (1989).

The Path Integral Formulation
of Quantum Mechanics

4.1 Introduction

In the 1940s, Feynman introduced a new formulation of quantum mechanics based on previous insights of Dirac's. In Feynman's formulation, the basic quantity is the quantum-mechanical propagator (1.2.9). Feynman discovered that such an amplitude can be described by an integration over all possible "paths" interpolating between two events in spacetime, with a weight $\exp(iS/\hbar)$, where S is the classical action of the path. In the classical limit, in which the action is large as compared to Planck's constant, one recovers the principle of least action of classical mechanics.

It turns out that one can reformulate quantum mechanics by taking, as the basic object, the quantum propagator and Feynman's path integral formula for it. Starting from this postulate, one can reconstruct nonrelativistic quantum mechanics, including the time-dependent Schrödinger equation. This was shown in detail by Feynman in his 1965 book with Hibbs (republished in 2010). Therefore, Feynman's approach provides a logically independent formulation of quantum mechanics. Certain aspects of quantum mechanics are more transparent in this formulation. For example, the fact that probability amplitudes of quadratic Lagrangians involve the classical action, as we showed in Examples 1.2.1 and 1.2.2, is an immediate consequence of the path integral formulation. Conversely, deriving the energy spectrum of even simple Hamiltonians turns out to be more difficult, since one has to first calculate the propagator, and then extract the energy eigenvalues from the spectral decomposition (1.2.13).

The path integral formulation has become of fundamental importance to modern theoretical physics, since it turns out to be much more convenient than traditional methods for the quantization of classical field theories. It is also extremely useful in condensed matter physics. In particular, it leads to a diagrammatic approach to perturbation theory (the famous Feynman diagrams) that has become the dominant language in many areas of physics. Perhaps the best introduction to the path integral formulation and Feynman diagrams is to study them in the context of quantum mechanics, where all results derived in this way can be carefully tested and checked against traditional methods. This is our purpose in this chapter.

4.2 Path Integral Representation of the Quantum Propagator

We will now see that the propagator (1.2.9) can be formulated in terms of a sum over classical paths. This is the famous path integral formulation of quantum mechanics due to Feynman, which can be regarded as a self-contained approach to quantum mechanics.

To calculate the QM propagator, let us divide the interval (t_0, t_f) into $N + 1$ small intervals

$$(t_0, t_1), (t_1, t_2), \ldots, (t_N, t_f), \tag{4.2.1}$$

and let us denote

$$\Delta t_k = t_{k+1} - t_k. \tag{4.2.2}$$

Therefore, we have

$$\mathsf{U}(t_f, t_0) = \mathsf{U}(t_f, t_N)\mathsf{U}(t_N, t_{N-1}) \cdots \mathsf{U}(t_1, t_0). \tag{4.2.3}$$

We define $t_{N+1} = t_f$. We can now introduce N resolutions of the identity,

$$\int_{\mathbb{R}} |q_i\rangle\langle q_i| \, dq_i = 1, \tag{4.2.4}$$

to obtain

$$K(q_f, q_0; t_f, t_0) = \int_{\mathbb{R}^N} \prod_{k=0}^{N} K(q_{k+1}, q_k; t_{k+1}, t_k) \, dq_N dq_{N-1} \cdots dq_1. \tag{4.2.5}$$

We are interested in taking the limit $N \to \infty$, $\Delta t_k \to 0$, as in the calculation of Riemann integrals. In this limit, we expect

$$K(q_{k+1}, q_k; t_{k+1}, t_k) = \left\langle q_{k+1} \left| e^{-\frac{i\Delta t_k}{\hbar}\mathsf{H}} \right| q_k \right\rangle \approx \left\langle q_{k+1} \left| 1 - \frac{i\Delta t_k}{\hbar}\mathsf{H} \right| q_k \right\rangle. \tag{4.2.6}$$

We calculate now

$$\langle q_{k+1}|\mathsf{H}|q_k\rangle = \int_{\mathbb{R}} \langle q_{k+1}|p_k\rangle\langle p_k|\mathsf{H}|q_k\rangle \, dp_k = \frac{1}{\sqrt{2\pi\hbar}} \int_{\mathbb{R}} e^{\frac{i}{\hbar}p_k q_{k+1}} \langle p_k|\mathsf{H}|q_k\rangle \, dp_k. \tag{4.2.7}$$

If we now assume that H is the standard nonrelativistic, time-independent Hamiltonian,

$$\mathsf{H} = \frac{\mathsf{p}^2}{2m} + V(\mathsf{q}), \tag{4.2.8}$$

we have

$$\langle p_k|\mathsf{H}|q_k\rangle = \left(\frac{p_k^2}{2m} + V(q_k)\right)\langle p_k|q_k\rangle = \frac{1}{\sqrt{2\pi\hbar}} \left(\frac{p_k^2}{2m} + V(q_k)\right) e^{-\frac{i}{\hbar}p_k q_k}$$

$$= \frac{1}{\sqrt{2\pi\hbar}} e^{-\frac{i}{\hbar}p_k q_k} H(q_k, p_k). \tag{4.2.9}$$

Therefore,

$$\langle q_{k+1}|\mathsf{H}|q_k\rangle = \int_{\mathbb{R}} \frac{dp_k}{2\pi\hbar} e^{\frac{i}{\hbar}p_k(q_{k+1}-q_k)} H(q_k, p_k), \tag{4.2.10}$$

and by using that

$$\langle q_{k+1}|q_k\rangle = \delta(q_{k+1} - q_k) = \int_{\mathbb{R}} \frac{dp_k}{2\pi\hbar} e^{\frac{i}{\hbar}p_k(q_{k+1}-q_k)}, \tag{4.2.11}$$

we find,

$$
\left\langle q_{k+1} \left| 1 - \frac{i\Delta t_k}{\hbar} H \right| q_k \right\rangle = \int_{\mathbb{R}} \frac{dp_k}{2\pi\hbar} e^{\frac{i}{\hbar} p_k (q_{k+1} - q_k)} \left(1 - \frac{i\Delta t_k}{\hbar} H(q_k, p_k) \right)
$$
$$
\approx \int_{\mathbb{R}} \frac{dp_k}{2\pi\hbar} \exp\left[\frac{i\Delta t_k}{\hbar} \left(p_k \frac{q_{k+1} - q_k}{\Delta t_k} - H(q_k, p_k) \right) \right].
$$
(4.2.12)

We obtain,

$$
K(q_f, q_0; t_f, t_0)
$$
$$
\approx \int_{\mathbb{R}^{2N+1}} \left[\prod_{k=1}^{N} \frac{dp_k \, dq_k}{2\pi\hbar} \right] \frac{dp_0}{2\pi\hbar} \exp\left[\sum_{k=0}^{N} \frac{i\Delta t_k}{\hbar} \left(p_k \frac{q_{k+1} - q_k}{\Delta t_k} - H(q_k, p_k) \right) \right].
$$
(4.2.13)

Let us now assume for simplicity that $\Delta t_k = \Delta t$ for all k. The approximation for the propagator in (4.2.13) becomes increasingly good in the limit $\Delta t \to 0$. Since

$$
(N + 1)\Delta t = t_f - t_0,
$$
(4.2.14)

is fixed, N has to go to infinity simultaneously. The appropriate limit is then

$$
N \to \infty, \qquad \Delta t \to 0,
$$
(4.2.15)

so that the product (4.2.14) is fixed. This type of limit is sometimes called a double-scaling limit, since we have to consider the limit of two different variables that are correlated. We then obtain the following formula for the propagator:

$$
K(q_f, q_0; t_f, t_0)
$$
$$
= \lim_{N \to \infty} \int_{\mathbb{R}^{2N+1}} \left[\prod_{k=1}^{N} \frac{dp_k \, dq_k}{2\pi\hbar} \right] \frac{dp_0}{2\pi\hbar} \exp\left[\sum_{k=0}^{N} \frac{i\Delta t}{\hbar} \left(p_k \frac{q_{k+1} - q_k}{\Delta t_k} - H(q_k, p_k) \right) \right],
$$
(4.2.16)

where it is understood that we have to consider the double-scaling limit (4.2.15). Although our argument for the validity of (4.2.16) has been heuristic, it can be justified rigorously by using the product formula of Lie–Kato–Trotter for the exponential of an operator.

It is possible to obtain an even more useful formula for the propagator by integrating over the momenta. Indeed, the integral over p_k in

$$
K(q_{k+1}, q_k; t_{k+1}, t_k) \approx \int_{\mathbb{R}} \frac{dp_k}{2\pi\hbar} \exp\left[\frac{i\Delta t_k}{\hbar} \left(p_k \frac{q_{k+1} - q_k}{\Delta t_k} - \frac{p_k^2}{2m} - V(q_k) \right) \right],
$$
(4.2.17)

is a Gaussian, and we can use (C.1) with

$$
A = -\frac{\Delta t_k}{m\hbar}.
$$
(4.2.18)

One finds,

$$
K(q_{k+1}, q_k; t_{k+1}, t_k) = \sqrt{\frac{m}{2\pi i \hbar \Delta t_k}} \exp\left[\frac{i\Delta t_k}{\hbar} \left(\frac{m}{2} \left(\frac{q_{k+1} - q_k}{\Delta t_k} \right)^2 - V(q_k) \right) \right].
$$
(4.2.19)

We can now set $\Delta t_k = \Delta t$ for all k, as before, and take the double-scaling limit $N \to \infty$ and $\Delta t \to 0$. We obtain

$$K(q_f, q_0; t_f, t_0)$$

$$= \lim_{N \to \infty} \left(\frac{m}{2\pi i \hbar \Delta t} \right)^{\frac{N+1}{2}} \int_{\mathbb{R}^N} \prod_{k=1}^{N} dq_k \, \exp \left[\sum_{k=0}^{N} \frac{i \Delta t}{\hbar} \left(\frac{m}{2} \left(\frac{q_{k+1} - q_k}{\Delta t} \right)^2 - V(q_k) \right) \right].$$

(4.2.20)

This formula has the following heuristic interpretation. We can regard the coordinates q_k as the discretization of a trajectory $q(t)$ between the points q_0 and q_f, as shown in Figure 4.1. The exponent appearing in the r.h.s. of (4.2.20) is then a discretization of the classical action of the mechanical system (1.2.58) for the trajectory $q(t)$. The integration over all possible intermediate coordinates q_k can be interpreted, in the limit $N \to \infty$, as an integration over all possible "paths" or "particle stories" $q(t)$, with the boundary conditions

$$q(t_0) = q_0, \qquad q(t_f) = q_f.$$

(4.2.21)

The "weight" of each path $q(t)$ in this integration is given by

$$e^{\frac{i}{\hbar} S(q(t))}.$$

(4.2.22)

According to this heuristic interpretation, we can write the quantum propagator as a *path integral* in position space,

$$K(q_f, q_0; t_f, t_0) = \int \mathcal{D}q(t) \, e^{\frac{i}{\hbar} S(q(t))},$$

(4.2.23)

where the "integration measure" $\mathcal{D}q(t)$ can be regarded as the limit

$$\lim_{N \to \infty} \left(\frac{m}{2\pi i \hbar \Delta t} \right)^{\frac{N+1}{2}} \int_{\mathbb{R}^N} \prod_{k=1}^{N} dq_k.$$

(4.2.24)

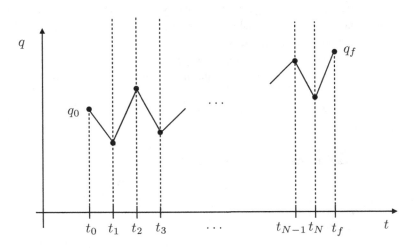

Figure 4.1   We can regard (4.2.20) as a sum over discretized "paths," which are obtained by giving a sequence of coordinates $q_1, \ldots, q_N$ associated to the intermediate times $t_1, \ldots, t_N$.

This formula is the basis for the path integral formulation of quantum mechanics pioneered by Feynman. However, the integral (4.2.23) should not be regarded as an integral, in the sense of the mathematical theory of integration, and has to be handled with care. When in doubt, one should always go back to the precise definition in terms of the limit in (4.2.20).

A similar heuristic interpretation holds for (4.2.16). We can regard the points (q_k, p_k) as discretizations of a trajectory $(q(t), p(t))$ in phase space, and the exponent appearing in the r.h.s. of (4.2.16) can be interpreted as a discretization of

$$\int_{t_0}^{t_f} [p(t)\dot{q}(t) - H(p(t), q(t))]\,dt. \tag{4.2.25}$$

We can also interpret the integration in (4.2.16), as $N \to \infty$, as an integration over all possible paths $q(t), p(t)$ in phase space going from $q(t_0) = q_0$ to $q(t_f) = q_f$. This leads to the following path integral formula in phase space:

$$K(q_f, q_0; t_f, t_0) = \int \mathcal{D}q(t)\mathcal{D}p(t)\, e^{\frac{i}{\hbar} \int_{t_0}^{t_f} (p\dot{q} - H(p,q))dt}. \tag{4.2.26}$$

Here, $\mathcal{D}q(t)\mathcal{D}p(t)$ is a representation of the measure of the path integral, which is defined more precisely in (4.2.16).

In the above derivation, we have assumed that our Hamiltonian is of the form (4.2.8). However, the expression (4.2.20) is also valid for a time-dependent Hamiltonian

$$\mathsf{H} = \frac{\mathsf{p}^2}{2m} + V(\mathsf{q}, t), \tag{4.2.27}$$

after making the replacement $V(q_k) \to V(q_k, t_k)$. The reason is that, for small times, the evolution operator is given by

$$\mathsf{U}(t_{k+1}, t_k) \approx 1 - \frac{i\Delta t_k}{\hbar}\mathsf{H}(t_k), \tag{4.2.28}$$

and the derivation of (4.2.20) with the replacement mentioned above remains valid.

The generalization of (4.2.20) to the d-dimensional propagator is straightforward. We simply have to take into account d copies of the one-dimensional result, and we obtain the following expression:

$$K(q_f, q_0; t_f, t_0)$$
$$= \lim_{N \to \infty} \left(\frac{m}{2\pi i \hbar \Delta t}\right)^{\frac{d(N+1)}{2}} \int_{\mathbb{R}^{dN}} \prod_{k=1}^{N} dq_k \, \exp\left[\sum_{k=0}^{N} \frac{i\Delta t}{\hbar}\left(\frac{m}{2}\left(\frac{q_{k+1} - q_k}{\Delta t}\right)^2 - V(q_k)\right)\right]. \tag{4.2.29}$$

4.3 Path Integral Representation of the Density Matrix

The QM propagator is closely related to the integral kernel of the canonical density matrix, as we saw in (1.2.20). Therefore, we should be able to compute $\rho(q_f, q_0; \beta)$ in a similar way. The analogue of $T = t_f - t_0$ is $u = \beta\hbar$, which should be regarded as an interval of "Euclidean" time. We will denote the density matrix by $\rho(q_f, q_0; u)$, since in this context u is a more appropriate variable than β. We will parametrize the interval of Euclidean time by the variable $0 \leq \tau \leq u$. To obtain a path integral representation of the density matrix, we split u into N intervals $\Delta \tau_k$. The analogue of formula (4.2.17) above is

$$\rho(q_{k+1}, q_k; \tau_{k+1}, \tau_k) \approx \int \frac{\mathrm{d}p_k}{2\pi\hbar} \exp\left[\frac{\mathrm{i}\Delta\tau_k}{\hbar} p_k \frac{q_{k+1} - q_k}{\Delta\tau_k} - \frac{\Delta\tau_k}{\hbar} \left(\frac{p_k^2}{2m} + V(q_k)\right)\right],$$
(4.3.30)

which can be integrated over p_k, this time with a *bone fide* Gaussian. One finds that

$$\rho(q_{k+1}, q_k; \tau_{k+1}, \tau_k) \approx \sqrt{\frac{m}{2\pi\hbar\Delta\tau_k}} \exp\left[-\frac{\Delta\tau_k}{\hbar} \left(\frac{m}{2} \left(\frac{q_{k+1} - q_k}{\Delta\tau_k}\right)^2 + V(q_k)\right)\right],$$
(4.3.31)

and, due to the change of relative sign between the kinetic and potential terms, we find a path integral representation for the density matrix

$$\rho(q_f, q_0; \beta) = \int \mathcal{D}q(t) \mathrm{e}^{-\frac{1}{\hbar}S_E(q(\tau))},$$
(4.3.32)

in terms of the *Euclidean* action:

$$S_E = \int_0^u \left(\frac{m}{2}\dot{q}^2 + V(q)\right) \mathrm{d}\tau, \qquad u = \beta\hbar.$$
(4.3.33)

This is an integral over paths satisfying the boundary conditions

$$q(0) = q_0, \qquad q(u) = q_f.$$
(4.3.34)

The calculation of the density matrix is formally equivalent to the one of the propagator, after performing a Wick rotation of the time variable,

$$t = -\mathrm{i}\tau.$$
(4.3.35)

We can also obtain from this a path-integral representation of the canonical partition function of the system at inverse temperature β. Since we are taking a trace, paths have to finish and end at the same point, i.e. we have *periodic* boundary conditions

$$q(0) = q(u) = q,$$
(4.3.36)

and in addition we have to integrate over all possible points q. This means that the thermal partition function can be written as a path integral with periodic boundary conditions:

$$Z(\beta) = \int_{q(0)=q(u)} \mathcal{D}q(\tau) \, \mathrm{e}^{-\frac{1}{\hbar}S_E(q(\tau))}.$$
(4.3.37)

In particular, we have the following representation in terms of a limit of integrations,

$$Z(\beta) = \lim_{N \to \infty} \left(\frac{m}{2\pi\hbar\Delta\tau}\right)^{\frac{N+1}{2}} \int \prod_{k=1}^{N+1} \mathrm{d}q_k \, \exp\left[-\sum_{k=0}^{N} \frac{\Delta\tau}{\hbar} \left(\frac{m}{2} \left(\frac{q_{k+1} - q_k}{\Delta\tau}\right)^2 + V(q_k)\right)\right],$$
(4.3.38)

where $q_0 = q_{N+1}$ due to the periodic boundary conditions. Note that there is an additional integration to perform, due to the trace.

4.4 The Free Particle

There are some simple examples where the path integral can be computed *exactly*. The free particle is of course one of them. Let us now calculate the path integral in the case $V(q) = 0$, by starting from the definition given in (4.2.20). In addition to providing a check of the formula, this will also introduce some of the techniques we will need in more complicated cases.

First, we write the sum appearing in the exponent of (4.2.20) as

$$\sum_{k=0}^{N} (q_{k+1} - q_k)^2 = q^T P_N q - 2v \cdot q + q_0^2 + q_f^2, \qquad (4.4.39)$$

where the $N \times N$ matrix P_N is given by[1]

$$P_N = \begin{pmatrix} 2 & -1 & 0 & \cdots & 0 & 0 \\ -1 & 2 & -1 & \cdots & 0 & 0 \\ 0 & -1 & 2 & \cdots & 0 & 0 \\ \vdots & \vdots & \vdots & \ddots & \vdots & \vdots \\ 0 & 0 & 0 & \cdots & 2 & -1 \\ 0 & 0 & 0 & \cdots & -1 & 2 \end{pmatrix}, \qquad (4.4.40)$$

and the N-dimensional vectors q, v are given by

$$q = (q_1, \ldots, q_N), \qquad v = (q_0, 0, \ldots, 0, q_f). \qquad (4.4.41)$$

Let us now use the basic formula of Gaussian integration (C.1), with

$$A = \frac{m}{\hbar \Delta t} P_N, \qquad p = -\frac{m}{\hbar \Delta t} v. \qquad (4.4.42)$$

To write down the result of this integration, we need various properties of the matrix P_N, such as the number of its positive and negative eigenvalues. It turns out that it is possible to calculate the eigenvalues of P_N explicitly. More generally, let us consider an $N \times N$ matrix $C^{(\alpha,\beta)}$ with entries

$$C_{jk}^{(\alpha,\beta)} = \alpha \delta_{jk} - \beta \left(\delta_{j,k+1} + \delta_{j,k-1} \right), \qquad j,k = 1, \ldots, N. \qquad (4.4.43)$$

Then, it is easy to check that the N-dimensional vectors $\chi^{(p)}$, $p = 1, \ldots, N$, with entries

$$\chi_j^{(p)} = \sqrt{\frac{2}{N+1}} \sin \frac{\pi p j}{N+1}, \qquad j = 1, \ldots, N, \qquad (4.4.44)$$

are eigenvectors of $C^{(\alpha,\beta)}$ with eigenvalues

$$\alpha - 2\beta \cos \frac{\pi p}{N+1}, \qquad p = 1, \ldots, N. \qquad (4.4.45)$$

The matrix P_N corresponds to $\alpha = 2$, $\beta = 1$, so that its eigenvalues are of the form

$$2 \left(1 - \cos \frac{\pi p}{N+1} \right), \qquad p = 1, \ldots, N, \qquad (4.4.46)$$

---

[1] This turns out to be the Cartan matrix of the Lie algebra A_{N+1}.

which are all positive. We then find,

$$\left(\frac{m}{2\pi i\hbar\Delta t}\right)^{\frac{N+1}{2}} \int \prod_{k=1}^{N} dq_k \, \exp\left[\frac{im\Delta t}{2\hbar} \sum_{k=0}^{N}\left(\frac{q_{k+1}-q_k}{\Delta t}\right)^2\right]$$

$$= \left(\frac{m}{2\pi i\hbar\Delta t}\right)^{\frac{N+1}{2}} \left(\frac{m}{\hbar\Delta t}\right)^{-N/2} \frac{(2\pi i)^{N/2}}{\sqrt{\det(P_N)}} \exp\left[\frac{im}{2\hbar\Delta t}\left(q_0^2 + q_f^2 - v^T P_N^{-1} v\right)\right]$$

$$= \sqrt{\frac{m}{2\pi i\hbar\Delta t \, \det(P_N)}} \exp\left[\frac{im}{2\hbar\Delta t}\left(q_0^2 + q_f^2 - v^T P_N^{-1} v\right)\right].$$

$$(4.4.47)$$

Note that, if we denote $B = P_N^{-1}$, we have

$$v^T P_N^{-1} v = B_{11}q_0^2 + B_{NN}q_f^2 + q_0 q_f (B_{1N} + B_{N1}). \qquad (4.4.48)$$

Therefore, we have to compute these matrix entries as well as $p_N = \det(P_N)$. By developing the determinant with respect to the last row, we easily find the recurrence relation

$$p_N = 2p_{N-1} + \det\begin{pmatrix} 2 & -1 & 0 & \cdots & 0 & 0 \\ -1 & 2 & -1 & \cdots & 0 & 0 \\ 0 & -1 & 2 & \cdots & 0 & 0 \\ \vdots & \vdots & \vdots & \ddots & \vdots & \vdots \\ 0 & 0 & 0 & \cdots & 2 & 0 \\ 0 & 0 & 0 & \cdots & 0 & -1 \end{pmatrix} = 2p_{N-1} - p_{N-2}, \qquad N \geq 3.$$

$$(4.4.49)$$

We can regard this relation as a second-order difference equation, with initial conditions

$$p_1 = 2, \qquad p_2 = 3. \qquad (4.4.50)$$

In order to find the solution to the recursion (4.4.49), we can use the following trick. First of all, although p_0 is not defined as a determinant, we can define it by extending the recursion relation to $N = 2$. One finds in this way that $p_0 = 1$. Let us now consider the auxiliary function

$$p(x) = \sum_{N \geq 0} p_N x^N. \qquad (4.4.51)$$

It is easy to see that the recursion (4.4.49) is equivalent to the following algebraic equation for $p(x)$:

$$p(x) - p_1 x - p_0 = 2x(p(x) - p_1) - x^2 p(x), \qquad (4.4.52)$$

whose solution is

$$p(x) = \frac{1}{(1-x)^2} = \sum_{N \geq 0}(N+1)x^N. \qquad (4.4.53)$$

We conclude that

$$p_N = N + 1. \qquad (4.4.54)$$

In order to calculate the inverse matrix, we use the minors of P_N. We find,

$$p_N B_{11} = p_N B_{NN} = p_{N-1},\qquad(4.4.55)$$

while for the other entry,

$$p_N B_{1N} = (-1)^{N+1}\det\begin{pmatrix} -1 & 2 & 0 & \cdots & 0 & 0 \\ 0 & -1 & 2 & \cdots & 0 & 0 \\ 0 & 0 & -1 & \cdots & 0 & 0 \\ \vdots & \vdots & \vdots & \ddots & \vdots & \vdots \\ 0 & 0 & 0 & \cdots & 0 & -1 \end{pmatrix} = 1.\qquad(4.4.56)$$

By symmetry, we have $B_{N1} = B_{1N}$. We then obtain,

$$B_{11} = B_{NN} = \frac{p_{N-1}}{p_N} = \frac{N}{N+1},\qquad(4.4.57)$$

and

$$B_{1N} = B_{N1} = \frac{1}{p_N} = \frac{1}{N+1}.\qquad(4.4.58)$$

It follows that

$$q_0^2 + q_f^2 - v^T P_N^{-1} v = q_0^2 + q_f^2 - B_{11}q_0^2 - B_{NN}q_f^2 - q_0 q_f (B_{1N} + B_{N1})$$

$$= q_0^2 + q_f^2 - \frac{N}{N+1}(q_0^2 + q_f^2) - \frac{2q_0 q_f}{N+1} = \frac{1}{N+1}(q_0 - q_f)^2.$$
$$(4.4.59)$$

Therefore, we find that (4.4.47) is given by

$$K(q_f, q_0; t_f, t_0) = \sqrt{\frac{m}{2\pi i\hbar T}}\exp\left[\frac{im}{2\hbar T}\left(q_0 - q_f\right)^2\right],\qquad(4.4.60)$$

where

$$T = (N+1)\Delta t = t_f - t_0.\qquad(4.4.61)$$

This is in agreement with the result obtained in (1.2.26).

4.5 Gaussian Theories

The main reason that we were able to calculate (4.2.20) explicitly in the case of the free particle is that all the integrals were Gaussian. In order to find solvable models, we should then consider theories defined by a quadratic Lagrangian of the form

$$L = \frac{m}{2}\dot{q}^2 - \frac{c(t)}{2}q^2 + f(t)q.\qquad(4.5.62)$$

We will now use the heuristic approach to the path integral based on (4.2.23), although it is easy to check that our results can be also derived from the basic formula (4.2.20). Let us denote by $q_c(t)$ the solution to the Lagrange EOM, with boundary conditions

$$q_c(t_0) = q_0,\qquad q_c(t_f) = q_f.\qquad(4.5.63)$$

An arbitrary path contributing to the path integral expression for $K(q_f, q_0; t_f, t_0)$ can be written as

$$q(t) = q_c(t) + y(t), \tag{4.5.64}$$

where

$$y(t_0) = y(t_f) = 0. \tag{4.5.65}$$

We can now expand the action around $q_c(t)$ as

$$S(q(t)) = S_c + \frac{1}{2} \int_{t_0}^{t_f} \int_{t_0}^{t_f} \frac{\delta^2 S}{\delta q(t) \delta q(t')}\Big|_{q(t) = q_c(t)} y(t) y(t') \, \mathrm{d}t \, \mathrm{d}t', \tag{4.5.66}$$

where S_c, introduced in (1.2.61), is the action evaluated at $q_c(t)$. We note that, in the functional Taylor expansion in (4.5.66), the term involving the first functional derivative vanishes, since $q_c(t)$ is an extremum of the action and one has (1.2.60). Since S is quadratic, the expansion in (4.5.66) stops at second order (in particular, the dependence on the linear term in (4.5.62) will only enter through the classical action S_c). By using (1.2.59), we obtain

$$\frac{\delta S}{\delta q(t)} = -m\ddot{q}(t) - c(t)q(t) + f(t), \tag{4.5.67}$$

therefore,

$$\frac{\delta^2 S}{\delta q(t) \delta q(t')} = \left[-m \left(\frac{\mathrm{d}}{\mathrm{d}t} \right)^2 - c(t) \right] \delta(t - t'). \tag{4.5.68}$$

The double integral involving the second derivative of the delta function can be simplified as follows:

$$\int_{t_0}^{t_f} \left(\int_{t_0}^{t_f} \frac{\mathrm{d}^2}{\mathrm{d}t^2} \delta(t - t') y(t) \mathrm{d}t \right) y(t') \mathrm{d}t' = - \int_{t_0}^{t_f} \left(\int_{t_0}^{t_f} \frac{\mathrm{d}}{\mathrm{d}t} \delta(t - t') \dot{y}(t) \mathrm{d}t \right) y(t') \mathrm{d}t'$$

$$= \int_{t_0}^{t_f} \left(\int_{t_0}^{t_f} \frac{\mathrm{d}}{\mathrm{d}t'} \delta(t - t') y(t') \mathrm{d}t' \right) \dot{y}(t) \mathrm{d}t$$

$$= - \int_{t_0}^{t_f} (\dot{y}(t))^2 \mathrm{d}t. \tag{4.5.69}$$

In the first line we integrated by parts w.r.t. t, and in going from the second to the third line, we integrated by parts w.r.t. t'. In both cases we took into account the boundary conditions (4.5.65), which set to zero the boundary terms. We conclude that

$$S(q(t)) = S_c + \mathcal{S}(y(t)), \tag{4.5.70}$$

where

$$\mathcal{S}(y) = \frac{1}{2} \int_{t_0}^{t_f} \left(m\dot{y}^2 - c(t)y^2 \right) \mathrm{d}t. \tag{4.5.71}$$

The propagator is then given by

$$K(q_f, q_0; t_f, t_0) = e^{\frac{i}{\hbar} S_c(q_f, q_0; t_f, t_0)} \int \mathcal{D}y(t) e^{\frac{i}{\hbar} \mathcal{S}(y(t))}, \tag{4.5.72}$$

where we have explicitly indicated that the action evaluated at the classical trajectory, S_c, is a function of the boundary data q_f, q_0, t_f, t_0. Note that, in doing the integration over paths, the decomposition (4.5.64) can be regarded as a change of variables, which is a translation by a "constant" (i.e. fixed) path. Therefore, the Jacobian of the change of variables is just one. The path integral over the path $y(t)$ is defined as above, by the limit

$$\int \mathcal{D}y(t) e^{\frac{i}{\hbar}S(y(t))}$$

$$= \lim_{N \to \infty} \left(\frac{m}{2\pi i \hbar \Delta t}\right)^{\frac{N+1}{2}} \int \prod_{k=1}^{N} dy_k \, \exp\left[\frac{im}{2\hbar \Delta t} \sum_{k=0}^{N} \left((y_{k+1} - y_k)^2 - \frac{(\Delta t)^2 c_k}{m} y_k^2\right)\right],$$
$$(4.5.73)$$

where $c_k = c(t_k)$, and we recall that in this case the boundary conditions are $y_0 = y_{N+1} = 0$. We can now generalize the calculation in the case of the free particle. The sum in the exponent of (4.5.73) can be written as

$$\mathbf{y}^T P_N \mathbf{y}, \qquad \mathbf{y}^T = (y_1, \ldots, y_N), \qquad (4.5.74)$$

where the $N \times N$ matrix P_N is now given by

$$P_N = \begin{pmatrix} 2 - \rho_1 & -1 & 0 & \cdots & 0 & 0 \\ -1 & 2 - \rho_2 & -1 & \cdots & 0 & 0 \\ 0 & -1 & 2 - \rho_3 & \cdots & 0 & 0 \\ \vdots & \vdots & \vdots & \ddots & \vdots & \vdots \\ 0 & 0 & 0 & \cdots & 2 - \rho_{N-1} & -1 \\ 0 & 0 & 0 & \cdots & -1 & 2 - \rho_N \end{pmatrix}, \qquad (4.5.75)$$

and

$$\rho_j = \frac{(\Delta t)^2 c_j}{m}. \qquad (4.5.76)$$

For any N, the integral (4.5.73) is then a Gaussian, and it can be evaluated explicitly by using (C.1). We obtain,

$$\left(\frac{m}{2\pi i \hbar \Delta t}\right)^{\frac{N+1}{2}} \int \prod_{k=1}^{N} dy_k \, \exp\left[\frac{im}{2\hbar \Delta t} \sum_{k=0}^{N} \left((y_{k+1} - y_k)^2 - \frac{(\Delta t)^2 c_k}{m} y_k^2\right)\right]$$
$$(4.5.77)$$
$$= \sqrt{\frac{m}{2\pi i \hbar \Delta t \, \det(P_N)}}.$$

In this formula, it has to be understood that

$$\frac{1}{\sqrt{\det(P_N)}} = e^{-\pi i \nu_N/2} \frac{1}{\sqrt{|\det(P_N)|}}, \qquad (4.5.78)$$

where ν_N is the number of negative eigenvalues of the matrix P_N. In some simple cases, such as the harmonic oscillator, ν_N, this can be calculated in detail, as we will see below. It is easy to find the recursion relation satisfied by the determinant $p_N = \det(P_N)$ by generalizing (4.4.49) for the free particle:

$$p_N = (2 - \rho_N) p_{N-1} - p_{N-2}, \qquad N \geq 3. \qquad (4.5.79)$$

As in Section 4.4, this can be extended to $N = 2$ by setting $p_0 = 1$. The recursion (4.5.79) can be also written as

$$\frac{p_{j+1} + p_{j-1} - 2p_j}{(\Delta t)^2} = -\frac{c_{j+1}p_j}{m}. \tag{4.5.80}$$

We want to regard p_j as the discretization of a function $\phi(t)$, defined as follows

$$\phi(t_j) = (\Delta t)\, p_j. \tag{4.5.81}$$

In this way, the recursion (4.5.80) becomes the discretization of a second-order differential equation satisfied by $\phi(t)$, namely

$$m\frac{d^2\phi}{dt^2} + c(t)\phi(t) = 0. \tag{4.5.82}$$

Indeed, let us multiply both sides of (4.5.80) by Δt. Then,

$$\lim_{\Delta t \to 0} \frac{\phi(t_{j+1}) + \phi(t_{j-1}) - 2\phi(t_j)}{(\Delta t)^2} = \frac{d^2\phi}{dt^2}, \tag{4.5.83}$$

where we identify $t_j = t$, and

$$\lim_{\Delta t \to 0} c(t_{j+1})\phi(t_j) = \lim_{\Delta t \to 0} c(t + \Delta t)\phi(t) = c(t)\phi(t). \tag{4.5.84}$$

The ODE (4.5.82) is the EOM for the action (4.5.71). The initial conditions for this ODE are as follows: in the continuum limit

$$\phi(t_0) = \lim_{\Delta t \to 0} (\Delta t)\, p_0 = 0. \tag{4.5.85}$$

The first derivative is given by

$$\frac{d\phi}{dt}\bigg|_{t=t_0} = \lim_{\Delta t \to 0} \frac{\Delta t(p_1 - p_0)}{\Delta t} = \lim_{\Delta t \to 0} \left(1 - \frac{(\Delta t)^2 c_1}{m}\right) = 1. \tag{4.5.86}$$

Therefore, we find that

$$\lim_{\Delta t \to 0} (\Delta t)\, p_N = \lim_{\Delta t \to 0} \phi(t_{N+1} - \Delta t) = \phi(t_f). \tag{4.5.87}$$

In order to keep track of the dependence on t_0, we will denote

$$f(t; t_0) = \phi(t), \tag{4.5.88}$$

which satisfies the ODE

$$m\frac{d^2 f(t; t_0)}{dt^2} + c(t)f(t; t_0) = 0, \tag{4.5.89}$$

with the boundary conditions

$$f(t_0; t_0) = 0, \qquad \dot{f}(t_0; t_0) = 1. \tag{4.5.90}$$

We conclude that

$$\int_{y(t_0)=y(t_f)=0} \mathcal{D}y(t) e^{\frac{i}{\hbar}S(y(t))} = \sqrt{\frac{m}{2\pi i\hbar f(t_f; t_0)}}, \tag{4.5.91}$$

so that the propagator is given by

$$K(q_f, q_0; t_f, t_0) = \sqrt{\frac{m}{2\pi i\hbar f(t_f; t_0)}} e^{iS_c(q_f,q_0;t_f,t_0)/\hbar}. \tag{4.5.92}$$

This is a remarkable formula. It states that, in a theory with a quadratic Lagrangian, the propagator has an interpretation in terms of classical paths: its phase is given by the classical action, and the prefactor is determined by a particular solution to the EOM. As a consequence of (4.5.78), we have to remember that

$$\frac{1}{\sqrt{f(t_f; t_0)}} = e^{-\pi i\nu/2} \frac{1}{\sqrt{|f(t_f; t_0)|}}, \tag{4.5.93}$$

where ν is the appropriate limit of the ν_N appearing in (4.5.78), and it depends on the value of t_f and t_0. It is sometimes called the *Maslov index*.

Example 4.5.1 *The free particle redux.* Let us verify that the prefactor in (4.4.60) is determined by the expression (4.5.92). The ODE (4.5.89) when $c(t) = 0$ is solved by

$$f(t; t_0) = at + b, \tag{4.5.94}$$

and the boundary conditions require

$$f(t_0; t_0) = at_0 + b = 0, \qquad \dot{f}(t_0; t_0) = a = 1, \tag{4.5.95}$$

so that

$$f(t; t_0) = t - t_0, \tag{4.5.96}$$

and

$$f(t_f; t_0) = T, \tag{4.5.97}$$

in agreement with (4.4.60). □

Example 4.5.2 *The harmonic oscillator.* Let us perform the same verification for the harmonic oscillator. We will set, as before, $T = t_f - t_0$. In this case, $c(t) = m\omega^2$, and

$$\rho_j = \rho = (\Delta t)^2 \omega^2, \tag{4.5.98}$$

is the same for all indices $j = 1, \ldots, N$. The matrix P_N is of the form (4.4.43) with $\beta = 1$ and

$$\alpha = 2 - (\Delta t)^2 \omega^2. \tag{4.5.99}$$

The eigenvalues of the matrix P_N are then of the form,

$$2\left(\cos\frac{\omega T}{N+1} - \cos\frac{\pi p}{N+1}\right), \qquad p = 1, \ldots, N+1, \tag{4.5.100}$$

for N sufficiently large. We conclude that, if

$$T_\nu < T < T_{\nu+1}, \tag{4.5.101}$$

where

$$T_\nu = \frac{\pi\nu}{\omega}, \tag{4.5.102}$$

there are exactly ν negative eigenvalues. This determines the Maslov index in the case of the harmonic oscillator. Let us now compute the function $f(t; t_0)$. It satisfies the EOM

$$\ddot{f}(t; t_0) + \omega^2 f(t; t_0) = 0. \tag{4.5.103}$$

Therefore

$$f(t; t_0) = A \cos \omega(t - t_0) + B \sin \omega(t - t_0). \tag{4.5.104}$$

Imposing the boundary conditions (4.5.90) leads to $A = 0$, $B = 1/\omega$, so that we find

$$f(t; t_0) = \frac{1}{\omega} \sin \omega(t - t_0), \tag{4.5.105}$$

and we conclude that

$$f(t_f; t_0) = \frac{1}{\omega} \sin \omega T. \tag{4.5.106}$$

In this case, it is easy to verify directly that the determinant of the matrix P_N behaves as

$$(\Delta t) p_N \approx f(t_f; t_0), \tag{4.5.107}$$

in the limit $N \to \infty$, $\Delta t \to 0$, $(N + 1)\Delta t = T$. To see this, we note that the recursion (4.5.79) reads in this case

$$p_N = \alpha p_{N-1} - p_{N-2}, \tag{4.5.108}$$

where α was defined in (4.5.99). As in the case of the free particle, we introduce a generating function

$$p(x) = \sum_{N \geq 0} p_N x^N, \tag{4.5.109}$$

and it is easy to see that the recursion relation is equivalent to the algebraic equation

$$p(x) - p_1 x - 1 = \alpha x (p(x) - 1) - x^2 p(x), \tag{4.5.110}$$

whose solution is

$$p(x) = \frac{1}{1 - \alpha x + x^2}. \tag{4.5.111}$$

The coefficient p_N can be obtained by the Cauchy formula

$$p_N = \oint \frac{dz}{2\pi i} \frac{1}{1 - \alpha z + z^2} \frac{1}{z^{N+1}}, \tag{4.5.112}$$

where we integrate around the origin in the complex plane. We can deform the integration contour and pick the residues of the poles at

$$z = z_\pm = \frac{\alpha \pm \sqrt{\alpha^2 - 4}}{2}. \tag{4.5.113}$$

We note that $z_+ z_- = 1$, and, since $\alpha < 2$, we have that $z_\pm = e^{\pm i\theta}$ for an angle θ. The residue calculation gives

$$p_N = \frac{z_+^{N+1} - z_-^{N+1}}{z_+ - z_-} = \frac{\sin(N+1)\theta}{\sin\theta}. \tag{4.5.114}$$

When Δt is very small, the angle θ is also very small. By comparing $\alpha = 2\cos\theta$ to its definition in (4.5.99), we find

$$\theta \approx \omega\Delta t = \frac{\omega T}{N+1}. \tag{4.5.115}$$

We conclude that

$$p_N \approx \frac{\sin\omega T}{\omega\Delta t}, \tag{4.5.116}$$

in accordance with (4.5.107). □

4.6 Semiclassical Approximation to the Propagator

As we have seen many times in this text, classical mechanics should emerge from quantum mechanics in the limit in which $\hbar$ is small as compared to the typical action involved in the problem. What happens to this limit in the path integral formulation? Let us consider the corresponding situation in the case of an ordinary integral with a similar structure,

$$\mathcal{I}(\hbar) = \int dz \, e^{if(z)/\hbar}. \tag{4.6.117}$$

When $\hbar$ is small, the integrand is a rapidly oscillating function, except at the points where the first order variation of $f(z)$ vanishes, i.e. at the critical points defined by $f'(z) = 0$. Let us assume for simplicity that there is a single critical point z_c. One expects that the most important contribution to $\mathcal{I}(z)$ comes from z_c and its neighborhood, and this is the basis of the so-called saddle-point approximation to oscillatory integrals of the form (4.6.117). More precisely, one expands $f(z)$ around z_c,

$$f(z) \approx f(z_c) + \frac{1}{2}f''(z_c)(z - z_c)^2, \tag{4.6.118}$$

and one obtains,

$$\mathcal{I}(\hbar) \approx \sqrt{\frac{2\pi i}{f''(z_c)}} e^{if(z_c)/\hbar}. \tag{4.6.119}$$

This approximation gives the two leading terms in an asymptotic expansion of the integral for small $\hbar$. In actual applications of the saddle-point approximation one has to consider subtle issues (e.g. how to deform the integration path into a steepest descent path that goes through the critical point). However, the basic result (4.6.119) will be enough for our purposes.

Let us apply a similar reasoning to the path integral formula (4.2.23) in the limit in which $\hbar$ is small when compared to the action of a typical path. As in the conventional saddle-point method, we expect that the most important contribution to the the path integral in this limit is due to the critical path for the functional $S(q(t))$, satisfying the equation

$$\frac{\delta S}{\delta q(t)} = 0, \qquad (4.6.120)$$

as well as the boundary conditions $q(t_0) = q_0$, $q(t_f) = q_f$. This defines the classical path, $q_c(t)$, precisely since (4.6.120) is nothing but the classical EOM.

We obtain in this way an intuitive and appealing picture of the semiclassical limit: in the path integral formulation, the probability amplitude of going from one point to another is an integral, i.e. a sum over all possible trajectories between the two points. The contribution of each path is

$$e^{iS(q(t))/\hbar}. \qquad (4.6.121)$$

When $\hbar$ is small, this sum is dominated by the path that makes $S(q(t))$ stationary, which is the classical trajectory, and we have at leading order

$$K(q_f, q_0; t_f, t_0) \approx e^{\frac{i}{\hbar} S_c(q_f, q_0; t_f, t_0)}, \qquad \hbar \ll 1. \qquad (4.6.122)$$

As in (4.6.119), we can expand around the classical trajectory to calculate the next-to-leading correction to this leading-order approximation. This is exactly what we did in (4.5.64), and we find

$$S(q(t)) = S_c + \frac{1}{2} \int dt \, dt' \, y(t) y(t') \frac{\delta^2 S}{\delta q(t) \delta q(t')}\bigg|_{q(t)=q_c(t)} + \cdots \qquad (4.6.123)$$

When the Lagrangian is of the standard form

$$L = \frac{m}{2} \dot{q}^2 - V(q), \qquad (4.6.124)$$

the second functional derivative reads

$$\frac{\delta^2 S}{\delta q(t) \delta q(t')}\bigg|_{q(t)=q_c(t)} = \left[-m \left(\frac{d}{dt} \right)^2 - V''(q_c(t)) \right] \delta(t - t'), \qquad (4.6.125)$$

and we conclude that

$$S(q(t)) = S_c + S_2(y(t)) + \cdots, \qquad (4.6.126)$$

where the quadratic functional is

$$S_2(y(t)) = \frac{1}{2} \int_{t_0}^{t_f} dt \left(m\dot{y}^2 - V''(q_c(t)) y^2 \right). \qquad (4.6.127)$$

The next correction to the leading-order result is given by evaluating the path integral over the path $y(t)$ with the action $S_2(y(t))$. This is exactly of the form considered in Section 4.5, where now

$$c(t) = V''(q_c(t)), \qquad (4.6.128)$$

so we find the answer

$$K(q_f, q_0; t_f, t_0) \approx \sqrt{\frac{m}{2\pi i \hbar f(t_f; t_0)}} e^{\frac{i}{\hbar} S_c(q_f, q_0; t_f, t_0)}, \qquad (4.6.129)$$

where $f(t; t_0)$ satisfies (4.5.89) with the boundary conditions (4.5.90). It turns out that this function can be written in terms of the action evaluated at the classical trajectory. To understand this, we consider a family of classical trajectories $q(t; q_0, p)$, parametrized by the initial position, q_0, and the initial momentum, p, so that they verify

$$q(t_0; q_0, p) = q_0, \qquad \dot{q}(t_0; q_0, p) = \frac{p}{m}. \tag{4.6.130}$$

They also satisfy Newton's equation

$$m\ddot{q}(t; q_0, p) + V'\left(q(t; q_0, p)\right) = 0. \tag{4.6.131}$$

Let us consider now the function

$$\xi(t; q_0, p) = m\frac{\partial q}{\partial p}. \tag{4.6.132}$$

Then, by taking a derivative w.r.t. p in Newton's equation, we find

$$m\ddot{\xi}(t; q_0, p) + V''\left(q(t; q_0, p)\right)\xi(t; q_0, p) = 0. \tag{4.6.133}$$

At the same time, by taking derivatives w.r.t. p in the initial conditions (4.6.130) for the family of paths, we find

$$\xi(t_0; q_0, p) = 0, \qquad \dot{\xi}(t_0; q_0, p) = 1. \tag{4.6.134}$$

We conclude that

$$f(t; t_0) = \xi(t; q_0, p). \tag{4.6.135}$$

To calculate $f(t_f; t_0)$, we first note that

$$q_f = q(t_f; q_0, p). \tag{4.6.136}$$

This gives p, the initial momentum, implicitly as a function of the final position q_f. Using the chain rule, we have that

$$\frac{\partial q(t_f; q_0, p)}{\partial p}\frac{\partial p}{\partial q_f} = 1. \tag{4.6.137}$$

But

$$f(t_f; t_0) = m\frac{\partial q(t_f; q_0, p)}{\partial p} = m\left(\frac{\partial p}{\partial q_f}\right)^{-1}. \tag{4.6.138}$$

As it is well known, the initial momentum can be obtained from the value of the action at the classical trajectory as

$$p = -\frac{\partial S_c}{\partial q_0}. \tag{4.6.139}$$

We conclude that

$$f(t_f; t_0) = -m\left(\frac{\partial^2 S_c}{\partial q_f \partial q_0}\right)^{-1}. \tag{4.6.140}$$

When we put all of this together, we end up with the following formula for the semiclassical quantum propagator:

$$K(q_f, q_0; t_f, t_0) \approx \frac{1}{\sqrt{2\pi i \hbar}} \sqrt{-\frac{\partial^2 S_c}{\partial q_f \partial q_0}} e^{\frac{i}{\hbar} S_c(q_f, q_0; t_f, t_0)}. \qquad (4.6.141)$$

We have then rederived the Van Vleck formula (2.11.443) in the one-dimensional case, using path integral techniques.

It follows from (4.5.92) that, in the case of a theory in which the potential is at most quadratic, like the ones analyzed in Section 4.5, the expression (4.6.141) is exact. This finally explains why we found (1.2.65) in the calculation of the QM propagator for such theories.

4.7 Semiclassical Approximation to the Partition Function

When we study the canonical partition function, there are two different semiclassical regimes that one can consider. In the first one, we take

$$\beta \hbar \to 0, \qquad (4.7.142)$$

which corresponds to either $\hbar$ small or high temperature. In this case, the Euclidean time interval u is very small. Let us consider a periodic trajectory satisfying $q(u) = q(0) = q$. Since the evolution time u is very short, the only way for trajectories to deviate significantly from the endpoints is to have a large kinetic energy. However, these trajectories are very much suppressed in the path integral due to the kinetic energy term in the Euclidean action. Therefore, we expect that the trajectories that contribute the most are the ones in which $q(\tau)$ deviates little from the endpoint q. If this is the case, we can approximate $V(q(\tau))$ by $V(q)$, so that

$$S_E(q(\tau)) \approx u V(q) + \frac{m}{2} \int_0^u \dot{q}^2(\tau) d\tau. \qquad (4.7.143)$$

The path integral computing the thermal partition function can be written as

$$\begin{aligned} Z(\beta) &= \int_{\mathbb{R}} dq \int_{q(0)=q(u)=q} \mathcal{D}q(\tau) e^{-\frac{1}{\hbar} S_E(q(\tau))} \\ &\approx \int_{\mathbb{R}} dq \, e^{-\beta V(q)} \int_{q(0)=q(u)=q} \mathcal{D}q(\tau) e^{-\frac{m}{2\hbar} \int_0^u \dot{q}^2(\tau) d\tau} \qquad (4.7.144) \\ &= \sqrt{\frac{m}{2\pi \hbar^2 \beta}} \int_{\mathbb{R}} dq \, e^{-\beta V(q)}. \end{aligned}$$

This is precisely the *classical* canonical partition function, displaying the Boltzmann factor.

Another limit occurs when taking $\hbar$ small, but keeping $\beta \hbar$ finite. This is exactly like the semiclassical limit studied in Section 4.6, and the analysis proceeds in the same way. First of all, we consider a stationary, periodic path for the Euclidean action, which we denote by $q_c(\tau)$. This is a classical solution of the EOM for an *inverted* potential:

$$m\ddot{q}_c(\tau) - V'(q_c(\tau)) = 0, \tag{4.7.145}$$

since the Euclidean action differs from the original action in the sign of the potential. The equation (4.7.145) is sometimes called the Euclidean EOM. We then expand the action around this path as

$$q(\tau) = q_c(\tau) + y(\tau). \tag{4.7.146}$$

Note that $y(\tau)$ is also a periodic path: $y(0) = y(u)$. We find,

$$S_E(q(\tau)) = S_E(q_c(\tau)) + \mathcal{S}_E(y(\tau)) + \cdots, \tag{4.7.147}$$

where the quadratic functional is

$$\mathcal{S}_E(y(\tau)) = \frac{1}{2} \int_0^u d\tau \left(m\dot{y}^2 + V''(q_c(\tau))y^2 \right). \tag{4.7.148}$$

We conclude that, in this approximation,

$$Z(\beta) \approx e^{-\frac{1}{\hbar}S_E(q_c(\tau))} \int_{y(0)=y(u)} \mathcal{D}y(\tau) \, e^{-\frac{1}{\hbar}\mathcal{S}_E(y(\tau))}. \tag{4.7.149}$$

In order to evaluate (4.7.149), we will write the quadratic functional (4.7.148) in a slightly more general form, as

$$S_E(y) = \frac{1}{2} \int_0^u d\tau \left(m\dot{y}^2 + c(\tau)y^2 \right). \tag{4.7.150}$$

We want to evaluate the Euclidean path integral appearing in (4.7.149) over all possible periodic paths $y(t)$. This path integral is defined as

$$\int_{y(0)=y(u)} \mathcal{D}y(\tau) e^{-\frac{1}{\hbar}S_E(y(\tau))}$$

$$= \lim_{N\to\infty} \left(\frac{m}{2\pi\hbar\Delta\tau} \right)^{\frac{N+1}{2}} \int \prod_{k=1}^{N+1} dy_k \, \exp\left[-\frac{m}{2\hbar\Delta\tau} \sum_{k=0}^{N} \left((y_{k+1} - y_k)^2 + \frac{(\Delta\tau)^2 c_k}{m} y_k^2 \right) \right], \tag{4.7.151}$$

where $c_k = c(\tau_k)$, and periodicity implies that

$$y_0 = y_{N+1}. \tag{4.7.152}$$

The calculation is similar to the one in Section 4.5. The exponent in (4.7.151) can be written as

$$-\frac{m}{2\hbar\Delta\tau} \mathbf{y}^T A_{N+1} \mathbf{y}, \tag{4.7.153}$$

where

$$\mathbf{y}^T = (y_0, y_1, \dots, y_N). \tag{4.7.154}$$

The matrix A_{N+1} is given by

$$A_{N+1} = \begin{pmatrix} 2+\rho_0 & -1 & 0 & \cdots & 0 & -1 \\ -1 & 2+\rho_1 & -1 & \cdots & 0 & 0 \\ 0 & -1 & 2+\rho_2 & \cdots & 0 & 0 \\ \vdots & \vdots & \vdots & \ddots & \vdots & \vdots \\ 0 & 0 & 0 & \cdots & 2+\rho_{N-1} & -1 \\ -1 & 0 & 0 & \cdots & -1 & 2+\rho_N \end{pmatrix}, \qquad (4.7.155)$$

where

$$\rho_j = \frac{(\Delta\tau)^2 c_j}{m}. \qquad (4.7.156)$$

The N th approximation to the path integral is now given by

$$\left(\frac{m}{2\pi\hbar\Delta\tau}\right)^{\frac{N+1}{2}} \int \prod_{k=1}^{N+1} dy_k \, \exp\left[-\frac{m}{2\hbar\Delta\tau} \sum_{k=0}^{N} \left((y_{k+1}-y_k)^2 + \rho_k y_k^2\right)\right] = \sqrt{\frac{1}{\det(A_{N+1})}}. \qquad (4.7.157)$$

To calculate the large N limit of this determinant, we consider the eigenvalue problem for the matrix A_{N+1}. A vector $\mathbf{y}$ will be an eigenvector with eigenvalue λ if

$$a_k y_k - y_{k+1} - y_{k-1} = \lambda y_k, \qquad (4.7.158)$$

where we set

$$a_k = 2 + \rho_k, \qquad (4.7.159)$$

and in addition we have the periodicity condition (4.7.152). We can think about (4.7.158) as the discretization of the second-order differential equation

$$-y'' + \frac{c(\tau)}{m} y = \lambda y. \qquad (4.7.160)$$

In addition, we can regard the y_i as variables associated to a periodic lattice. We will write the eigenvalue equations in the form

$$\begin{pmatrix} y_i \\ y_{i+1} \end{pmatrix} = t_i(\lambda) \begin{pmatrix} y_{i-1} \\ y_i \end{pmatrix}, \qquad (4.7.161)$$

where

$$t_i(\lambda) = \begin{pmatrix} 0 & 1 \\ -1 & a_i - \lambda \end{pmatrix}. \qquad (4.7.162)$$

In other words, to each site of the lattice we associate a vector with two coordinates y_{i-1}, y_i. Two consecutive sites are related by the "transfer matrix" $t_i(\lambda)$. Note that

$$\det(t_i(\lambda)) = 1. \qquad (4.7.163)$$

The matrices $t_i(\lambda)$ play the role of the Lax matrices (2.12.501) of the periodic Toda lattice. Given an initial condition y_{-1}, y_0, we can find a solution to (4.7.158) at all discretized times by acting with products of the $t_i(\lambda)$:

$$\begin{pmatrix} y_{i-1} \\ y_i \end{pmatrix} = t_{i-1}(\lambda) \cdots t_0(\lambda) \begin{pmatrix} y_{-1} \\ y_0 \end{pmatrix}. \tag{4.7.164}$$

In analogy to (2.12.503), the monodromy matrix is defined by

$$T_{N+1}(\lambda) = t_N(\lambda) \cdots t_0(\lambda), \tag{4.7.165}$$

and we have

$$\begin{pmatrix} y_N \\ y_{N+1} \end{pmatrix} = T_{N+1}(\lambda) \begin{pmatrix} y_{-1} \\ y_0 \end{pmatrix}. \tag{4.7.166}$$

The existence of a periodic solution to the discretized equation (4.7.158) is equivalent to the existence of an eigenvector of the monodromy matrix of eigenvalue 1. Therefore, λ is an eigenvalue of A_{N+1} if and only if it satisfies

$$\det\left(T_{N+1}(\lambda) - \mathbf{1}\right) = 0. \tag{4.7.167}$$

If we write

$$T_{N+1}(\lambda) = \begin{pmatrix} a(\lambda) & b(\lambda) \\ c(\lambda) & d(\lambda) \end{pmatrix}, \tag{4.7.168}$$

we find that

$$p(\lambda) \equiv \det\left(T_{N+1}(\lambda) - \mathbf{1}\right) = 2 - \operatorname{Tr} T_{N+1}, \tag{4.7.169}$$

since $\det(T_{N+1}) = 1$. It is easy to see that the entry $d(\lambda)$ in (4.7.168) is a polynomial of degree $N + 1$ in λ, which goes like

$$d(\lambda) \approx (-\lambda)^{N+1}, \qquad \lambda \gg 1. \tag{4.7.170}$$

This is similar to the result (2.12.507). One can also see that $a(\lambda)$ is a polynomial of degree $N - 1$ in λ, for $N \geq 2$. It follows that $p(\lambda)$ is a polynomial in λ that has the same degree as the characteristic polynomial of A_{N+1}, and the same roots. In addition, its leading coefficient $-(-\lambda)^{N+1}$ is minus the leading coefficient of the characteristic polynomial. We conclude that

$$\det(A_{N+1} - \lambda) = -\det\left(T_{N+1}(\lambda) - \mathbf{1}\right). \tag{4.7.171}$$

Let us now compute the diagonal entries of $T_{N+1}(\lambda)$. To do this, we take the following approach. Let us consider a basis for the vector at the initial site:

$$\begin{pmatrix} v_{-1}^{(1)} \\ v_0^{(1)} \end{pmatrix} = \begin{pmatrix} 1 \\ 0 \end{pmatrix}, \qquad \begin{pmatrix} v_{-1}^{(2)} \\ v_0^{(2)} \end{pmatrix} = \begin{pmatrix} 0 \\ 1 \end{pmatrix}. \tag{4.7.172}$$

Then,

$$\begin{pmatrix} v_N^{(1)} \\ v_{N+1}^{(1)} \end{pmatrix} = T_{N+1}(\lambda) \begin{pmatrix} v_{-1}^{(1)} \\ v_0^{(1)} \end{pmatrix}, \qquad \begin{pmatrix} v_N^{(2)} \\ v_{N+1}^{(2)} \end{pmatrix} = T_{N+1}(\lambda) \begin{pmatrix} v_{-1}^{(2)} \\ v_0^{(2)} \end{pmatrix}. \tag{4.7.173}$$

It follows that

$$T_{N+1}(\lambda) = \begin{pmatrix} v_N^{(1)} & v_N^{(2)} \\ v_{N+1}^{(1)} & v_{N+1}^{(2)} \end{pmatrix}. \tag{4.7.174}$$

Let us now consider a solution to (4.7.158) with initial conditions

$$y_{-1}^{(1)} = y_0^{(1)} = 1, \tag{4.7.175}$$

and another one with

$$y_{-1}^{(2)} = 0, \qquad y_0^{(2)} = \Delta t. \tag{4.7.176}$$

This leads to another basis of solutions, so the previous basis can be expressed as a linear combination of this one. One quickly finds,

$$v_k^{(1)} = y_k^{(1)} - \frac{1}{\Delta t} y_k^{(2)}, \qquad v_k^{(2)} = \frac{1}{\Delta t} y_k^{(2)}. \tag{4.7.177}$$

Therefore,

$$\det(A_{N+1} - \lambda) = v_N^{(1)} + v_{N+1}^{(2)} - 2 = y_N^{(1)} + \frac{y_{N+1}^{(2)} - y_N^{(2)}}{\Delta t} - 2. \tag{4.7.178}$$

In the continuum limit $N \to \infty$, $y_k^{(1)}$ becomes a solution $y_1(\tau; \lambda)$ of (4.7.160) with boundary conditions

$$y_1(0; \lambda) = 1, \qquad y_1'(0; \lambda) = 0, \tag{4.7.179}$$

while $y_k^{(2)}$ becomes a solution $y_2(\tau; \lambda)$ with boundary conditions

$$y_2(0; \lambda) = 0, \qquad y_2'(0; \lambda) = 1. \tag{4.7.180}$$

Given such a solution, we have that

$$\lim_{N \to \infty} \det(A_{N+1} - \lambda) = y_1(u; \lambda) + y_2'(u; \lambda) - 2. \tag{4.7.181}$$

This makes it possible to compute the path integral from a basis of solutions to the EOM (4.7.160), and we conclude in particular that

$$\int_{y(\beta\hbar)=y(0)} e^{-\frac{1}{2\hbar} \int_0^{\beta\hbar} (\dot{y}^2 + (c(\tau) - \lambda)y^2) d\tau} \, \mathcal{D}y(\tau) = \frac{1}{\sqrt{y_1(u; \lambda) + y_2'(u; \lambda) - 2}}. \tag{4.7.182}$$

Example 4.7.1 Let us consider the (Euclidean) harmonic oscillator. In this case, the function $c(\tau)$ in (4.7.150) is given by

$$c(\tau) = \omega^2. \tag{4.7.183}$$

We will also set $m = 1$. The path integral (4.7.151) computes the canonical partition function of the harmonic oscillator, which we will denote by $Z_G(\beta)$ (here, G stands for Gaussian). The differential equation (4.7.160) is in this case,

$$-y'' + \omega^2 y = 0. \tag{4.7.184}$$

We have already set $\lambda = 0$. The solutions $y_1(\tau)$, $y_2(\tau)$ with the above boundary conditions are

$$y_1(\tau) = \cosh(\omega\tau), \qquad y_2(\tau) = \frac{\sinh(\omega\tau)}{\omega}. \tag{4.7.185}$$

Then,

$$y_1(u) + y_2'(u) - 2 = 2\cosh(\beta\hbar\omega) - 2 = 4\sinh^2\left(\frac{\beta\omega\hbar}{2}\right). \tag{4.7.186}$$

We conclude that

$$Z_G(\beta) = \frac{1}{2\sinh\left(\frac{\beta\omega\hbar}{2}\right)}, \tag{4.7.187}$$

which is the standard partition function for the harmonic oscillator at finite temperature.   □

Example 4.7.2 *Anti-periodic boundary conditions.* Often we are interested in calculating partition functions with the insertion of an operator that commutes with the Hamiltonian. A typical example is the parity operator P, which acts on wavefunctions as

$$P\psi(q) = \psi(-q). \tag{4.7.188}$$

This operator has eigenvalues $(-1)^P$, where P is the parity of the wavefunctions: $P = 0, 1$ for wavefunctions that are even (respectively odd) under reflection. If P commutes with H, both operators can be diagonalized simultaneously. One then defines the "twisted" partition function as

$$Z_a(\beta) = \text{Tr}\left(P\, e^{-\beta H}\right). \tag{4.7.189}$$

Assuming for simplicity that the spectrum of H is discrete, and labelled by an integer $n = 0, 1, \ldots$, we can calculate the trace in (4.7.189) in a basis of eigenstates of H and P, as

$$Z_a(\beta) = \sum_{n \geq 0} (-1)^{P_n} e^{-\beta E_n}, \tag{4.7.190}$$

where P_n is the parity of the nth state.

An important example of this situation is that of Hamiltonians in one dimension of the form (1.3.83), where $V(q)$ is an even, confining potential. In this case, the spectrum is non-degenerate, and the eigenfunctions of the Hamiltonian have a well-defined parity: if we label them with an integer $n = 0, 1, \ldots$, according to increasing energy, we have $P_n = n \bmod 2$. For example, for the harmonic oscillator we have

$$Z_a^G(\beta) = \sum_{n=0}^{\infty} (-1)^n e^{-\beta\hbar\omega(n+1/2)} = \frac{1}{2\cosh\left(\frac{\beta\omega\hbar}{2}\right)}. \tag{4.7.191}$$

It is easy to see that $Z_a(\beta)$ can be written in terms of a path integral, as we did in Section 4.3 for the thermal partition function. The only difference with the expression (4.3.37) is that, due to the insertion of the parity operator P, which acts at the very end, one has to consider *anti-periodic* boundary conditions

$$q(u) = -q(0), \tag{4.7.192}$$

and we can write

$$Z_a(\beta) = \int_{q(u)=-q(0)} \mathcal{D}q(t) \, \exp\left[-S_E\left(q(\tau)\right)\right]. \tag{4.7.193}$$

As an example of this, we can consider the path integral (4.7.151) with anti-periodic boundary conditions $y(0) = -y(u)$. By following the arguments above, one finds, instead of (4.7.171),

$$\det(A_{N+1} - \lambda) = \det\left(T_{N+1}(\lambda) + 1\right). \tag{4.7.194}$$

We then obtain the formula

$$\int_{y(\beta\hbar)=-y(0)} e^{-\frac{1}{2\hbar}\int_0^{\beta\hbar}(\dot{y}^2 + (c(\tau)-\lambda)y^2)d\tau} \, \mathcal{D}y(\tau) = \frac{1}{\sqrt{y_1(u;\lambda) + y_2'(u;\lambda) + 2}}. \tag{4.7.195}$$

It is easy to check that this reproduces the result (4.7.191) for the harmonic oscillator.

$$\square$$

4.8 Functional Determinants

In the Gaussian approximation, the calculation of path integrals involves taking the limit as $N \to \infty$ of the determinant of a matrix of size $N \times N$. It is natural to interpret this limit as the *determinant of an operator*. As an example, let us reconsider the case of the free particle, as an example of a Gaussian integration. The propagator takes the form in (4.5.72),

$$K(q_f, q_0; t_f, t_0) = e^{\frac{i}{\hbar}S_c(q_f,q_0;t_f,t_0)} \int e^{\frac{im}{2\hbar}\int_{t_0}^{t_f} \dot{y}^2 dt} \, \mathcal{D}y(t). \tag{4.8.196}$$

Let us focus on the path integral over $y(t)$. Let

$$A = -\frac{d^2}{dt^2}, \tag{4.8.197}$$

be the second-order differential operator acting on square integrable functions defined on the interval $[t_0, t_f]$, and satisfying Dirichlet boundary conditions $y(t_0) = y(t_f) = 0$. Then,

$$\langle y|A|y\rangle = -\int_{t_0}^{t_f} y(t)\ddot{y}(t)dt = \int_{t_0}^{t_f} \dot{y}^2(t)dt. \tag{4.8.198}$$

Therefore, the path integral can be written as

$$\int e^{\frac{im}{2\hbar}\langle y|A|y\rangle} \, \mathcal{D}y(t). \tag{4.8.199}$$

We can think about this integral as an infinite-dimensional version of a Gaussian integral, in which the operator is regarded as an infinite-dimensional version of a matrix. We would expect the answer to be proportional to

$$\frac{1}{\sqrt{\det(A)}},\qquad(4.8.200)$$

where $\det(A)$ is an appropriate definition of the determinant of the operator A. How do we define it? One possibility is to use the eigenvalues of A, λ_n, $n = 1, 2, \ldots$, and set

$$\det(A) = \prod_{n=1}^{\infty} \lambda_n.\qquad(4.8.201)$$

The spectrum of A is easy to find. The eigenfunctions are

$$\psi_n(t) = \sin\left(\frac{n\pi t}{T}\right),\qquad n = 1, 2, \ldots,\qquad(4.8.202)$$

where we have set $t_0 = 0$ for simplicity, and the corresponding eigenvalues are

$$\lambda_n = \left(\frac{\pi n}{T}\right)^2,\qquad n = 1, 2, \ldots.\qquad(4.8.203)$$

However, the product over the eigenvalues in (4.8.201) is clearly divergent, and so this definition does not make sense.

There is however a natural way to define the *regularized* determinant of an operator A. Let us assume that A is an operator with a discrete and positive spectrum $\lambda_n > 0$, $n = 0, 1, \ldots$. The *zeta function* of the operator A is defined by

$$\zeta_A(s) = \sum_{n\geq 0} \frac{1}{\lambda_n^s}.\qquad(4.8.204)$$

If s is real and sufficiently large, this sum converges. For example, in the case of the operator (4.8.197), the sum converges if $\mathrm{Re}(s) > 1/2$. Let us now suppose that $\zeta_A(s)$ can be analytically continued to a larger domain containing $s = 0$, where it is regular. Then, the regularized determinant of A is defined as

$$\det A = \exp\left(-\zeta_A'(0)\right).\qquad(4.8.205)$$

The rationale for this definition is that, if the infinite product (4.8.201) converges, one has

$$\zeta_A'(0) = \sum_{n\geq 0} \frac{d}{ds}\left(e^{-s\log\lambda_n}\right)_{s=0} = -\sum_{n\geq 0} \log(\lambda_n).\qquad(4.8.206)$$

The classical example for this procedure is Riemann's zeta function, which is defined as

$$\zeta(s) = \sum_{n=1}^{\infty} \frac{1}{n^s}.\qquad(4.8.207)$$

Clearly, this converges when $\mathrm{Re}(s) > 1$. It can be shown that $\zeta(s)$ can be extended to the whole complex plane, and the only singularity in this extension is a simple pole at $s = 1$.

In the case of the operator (4.8.197), its zeta function is closely related to Riemann's zeta function, since

$$\zeta_A(s) = \left(\frac{T}{\pi}\right)^{2s} \sum_{n=1}^{\infty} \frac{1}{n^{2s}} = \left(\frac{T}{\pi}\right)^{2s} \zeta(2s). \tag{4.8.208}$$

In this case, the fact that Riemann's zeta function has an analytic extension to $s = 0$ makes it possible to define and calculate the regularized determinant of A. One finds,

$$\zeta_A'(0) = 2 \log\left(\frac{T}{\pi}\right) \zeta(0) + 2\zeta'(0). \tag{4.8.209}$$

We can now use the values

$$\zeta(0) = -\frac{1}{2}, \qquad \zeta'(0) = -\frac{1}{2} \log(2\pi), \tag{4.8.210}$$

to obtain

$$\zeta_A'(0) = -\log(2T), \tag{4.8.211}$$

therefore

$$\det A = 2T. \tag{4.8.212}$$

We have found in the analysis of the free particle that (we set $m = 1$ for convenience)

$$\int e^{\frac{i}{2\hbar} \int_{t_0}^{t_f} \dot{y}^2 dt} \mathcal{D}y(t) = \frac{1}{\sqrt{2\pi i \hbar T}}, \tag{4.8.213}$$

so the path integral is indeed proportional to $1/\sqrt{\det(A)}$:

$$\int e^{\frac{i}{2\hbar} \int_{t_0}^{t_f} \dot{y}^2 dt} \mathcal{D}y(t) = \frac{1}{\sqrt{\pi i \hbar \det(A)}}. \tag{4.8.214}$$

The result (4.8.212) can be generalized to operators that appear naturally in path integrals. Let us first consider the operator

$$A = -\frac{d^2}{dt^2} - c(t), \tag{4.8.215}$$

with Dirichlet boundary conditions in the interval $[t_0, t_f]$. Let us assume that this operator does not have a zero eigenvalue (one usually says that it has no zero modes). By using the zeta function regularization defined above, one can show that

$$\det A = 2f(t_f; t_0), \tag{4.8.216}$$

where the function $f(t_f; t_0)$ is defined by the ODE (4.5.89) with boundary conditions (4.5.90) (and we again set $m = 1$). Therefore, the result (4.5.91) can be written as

$$\int_{y(t_0)=y(t_f)=0} e^{\frac{i}{2\hbar} \int_{t_0}^{t_f} (\dot{y}^2 - c(t)y^2) dt} \mathcal{D}y(t) = \frac{1}{\sqrt{\pi i \hbar \det(A)}}. \tag{4.8.217}$$

We also have the Euclidean version,

$$\int_{y(0)=y(\beta\hbar)=0} e^{-\frac{1}{2\hbar} \int_0^{\beta\hbar} (\dot{y}^2 + c(\tau)y^2) d\tau} \mathcal{D}y(\tau) = \frac{1}{\sqrt{\pi \hbar \det(A)}}, \tag{4.8.218}$$

where the operator A is now given by

$$A = -\frac{d^2}{d\tau^2} + c(\tau). \tag{4.8.219}$$

When the operator A in (4.8.215) has zero modes one has to be more careful, since the standard determinant vanishes. To take into account this more general case, it is useful to consider the operator

$$A_\lambda = A - \lambda, \tag{4.8.220}$$

associated to the eigenvalue problem

$$-y''(t) - c(t)y(t) = \lambda y(t). \tag{4.8.221}$$

One has in this case the following result. Let $y(t; \lambda)$ a solution of (4.8.221) with the boundary conditions

$$y(t_0; \lambda) = 0, \qquad y'(t_0; \lambda) = 1. \tag{4.8.222}$$

Then,

$$\det (A - \lambda) = 2y(t_f; \lambda). \tag{4.8.223}$$

When A has no zero mode, we recover (4.8.216) by setting $\lambda = 0$. When A has a zero mode, we can extract it as follows. The effect of subtracting $-\lambda$ in (4.8.220) is to shift all eigenvalues by $-\lambda$. We can then define

$$\det'(A) = -\frac{\partial}{\partial \lambda} \det (A - \lambda) \Big|_{\lambda=0}. \tag{4.8.224}$$

If A has a discrete spectrum of eigenvalues $\lambda_n, n = 0, 1, \ldots$, with $0 = \lambda_0 < \lambda_1 < \ldots$, the primed determinant (4.8.224) is the regularized version of

$$\prod_{n \geq 1} \lambda_n. \tag{4.8.225}$$

It is also possible to consider the operator (4.8.215) acting on the space of *periodic* functions in the interval $[0, u]$, as we had to do in the evaluation of the path integral (4.7.151). In this case we have the result

$$\det (A - \lambda) = y_1(u; \lambda) + y_2'(u; \lambda) - 2, \tag{4.8.226}$$

where $y_{1,2}(t; \lambda)$ are solutions of (4.8.221) with the boundary conditions (4.7.179) and (4.7.180), respectively. This can also be proved using zeta function techniques. In particular, the result (4.7.182) of Section 4.7 can be written as

$$\int_{y(\beta\hbar)=y(0)} e^{-\frac{1}{2\hbar} \int_0^{\beta\hbar} (\dot{y}^2 + (c(\tau)-\lambda)y^2) d\tau} \mathcal{D}y(\tau) = \frac{1}{\sqrt{\det(A - \lambda)}}, \tag{4.8.227}$$

where we again assumed that the operator A has no zero modes.

There is one special class of operators appearing in one-dimensional quantum mechanics where the determinant can be computed in closed form by an explicit study of the spectrum: the *Pöschl–Teller operators*. These operators are labeled by two parameters $\ell \in \mathbb{Z}_{\geq 0}$, m, and they have the form

$$A_{\ell,m} = -\frac{d^2}{dt^2} + m^2 - \frac{\ell(\ell+1)}{\cosh^2(t)}. \tag{4.8.228}$$

They can be regarded as the Schrödinger operators associated to the Pöschl–Teller potential studied in Example 1.5.3. They are obtained from (1.5.234) by setting $\alpha = 1$ and $\lambda = \ell + 1$. We have a spectrum of bound states in the well, which was obtained from the poles in the transition coefficient $a(k)$. The result is written down in (1.5.250) and (1.5.251). From this result we obtain the energies of the bound states as

$$E_{\ell,m}^{(j)} = m^2 - (\ell - j)^2, \qquad j = 0, \ldots, \ell - 1. \tag{4.8.229}$$

In addition, we have a continuum of scattering states, labeled by the momentum, k, and with energy

$$E_{k,m} = k^2 + m^2. \tag{4.8.230}$$

Since the spectrum of the operator $A_{\ell,m}$ is known, we should be able to compute its determinant. In our considerations above we have assumed that we work in a finite interval and the spectrum is discrete. In the case of the Pöschl–Teller operator, we work on the real line and part of the spectrum lies in the continuum. We then have to be more precise about what we mean by the determinant in this more general case. If an operator A has a discrete spectrum $\{\lambda_n\}_{n \in \mathcal{I}}$ and a continuum spectrum $\lambda(k)$, the determinant should be understood as

$$\log \det A = \sum_{n \in \mathcal{I}} \log(\lambda_n) + \int_{\mathbb{R}} dk \, \rho(k) \log(\lambda(k)), \tag{4.8.231}$$

where $\rho(k)$ is the density of states for the continuum part. This density is easily determined by introducing an IR regulator and putting the system in a box. In the case of the Pöschl–Teller potential, the calculation of $\rho(k)$ goes as follows. A scattering state will experience phase shifts. A plane wave e^{ikt} at $t \to -\infty$ will acquire a phase $\theta(k)$ at $t \to \infty$, given by

$$a(k) = |a(k)| e^{-i\theta(k)}. \tag{4.8.232}$$

This follows from the definition of $a(k)$ in (1.5.163). The quantization condition, once we put these scattering states in a box of length β, is simply

$$k\beta + \theta(k) = 2\pi n, \tag{4.8.233}$$

and the density of states is given by

$$\rho(k) = \frac{dn}{dk} = \rho_{\text{free}}(k) + \rho_\theta(k), \tag{4.8.234}$$

where

$$\rho_{\text{free}}(k) = \frac{\beta}{2\pi}, \qquad \rho_\theta(k) = \frac{1}{2\pi} \theta'(k). \tag{4.8.235}$$

By using the explicit expression for $a(k)$ in (1.5.248), one finds

$$\rho_\theta(k) = -\frac{1}{\pi} \sum_{j=1}^{\ell} \frac{j}{k^2 + j^2}. \tag{4.8.236}$$

Let us now compute $\log \det \mathsf{A}_{\ell,m}$ by using these results. In the expression (4.8.231), the sum over λ_n is over a finite set of discrete eigenvalues. The integration over the continuum part is divergent due to the contribution of $\rho_{\text{free}}(k)$. However, if we subtract $\log \det \mathsf{A}_{0,m}$, this part cancels and we find

$$\log \left(\frac{\det \mathsf{A}_{\ell,m}}{\det \mathsf{A}_{0,m}} \right) = \sum_{1 \le j \le \ell, j} \log(m^2 - j^2) + \int_{-\infty}^{\infty} dk \, \rho_\theta(k) \log(k^2 + m^2)$$

$$= \sum_{1 \le j \le \ell, j} \log(m^2 - j^2) - \frac{1}{\pi} \sum_{j=1}^{\ell} j \int_{-\infty}^{\infty} \frac{dk}{k^2 + j^2} \log(k^2 + m^2). \tag{4.8.237}$$

Since

$$\int_{-\infty}^{\infty} \frac{dk}{k^2 + j^2} \log(k^2 + m^2) = \frac{2\pi}{j} \log(j + m), \tag{4.8.238}$$

the end result is

$$\frac{\det \mathsf{A}_{\ell,m}}{\det \mathsf{A}_{0,m}} = \frac{\prod_{1 \le j \le \ell}(m^2 - j^2)}{\prod_{1 \le j \le \ell}(m + j)^2}. \tag{4.8.239}$$

If $m \in \{1, \ldots, \ell\}$, the operator $\mathsf{A}_{\ell,m}$ has a zero mode. By using (4.8.224) we find

$$\frac{\det' \mathsf{A}_{\ell,m}}{\det \mathsf{A}_{0,m}} = \frac{\prod_{1 \le j \le \ell, j \ne m}(m^2 - j^2)}{\prod_{1 \le j \le \ell}(m + j)^2}. \tag{4.8.240}$$

In some examples, we have to evaluate the determinant of an operator that is proportional to a Pöschl–Teller operator. The behavior of the determinant of an operator under rescaling requires a careful analysis. Let us suppose that we rescale

$$\mathsf{A}_{\ell,m} \to \xi \mathsf{A}_{\ell,m}, \qquad \mathsf{A}_{0,m} \to \xi \mathsf{A}_{0,m}. \tag{4.8.241}$$

Then,

$$\frac{\det' \mathsf{A}_{\ell,m}}{\det \mathsf{A}_{0,m}} \to \xi^{N'_{\ell,m} - N_{0,m}} \frac{\det' \mathsf{A}_{\ell,m}}{\det \mathsf{A}_{0,m}}, \tag{4.8.242}$$

where $N'_{\ell,m} - N_{0,m}$ is the number of nonzero modes of $\mathsf{A}_{\ell,m}$ minus the number of modes of $\mathsf{A}_{0,m}$. This can be computed by adapting the procedure in (4.8.237): $\mathsf{A}_{\ell,m}$ has $\ell - 1$ discrete nonzero modes for $m \le \ell$, plus a continuum. To calculate the difference between the zero modes in the continuum for $\mathsf{A}_{\ell,m}$ and $\mathsf{A}_{0,m}$ we can again use the spectral density. We find,

$$N'_{\ell,m} - N_{0,m} = \ell - 1 + \int_{-\infty}^{\infty} dk \, \rho(k) = \ell - 1 - \frac{1}{\pi} \sum_{j=1}^{\ell} j \int_{-\infty}^{\infty} \frac{dk}{k^2 + j^2} \tag{4.8.243}$$

$$= -1.$$

Therefore, we conclude that after a rescaling, the quotient of determinants behaves as

$$\frac{\det' A_{\ell,m}}{\det A_{0,m}} \rightarrow \frac{1}{\xi} \frac{\det' A_{\ell,m}}{\det A_{0,m}}. \tag{4.8.244}$$

4.9 Correlation Functions and Wick's Theorem

So far we have used the Feynman path integral to calculate propagators, but it can be also used to compute correlation functions. To do this, we work in the Heisenberg picture. The Heisenberg operator associated to q is given in (1.2.32). The eigenvector of this operator is

$$|q, t\rangle = e^{iHt/\hbar}|q\rangle, \tag{4.9.245}$$

since

$$q_H(t)|q, t\rangle = q|q, t\rangle. \tag{4.9.246}$$

We are interested in computing *correlation functions*, i.e. averages of operators in the Heisenberg picture, such as

$$\langle q_f, t_f | q_H(t) | q_0, t_0 \rangle. \tag{4.9.247}$$

In order to obtain a path integral representation for this function, we write it as

$$\begin{aligned}
\langle q_f, t_f | q_H(t) | q_0, t_0 \rangle &= \int_{\mathbb{R}} dq \langle q_f, t_f | q_H(t) e^{iHt/\hbar} | q \rangle \langle q | e^{-iHt/\hbar} | q_0, t_0 \rangle \\
&= \int_{\mathbb{R}} dq \langle q_f, t_f | q_H(t) | q, t \rangle \langle q, t | q_0, t_0 \rangle \\
&= \int_{\mathbb{R}} dq \langle q_f, t_f | q, t \rangle q \langle q, t | q_0, t_0 \rangle \\
&= \int_{\mathbb{R}} dq \, K(q_f, q; t_f, t) q \, K(q, q_0; t, t_0).
\end{aligned} \tag{4.9.248}$$

If we now use the path integral expression for the propagator, we obtain

$$\langle q_f, t_f | q_H(t) | q_0, t_0 \rangle = \int \mathcal{D}q(s) \, e^{\frac{i}{\hbar} S(q(s))} q(t). \tag{4.9.249}$$

To derive this equality, let us consider a $t \in [t_0, t_f]$. Any path $q(s)$ going from q_0 at $s = t_0$ to q_f at $s = t_f$ can be regarded as the union of two paths: the first which we will denote as $q_1(s)$ goes from q_0 at $s = t_0$, to q at $s = t$; the second one which we will denote as $q_2(s)$, goes from q at t, to q_f at $s = t_f$. The integration over all paths going from q_0 to q_f can then be written as

$$\int_{q(t_0)=q_0, q(t_f)=q_f} \mathcal{D}q(s) = \int_{\mathbb{R}} dq \int_{q_1(t_0)=q_0, q_1(t)=q} \mathcal{D}q_1(s) \int_{q_2(t)=q, q_2(t_f)=q_f} \mathcal{D}q_2(s). \tag{4.9.250}$$

Then, taking into account that

$$S(q(s)) = S(q_1(s)) + S(q_2(s)), \tag{4.9.251}$$

we find

$$\int \mathcal{D}q(s)\, e^{\frac{i}{\hbar}S(q(s))}q(t)$$

$$= \int_{\mathbb{R}} dq\, q \int_{q_1(t_0)=q_0,\, q_1(t)=q} \mathcal{D}q_1(s)e^{\frac{i}{\hbar}S(q_1(s))} \int_{q_2(t)=q,\, q_2(t_f)=q_f} \mathcal{D}q_2(s)e^{\frac{i}{\hbar}S(q_2(s))},$$

$$(4.9.252)$$

which is precisely the last line in (4.9.248).

Let us now consider

$$\int \mathcal{D}q(s)\, e^{\frac{i}{\hbar}S(q(s))}q(t_1)q(t_2). \qquad (4.9.253)$$

The result of this path integral depends on the time ordering: if $t_1 > t_2$, the integral can be written as

$$\int_{\mathbb{R}^2} dq_1\, dq_2\, K(q_f, q_1; t_f, t_1)q_1 K(q_1, q_2; t_1, t_2)q_2 K(q_2, q_0; t_2, t_0), \qquad (4.9.254)$$

and if $t_2 > t_1$, then

$$\int_{\mathbb{R}^2} dq_1\, dq_2\, K(q_f, q_2; t_f, t_2)q_2 K(q_2, q_1; t_2, t_1)q_1 K(q_1, q_0; t_2, t_0). \qquad (4.9.255)$$

We conclude that

$$\langle q_f, t_f | T\, (\mathsf{q}_H(t_1)\mathsf{q}_H(t_2))\, |q_0, t_0\rangle = \int \mathcal{D}q(s)\, e^{\frac{i}{\hbar}S(q(s))}q(t_1)q(t_2), \qquad (4.9.256)$$

where T denotes time-ordering, i.e.

$$T\, (\mathsf{q}_H(t_1)\mathsf{q}_H(t_2)) = \begin{cases} \mathsf{q}_H(t_1)\mathsf{q}_H(t_2), & \text{if } t_1 > t_2, \\ \mathsf{q}_H(t_2)\mathsf{q}_H(t_1), & \text{if } t_2 > t_1. \end{cases} \qquad (4.9.257)$$

In the Euclidean theory, "time" evolution is implemented by

$$\mathsf{q}_H(\tau) = e^{\tau \mathsf{H}/\hbar}\mathsf{q}\, e^{-\tau \mathsf{H}/\hbar}, \qquad (4.9.258)$$

and we define

$$|q, \tau\rangle = e^{\tau \mathsf{H}/\hbar}|q\rangle, \qquad \langle q, \tau| = \langle q|e^{-\tau \mathsf{H}/\hbar}. \qquad (4.9.259)$$

Note that $|q, \tau\rangle$ and $\langle q, \tau|$ are no longer related by Hermitian conjugation. In the Euclidean path integral with periodic boundary conditions, as in (4.3.37), the natural correlation functions are of the form

$$\int_{q(0)=q(u)} \mathcal{D}q(\tau)\, e^{-S_E(q(\tau))/\hbar}q(\tau_1)\cdots q(\tau_\ell), \qquad (4.9.260)$$

where $u = \beta\hbar$. This path integral is calculated as

$$\int_{\mathbb{R}} dq\langle q, u | T\, (\mathsf{q}_H(\tau_1)\cdots\mathsf{q}_H(\tau_\ell))\, |q, 0\rangle = \int_{\mathbb{R}} dq\langle q|e^{-\beta \mathsf{H}}T\, (\mathsf{q}_H(\tau_1)\cdots\mathsf{q}_H(\tau_\ell))\, |q\rangle$$

$$= \text{Tr}\left\{e^{-\beta \mathsf{H}}T\, (\mathsf{q}_H(\tau_1)\cdots\mathsf{q}_H(\tau_\ell))\right\}. \qquad (4.9.261)$$

The normalized correlator will be defined as

$$
\begin{aligned}
\langle q(\tau_1) \cdots q(\tau_\ell) \rangle &= \frac{1}{Z(\beta)} \int_{q(0)=q(u)} \mathcal{D}q(\tau) \, e^{-S_E(q(\tau))/\hbar} q(\tau_1) \cdots q(\tau_\ell) \\
&= \frac{1}{Z(\beta)} \mathrm{Tr} \left\{ e^{-\beta \mathsf{H}} T \left(\mathsf{q}_H(\tau_1) \cdots \mathsf{q}_H(\tau_\ell) \right) \right\}.
\end{aligned}
\tag{4.9.262}
$$

We will focus on the correlators of the Euclidean theory. They have the following important property. Let us suppose that the Hamiltonian H has a discrete spectrum, with eigenstates $|n\rangle$ and eigenvalues E_n, $n = 0, 1, 2, \ldots$, such that

$$
E_0 < E_1 < \ldots.
\tag{4.9.263}
$$

This means, in particular, that there is an energy gap between the ground state and the excited states. Let us now consider the low temperature limit

$$
\beta \to \infty.
\tag{4.9.264}
$$

We note that,

$$
\begin{aligned}
\mathrm{Tr} &\left\{ e^{-\beta \mathsf{H}} T \left(\mathsf{q}_H(\tau_1) \cdots \mathsf{q}_H(\tau_\ell) \right) \right\} \\
&= \sum_{n \geq 0} e^{-\beta E_n} \langle n| T \left(\mathsf{q}_H(\tau_1) \cdots \mathsf{q}_H(\tau_\ell) \right) |n\rangle \\
&= e^{-\beta E_0} \langle 0| T \left(\mathsf{q}_H(\tau_1) \cdots \mathsf{q}_H(\tau_\ell) \right) |0\rangle + \mathcal{O}\left(e^{-\beta(E_1 - E_0)} \right).
\end{aligned}
\tag{4.9.265}
$$

Similarly,

$$
Z(\beta) = e^{-\beta E_0} + \mathcal{O}\left(e^{-\beta(E_1 - E_0)} \right).
\tag{4.9.266}
$$

Therefore, in the limit (4.9.264) only the ground state survives, and we have

$$
\lim_{\beta \to \infty} \langle q(\tau_1) \cdots q(\tau_\ell) \rangle = \langle 0| T \left(\mathsf{q}_H(\tau_1) \cdots \mathsf{q}_H(\tau_\ell) \right) |0\rangle,
\tag{4.9.267}
$$

i.e. this limit computes correlation functions in the vacuum state.

Correlation functions can be easily obtained from the generating functional

$$
Z[j] = \int_{q(0)=q(u)} \mathcal{D}q(\tau) \, e^{-\frac{1}{\hbar} S_E(q(\tau)) + \int q(\tau) j(\tau) d\tau},
\tag{4.9.268}
$$

which is the partition function in a theory with a modified action depending on an arbitrary source $j(\tau)$. Note that $Z[0]$ is the standard partition function. Then, by functional differentiation we find

$$
\langle q(\tau_1) \cdots q(\tau_\ell) \rangle = \frac{1}{Z[0]} \frac{\delta^\ell Z[j]}{\delta j(\tau_1) \cdots \delta j(\tau_\ell)} \bigg|_{j=0}.
\tag{4.9.269}
$$

The path integral $Z[j]$ can be computed exactly in the case of a Gaussian theory. Let us consider the Euclidean action

$$
S_E(q; j) = \int_0^u \left(\frac{1}{2} \dot{q}^2 + \frac{\omega^2}{2} q^2 - \hbar j q \right) d\tau.
\tag{4.9.270}
$$

The classical EOM is given by

$$\left(-\partial_\tau^2 + \omega^2\right) q(\tau) = \hbar j(\tau), \tag{4.9.271}$$

with periodic boundary conditions. To solve this equation, let us introduce the Green's function or propagator satisfying

$$\left(-\partial_\tau^2 + \omega^2\right) G(\tau, \tau') = \delta(\tau - \tau'). \tag{4.9.272}$$

We impose periodic boundary conditions for both this function and its first derivative. It follows from time translation invariance that in this case

$$G(\tau, \tau') = G(\tau - \tau', 0) \equiv G(\tau - \tau'). \tag{4.9.273}$$

We also have, by time reversal symmetry, that

$$G(\tau, \tau') = G(\tau', \tau). \tag{4.9.274}$$

Then, the solution of (4.9.271) is given by

$$q_c(\tau) = \hbar \int_0^u G(\tau, \tau') j(\tau') \mathrm{d}\tau'. \tag{4.9.275}$$

Let us use bra and ket notation and write this as

$$|q_c\rangle = \hbar G |j\rangle, \tag{4.9.276}$$

where G is the operator whose integral kernel is the Green's function $G(\tau, \tau')$. It follows that the classical Euclidean action is given by

$$\frac{1}{\hbar} S_E(q_c, j) = \frac{1}{2\hbar} \langle q_c | G^{-1} | q_c \rangle - \langle j | q_c \rangle = -\frac{\hbar}{2} \langle j | G | j \rangle$$
$$= -\frac{\hbar}{2} \int j(\tau) G(\tau - \tau') j(\tau') \mathrm{d}\tau \mathrm{d}\tau'. \tag{4.9.277}$$

We can now expand the action around the classical trajectory, as we did in Section 4.6, to obtain

$$Z[j] = \mathrm{e}^{-S_E(q_c, j)/\hbar} \int \mathcal{D}y \, \mathrm{e}^{-\frac{1}{\hbar} S_G(y)}, \tag{4.9.278}$$

where

$$S_G(y) = \int_0^u \left(\frac{1}{2}\dot{y}^2 + \frac{\omega^2}{2} y^2\right) \mathrm{d}\tau. \tag{4.9.279}$$

In other words, we find that

$$Z[j] = Z[0] \exp\left(\frac{\hbar}{2} \int j(\tau) G(\tau - \tau') j(\tau') \mathrm{d}\tau \mathrm{d}\tau'\right). \tag{4.9.280}$$

We can now use this explicit formula for $Z[j]$, together with (4.9.269), to calculate correlation functions for the harmonic oscillator. For the correlation function of two variables or *two-point function* we find

$$\langle q(\tau) q(\tau') \rangle = \hbar G(\tau - \tau'). \tag{4.9.281}$$

To calculate a generic correlation function, let us expand the exponential in (4.9.280):

$$\exp\left(\frac{\hbar}{2}\int d\tau d\tau' j(\tau)G(\tau-\tau')j(\tau')\right)$$

$$=\sum_{n=0}^{\infty}\frac{\hbar^n}{2^n n!}\left(\int d\tau d\tau' j(\tau)G(\tau-\tau')j(\tau')\right)^n$$

$$=\sum_{n=0}^{\infty}\frac{\hbar^n}{2^n n!}\int d\tau_1\cdots d\tau_{2n}j(\tau_1)\cdots j(\tau_{2n})G(\tau_1,\tau_2)\cdots G(\tau_{2n-1},\tau_{2n})$$

$$=\sum_{n=0}^{\infty}\frac{\hbar^n}{2^n n!}\int d\tau_1\cdots d\tau_{2n}j(\tau_1)\cdots j(\tau_{2n})$$

$$\times\frac{1}{(2n)!}\sum_{\sigma\in S_{2n}}G(\tau_{\sigma(1)},\tau_{\sigma(2)})\cdots G(\tau_{\sigma(2n-1)},\tau_{\sigma(2n)}).$$

$$(4.9.282)$$

In going to the last line, we have used the fact that $G(\tau_1,\tau_2)\cdots G(\tau_{2n-1},\tau_{2n})$ can be symmetrized w.r.t. the indices $1,\ldots,2n$, since it multiplies a symmetric function and is integrated with the symmetric measure $d\tau_1\cdots d\tau_{2n}$. Therefore, we sum over all possible permutations σ of the indices in the permutation group of $2n$ elements S_{2n}. On the other hand, (4.9.282) is computing the generating functional (4.9.268),

$$Z[j]=Z[0]\sum_{M=0}^{\infty}\frac{1}{M!}\int d\tau_1\cdots d\tau_M j(\tau_1)\cdots j(\tau_M)\langle q(\tau_1)\cdots q(\tau_M)\rangle. \quad (4.9.283)$$

Since (4.9.282) with (4.9.283) are equal, and since the source $j(\tau)$ is arbitrary, we conclude first of all that the correlation functions with an odd number of terms vanish, as expected from the symmetry $q(\tau)\leftrightarrow -q(\tau)$ of the theory. When $M=2n$ is even, we find that

$$\langle q(\tau_1)\cdots q(\tau_{2n})\rangle=\frac{\hbar^n}{2^n n!}\sum_{\sigma\in S_{2n}}G(\tau_{\sigma(1)},\tau_{\sigma(2)})\cdots G(\tau_{\sigma(2n-1)},\tau_{\sigma(2n)}). \quad (4.9.284)$$

We notice, however, that in the r.h.s. many terms are equal, namely, terms that differ by a permutation of the two labels inside $G(\tau,\tau')$ (recall (4.9.274)), or by a permutations of the n groups of paired indices. For example, if $n=2$, the terms $G(\tau_1,\tau_2)G(\tau_3,\tau_4)$ and $G(\tau_2,\tau_1)G(\tau_3,\tau_4)$ give the same contribution in the sum, as does the term $G(\tau_3,\tau_4)G(\tau_1,\tau_2)$. We will say that two permutations of $2n$ elements are equivalent if they give the same contribution in the r.h.s. of (4.9.284). An equivalence class of permutations is called a *pairing*. Each pairing corresponds to $2^n n!$ different permutations, which is precisely the factor in front of the sum in (4.9.284). The different pairings are obtained by grouping the $2n$ labels in groups of 2, irrespective of the ordering of the groups and the order of the labels inside each pair. We conclude that (4.9.284) can be written as a sum over pairings, as follows:

$$\langle q(\tau_1)\cdots q(\tau_{2n})\rangle=\hbar^n\sum_{\text{pairings }P}G(\tau_{P(1)},\tau_{P(2)})\cdots G(\tau_{P(2n-1)},\tau_{P(2n)}). \quad (4.9.285)$$

This important result is *Wick's theorem*. Let us note that there are

$$\frac{(2n)!}{2^n n!}, \quad (4.9.286)$$

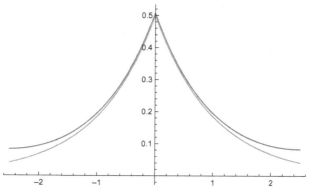

Figure 4.2 (Top) The propagator (4.9.288) for $u = 5$, $\omega = 1$; and (bottom) its large u limit (4.9.290).

inequivalent pairings contributing to the r.h.s. of (4.9.285). For example, when $n = 2$ there are $4!/(4 \cdot 2!) = 3$ different pairings, and one finds

$$\langle q(\tau_1)q(\tau_2)q(\tau_3)q(\tau_4)\rangle = \langle q(\tau_1)q(\tau_2)\rangle\langle q(\tau_3)q(\tau_4)\rangle + \langle q(\tau_1)q(\tau_3)\rangle\langle q(\tau_2)q(\tau_4)\rangle$$
$$+ \langle q(\tau_1)q(\tau_4)\rangle\langle q(\tau_2)q(\tau_3)\rangle.$$
$$(4.9.287)$$

Wick's theorem applies to any Gaussian probability measure, and it is the basis of the perturbation theory of path integrals, as we will see in Section 4.10.

Finally, it is useful to have an explicit expression for the propagator $G(\tau)$. This can be obtained in various ways, but it easy to check by direct substitution that

$$G(\tau) = \frac{1}{2\omega} \frac{\cosh\left(\omega\left(|\tau| - \frac{u}{2}\right)\right)}{\sinh\left(\frac{\omega u}{2}\right)}, \qquad 0 \le \tau \le u, \qquad (4.9.288)$$

satisfies (4.9.272). It is also possible to work in the interval

$$-\frac{u}{2} \le \tau \le \frac{u}{2}. \qquad (4.9.289)$$

In this interval, the propagator (4.9.288) can be approximated as $u \to \infty$ by

$$G_{\mathbb{R}}(\tau) = \frac{e^{-\omega|\tau|}}{2\omega}, \qquad (4.9.290)$$

as can be seen in Figure 4.2.

Example 4.9.1 According to the path integral arguments presented above, the quantity

$$\langle q(\tau)q(0)\rangle = \frac{1}{Z(\beta)} \text{Tr}\left\{e^{-\frac{u\mathsf{H}}{\hbar}} T\left(\mathsf{q}_H(\tau)\mathsf{q}_H(0)\right)\right\}, \qquad (4.9.291)$$

is given by $\hbar$, multiplied by the propagator (4.9.288). Let us verify this with a calculation in the conventional formulation of quantum mechanics. By taking into account the two possible cases for the time ordering, $\tau > 0$ or $\tau < 0$, we can write

$$\langle q(\tau)q(0)\rangle = \frac{1}{Z(\beta)} \text{Tr}\left\{e^{\left(\frac{|\tau|}{\hbar} - \beta\right)\mathsf{H}} \mathsf{q}\, e^{-\frac{|\tau|}{\hbar}\mathsf{H}} \mathsf{q}\right\}. \qquad (4.9.292)$$

We evaluate this trace in the basis of eigenvectors of H, which we denote by $|n\rangle$, $n = 0, 1, 2, \ldots$, which satisfy

$$H|n\rangle = \hbar\omega \left(n + \frac{1}{2}\right) |n\rangle. \tag{4.9.293}$$

In this way, we obtain

$$\langle q(\tau)q(0)\rangle = \frac{1}{Z(\beta)} \sum_{m,n=0}^{\infty} \exp\left(\left(|\tau| - \beta\hbar\right)\omega\left(n + \frac{1}{2}\right) - |\tau|\omega\left(m + \frac{1}{2}\right)\right) |\langle n|q|m\rangle|^2. \tag{4.9.294}$$

It is an elementary result in quantum mechanics that

$$|\langle n|q|m\rangle|^2 = \frac{\hbar}{2\omega}\left(m\delta_{n,m-1} + (m+1)\delta_{n,m+1}\right), \tag{4.9.295}$$

where we have set the mass equal to unity, as in (4.9.270). We conclude that

$$\begin{aligned}
\langle q(\tau)q(0)\rangle &= \frac{\hbar}{2\omega Z(\beta)} e^{-\beta\hbar\omega/2} \left\{ \sum_{n=0}^{\infty} (n+1)e^{-|\tau|\omega-\beta\hbar\omega n} + \sum_{m=0}^{\infty} (m+1)e^{|\tau|\omega-\beta\hbar\omega(m+1)} \right\} \\
&= \frac{\hbar}{2\omega Z(\beta)} e^{-\beta\hbar\omega/2} \left\{ \frac{e^{-\omega|\tau|}}{(1-e^{-\beta\hbar\omega})^2} + \frac{e^{\omega|\tau|-\beta\omega\hbar}}{(1-e^{-\beta\hbar\omega})^2} \right\} \\
&= \frac{\hbar}{2\omega} \frac{\cosh\left(\omega\left(|\tau| - \frac{\beta\hbar}{2}\right)\right)}{\sinh\left(\frac{\omega\beta\hbar}{2}\right)},
\end{aligned} \tag{4.9.296}$$

where we have used the standard expression for $Z(\beta)$ given in e.g. (4.7.187). This confirms the result obtained in (4.9.288) using path integral methods. □

4.10 Perturbation Theory and Feynman Diagrams

Most path integrals cannot be calculated exactly, so one is obliged to perform some type of approximation. One possibility is to develop the semiclassical approximation to the path integral started in Section 4.6, which is equivalent to the conventional WKB approximation in quantum mechanics. Another possibility is to develop a perturbative approach in a small parameter, as we do in stationary perturbation theory in quantum mechanics. We will now explore the analogue of this second approach from the path integral point of view.

Let us suppose that the action of the theory we are interested in can be written as

$$S(q(\tau)) = S_G(q(\tau)) + \lambda S_I(q(\tau)), \tag{4.10.297}$$

where S_G is a quadratic action, λ is a small parameter, and S_I is an interaction term (for simplicity, we will work in the Euclidean theory). A typical example is a non-relativistic particle in a onedimensional potential of the form

$$V(q) = \frac{\omega^2 q^2}{2} + V_1(q), \tag{4.10.298}$$

where $V_1(q) = \lambda q^M$. In this case, the quadratic action is

$$S_G(q(t)) = \int_0^u \left(\frac{1}{2}\dot{q}^2 + \frac{\omega^2}{2}q^2\right) d\tau, \tag{4.10.299}$$

and the interaction is

$$S_1(q(\tau)) = \int_0^u q^M(\tau)d\tau. \tag{4.10.300}$$

The perturbative evaluation of the partition function is found by expanding the interacting part in a formal power series in λ,

$$Z(\lambda) = \int_{q(0)=q(u)} \mathcal{D}q(\tau)\, e^{-\frac{1}{\hbar}S_G(q(\tau))-\frac{\lambda}{\hbar}S_1(q(\tau))}$$

$$= \sum_{N=0}^{\infty} \frac{(-\lambda/\hbar)^N}{N!} \int_{q(0)=q(u)} \mathcal{D}q(\tau)\, e^{-S_G(q(\tau))}\, (S_1(q(\tau)))^N \tag{4.10.301}$$

$$= Z(0) \sum_{N=0}^{\infty} \frac{(-\lambda/\hbar)^N}{N!} \langle (S_1(q(\tau)))^N \rangle.$$

In this equation, the argument of the partition function is the coupling constant λ, and not the inverse temperature β. If the interaction is polynomial in $q(\tau)$, the averages appearing here can be computed by using Wick's theorem, and we obtain in this way a formal power series in λ. If λ is small enough, this series might give us some understanding of the interacting theory. It turns out that the information carried by this series is limited, since the series is not even convergent. Nevertheless, this approximation scheme, known as perturbation theory, is widely used in quantum mechanics, quantum field theory, and many-body physics.

It turns out that the information contained in each term of the above series can be expressed in diagrammatic form. Let us assume for definiteness that we have a quartic interaction, i.e. we have (4.10.300) with $M = 4$. If this is the case, we have to evaluate terms of the form

$$\left\langle \left(\int d\tau q^4(\tau)\right)^N \right\rangle, \tag{4.10.302}$$

using Wick's theorem. Let us represent each power of the interaction term by a *vertex* with four edges, or quartic vertex, as shown in Figure 4.3. The Nth order term (4.10.302) leads to N vertices. According to Wick's theorem, we have to pair

Figure 4.3   Diagram representing the quartic vertex.

Figure 4.4   Feynman diagram obtained by pairing all the legs in a single quartic vertex.

the edges in all possible ways. Each of these pairings can be represented by a diagram called a *Feynman diagram*. Consider, for example, $N = 1$. We have to compute

$$\left\langle \int d\tau\, q^4(\tau) \right\rangle, \tag{4.10.303}$$

which leads to three possible pairings. They all give the same contribution, and we find

$$\left\langle \int d\tau\, q^4(\tau) \right\rangle = 3\hbar^2 G^2(0) \int_{-u/2}^{u/2} d\tau = 3u\hbar^2 G^2(0). \tag{4.10.304}$$

Each of the three equivalent pairings correspond to the Feynman diagram shown in Figure 4.4. We are now working on the interval $[-u/2, u/2]$ instead of the interval $[0, u]$ since we will eventually be interested in taking the limit $u \to \infty$, and in the first interval we can use (4.9.290) as an approximation to the propagator. Of course, the integrals for finite u are the same due to the symmetry properties of $G(\tau, \tau')$.

Life becomes more interesting when $N = 2$, and one has to compute

$$\left\langle \left(\int d\tau\, q^4(\tau) \right)^2 \right\rangle = \left\langle \left(\int d\tau\, q^4(\tau) \right) \left(\int d\tau\, q^4(\tau') \right) \right\rangle. \tag{4.10.305}$$

The first thing to note is that we now have two types of diagrams: we can consider diagrams in which the edges of the first vertex do not pair with edges in the second vertex. This gives a diagram with two separated pieces, which is a particular example of a *disconnected diagram*. In addition, when $N = 2$ there are diagrams in which at least one edge of one vertex pairs with an edge of the other vertex. This leads to diagrams that cannot be separated into two pieces without cutting an edge. They are examples of *connected* diagrams. It is easy to see that for $N = 2$ there are two different types of connected diagrams, which are shown in Figure 4.5.

There are many pairings that lead to the same Feynman diagram. Each component of the disconnected diagram is obtained from three different pairings, as we saw before, so for the disconnected diagram in Figure 4.5 there are $3^2 = 9$ pairings. This diagram contributes

$$9\hbar^4 G^4(0)u^2, \tag{4.10.306}$$

to the correlation function. The first connected diagram shown in Figure 4.5 is obtained from 72 different pairings. To calculate this number, note that in each vertex there are six ways to pair two legs, and then there are two ways to pair the remaining two legs of the two vertices. This gives in total $6^2 \cdot 2 = 72$ pairings. The contribution of this connected diagram to the correlation function is

$$72\hbar^4 G^2(0) \int_{-u/2}^{u/2} \int_{-u/2}^{u/2} G^2(\tau, \tau') d\tau d\tau'. \tag{4.10.307}$$

The second connected diagram comes from $4! = 24$ different pairings (the different permutations of the four connecting legs). Its contribution is

Disconnected   Connected

Figure 4.5 Feynman diagrams obtained at $N = 2$.

$$24\hbar^4 \int_{-u/2}^{u/2} \int_{-u/2}^{u/2} G^4(\tau, \tau') \mathrm{d}\tau \mathrm{d}\tau'. \tag{4.10.308}$$

Note that the counting of the possible pairings associated to a given Feynman diagram might be an intricate combinatorial problem. A useful constraint is that the total number of pairings must be equal to (4.9.286). For example, when $n = 4$, one verifies that

$$9 + 72 + 24 = \frac{8!}{2^4 \cdot 4!}. \tag{4.10.309}$$

Very often, we are interested in the computation of the free energy rather than in the computation of the partition function. This means that we want to calculate $\log Z(\lambda)$. It turns out that this quantity is simpler, from the diagrammatic point of view, since it only involves connected Feynman diagrams. This is proved as follows. Let us denote by

$$\langle S_I^\ell \rangle^{(c)}, \tag{4.10.310}$$

the contribution of connected diagrams to the correlation function involving ℓ vertices. It is clear that a general, not necessarily connected, diagram with N vertices, can be partitioned into connected diagrams. Let us denote by m_ℓ the number of connected diagrams with ℓ vertices appearing in a diagram with N vertices in total. Clearly, we must have

$$\sum \ell m_\ell = N. \tag{4.10.311}$$

We then have,

$$\langle S_I^N \rangle = \sum_{\{m_l\}}{}' c_{\{m_\ell\}} \prod_\ell \left(\langle S_I^N \rangle^{(c)} \right)^{m_\ell}, \tag{4.10.312}$$

where the sum is over all possible sets $\{m_\ell\}$, satisfying the constraint (4.10.311). The coefficient $c_{\{m_\ell\}}$ is a combinatorial factor that gives all possible, nonequivalent ways of grouping N vertices into m_ℓ connected groups of ℓ vertices. This factor is given by

$$c_{\{m_\ell\}} = \frac{N!}{\prod_\ell m_\ell! \, (\ell!)^{m_\ell}}. \tag{4.10.313}$$

In this expression, $N!$ gives all possible permutations of the vertices. However, we have to divide the $N!$ by the number of equivalent configurations. These are obtained in two ways: (a) by permuting the labels of the vertices *inside* each connected subdiagram, which gives $(\ell!)^{m_\ell}$ for each ℓ; (b) by permuting the m_ℓ connected diagrams among them, which gives $m_\ell!$ for each ℓ.

Let us now compute (4.10.301) by using (4.10.312) and (4.10.313). The sum over N can be written as a sum over all possible m_ℓ:

$$
\frac{Z(\lambda)}{Z(0)} = \sum_{\{m_l\}} \prod_{\ell=1}^{\infty} \frac{1}{m_\ell!} \left(\left(-\frac{\lambda}{\hbar} \right)^\ell \frac{\langle S_I^\ell \rangle^{(c)}}{\ell!} \right)^{m_\ell} = \prod_{\ell=1}^{\infty} \sum_{m_\ell=0}^{\infty} \frac{1}{m_\ell!} \left(\left(-\frac{\lambda}{\hbar} \right)^\ell \frac{\langle S_I^\ell \rangle^{(c)}}{\ell!} \right)^{m_\ell}
$$
$$
= \exp \left(\sum_{\ell=1}^{\infty} \left(-\frac{\lambda}{\hbar} \right)^\ell \frac{\langle S_I^\ell \rangle^{(c)}}{\ell!} \right).
$$

$$(4.10.314)$$

If we regard (4.10.301) as a generating function of averages in a probability distribution, the logarithm of $Z(\lambda)$ defines the *cumulants* of the distribution, which are given by the connected averages $\langle S_I^n \rangle^{(c)}$. It is easy to see that

$$
\langle S_I \rangle^{(c)} = \langle S_I \rangle,
$$
$$
\langle S_I^2 \rangle^{(c)} = \langle S_I^2 \rangle - \langle S_I \rangle^2.
$$

$$(4.10.315)$$

The result (4.10.314) is very useful when computing the perturbative expansion of the thermal free energy

$$
\beta F = -\log Z.
$$

$$(4.10.316)$$

We find,

$$
\beta F(\lambda) = \beta F(0) + 3\beta \lambda \hbar^2 G^2(0)
$$
$$
- \frac{\lambda^2 \hbar^2}{2} \left(72 G^2(0) \int_{-u/2}^{u/2} \int_{-u/2}^{u/2} G^2(\tau, \tau') \mathrm{d}\tau \mathrm{d}\tau' \right.
$$
$$
\left. + 24 \int_{-u/2}^{u/2} \int_{-u/2}^{u/2} G^4(\tau, \tau') \mathrm{d}\tau \mathrm{d}\tau' \right) + \mathcal{O}(\lambda^3).
$$

$$(4.10.317)$$

From this perturbative series, one can derive a simpler one to compute the ground state energy of the system. This is because

$$
E_0 = -\lim_{\beta \to \infty} \frac{1}{\beta} \log Z = \lim_{\beta \to \infty} F.
$$

$$(4.10.318)$$

In this limit we can replace the propagator $G(\tau, \tau')$ by (4.9.290), and the resulting integrals are much simpler. Note that these integrals give always a single, overall factor of u, due to time translation invariance. We have, for example,

$$
\lim_{u \to \infty} \frac{1}{u} \int_{-u/2}^{u/2} \int_{-u/2}^{u/2} G^2(\tau, \tau') \mathrm{d}\tau \mathrm{d}\tau' = \int_{-\infty}^{\infty} G_{\mathbb{R}}^2(\tau) \mathrm{d}\tau.
$$

$$(4.10.319)$$

Therefore, we find the following result for the perturbative expansion of the ground state energy,

$$
E_0(\lambda) = E_0(0) + 3\hbar^2 \lambda G_{\mathbb{R}}^2(0)
$$
$$
- \frac{\lambda^2 \hbar^3}{2} \left(72 G_{\mathbb{R}}^2(0) \int_{-\infty}^{\infty} G_{\mathbb{R}}^2(\tau) \mathrm{d}\tau + 24 \int_{-\infty}^{\infty} G_{\mathbb{R}}^4(\tau) \mathrm{d}\tau \right) + \mathcal{O}(\lambda^3).
$$

$$(4.10.320)$$

The integrals of the propagators are immediate,

$$\int_{-\infty}^{\infty} G_{\mathbb{R}}^n(\tau)\mathrm{d}\tau = \frac{4}{(2\omega)^{n+1}n}.$$

(4.10.321)

We finally obtain,

$$\frac{1}{\hbar\omega}E_0(\lambda) = \frac{1}{2} + \frac{3}{4}\widehat{\lambda} - \frac{21}{8}\widehat{\lambda}^2 + \cdots,$$

(4.10.322)

where

$$\widehat{\lambda} = \frac{\hbar\lambda}{m^2\omega^3},$$

(4.10.323)

is the dimensionless coupling, and we have restored the dependence on m. In this way, we have recovered the standard stationary perturbation theory expansion from the path integral.

4.11 Instantons in Quantum Mechanics

In Section 4.10 we saw how we can recover stationary perturbation theory in quantum mechanics by using the path integral formalism. However, we have seen in Chapter 2 that perturbation theory is typically insufficient to solve the spectral problems appearing in quantum mechanics, and one needs nonperturbative corrections, which are exponentially small in $\hbar$ or in the coupling constant. In this section we show that, in many cases, these corrections are due to nontrivial saddle points of the path integral. More precisely, we will study expansions of the path integral around solutions to the Euclidean EOM, called *instantons*, which lead to nonperturbative corrections.

Perhaps the simplest example of such a correction occurs in the double-well potential. As we saw in (2.8.339), the nonperturbative splitting between the ground state energy and the first excited state in the double-well potential is a purely nonperturbative effect, so it is an interesting question to ask how this can be derived in the context of the path integral formalism. Let us then consider the double-well potential with a potential of the form

$$V(x) = \frac{g^2}{2}\left(x^2 - \frac{1}{4g^2}\right)^2,$$

(4.11.324)

which is obtained from (2.8.308) by shifting x, so that the parity symmetry is implemented as $x \to -x$. In order to study the energy splitting, we have to find a natural quantity in the path integral formulation that gives access to it. One could consider the usual thermal partition function, but it proves much simpler to consider the "twisted" partition function (4.7.189). When β is large, the main contribution to $Z_a(\beta)$ comes from the first two energy levels of the double-well potential, the ground state E_0 and the first excited state E_1, and one has

$$Z_a(\beta) \approx \mathrm{e}^{-\beta E_0} - \mathrm{e}^{-\beta E_1} \approx -\beta\mathrm{e}^{-\beta/2}\left(E_0 - E_1\right),$$

(4.11.325)

where we have used that, for small g, perturbation theory gives $E_{0,1} \approx 1/2 + \mathcal{O}(g^2)$. In order to compute the energy splitting, we then have to compute the twisted partition function at large β and small g. We will use the path integral representation (4.7.193). As in (4.9.289), and for convenience, we will shift the Euclidean time so that the paths are defined in the interval $[-u/2, u/2]$. Let us now understand what kind of paths will contribute to the path integral in the limit of β large and g small. The Euclidean action is of the form

$$\frac{1}{\hbar} S_E(q(\tau)) = \frac{1}{\hbar} \int_{-u/2}^{u/2} \left\{ \frac{m}{2} \dot{q}^2(\tau) + \frac{g^2}{2} \left(q^2(\tau) - \frac{1}{4g^2} \right)^2 \right\} d\tau. \qquad (4.11.326)$$

We can rescale $q \to q/g$ to obtain

$$\frac{1}{\hbar} S_E(q(\tau)) = \frac{1}{\hbar g^2} \int_{-u/2}^{u/2} \left\{ \frac{m}{2} \dot{q}^2(\tau) + \frac{1}{2} \left(q^2(\tau) - \frac{1}{4} \right)^2 \right\} d\tau. \qquad (4.11.327)$$

Therefore, small g is like small $\hbar$, and the considerations of Sections 4.6 and 4.7 apply, namely, we should calculate the path integral by expanding around solutions of the (Euclidean) EOM. Taking β large means that configurations of large energy will be suppressed, and we should consider configurations with the smallest possible energy. We conclude that we have to expand around classical solutions of the EOM, with zero energy, and satisfying the boundary condition (4.7.192).

One way to investigate these solutions is to take into account that, as we mentioned in (4.7.145), the Euclidean EOM corresponds to the standard Lagrangian EOM but in an inverted potential. It is easy to see that solutions with zero energy in such a potential are solutions that start in the infinite past, $\tau \to -\infty$, in one of the minima of the potential, and then go to the other minimum as $\tau \to \infty$. We can calculate these solutions explicitly by using conservation of the energy in the inverted potential,

$$\frac{1}{2} \dot{q}^2 - V(q) = E, \qquad (4.11.328)$$

where we set $m = 1$. If the energy vanishes, (4.11.328) can be integrated immediately when $V(q)$ is of the form (4.11.324). One finds,

$$q_{\pm}(\tau; \tau_0) = \pm \frac{1}{2g} \tanh \left(\frac{\tau - \tau_0}{2} \right), \qquad (4.11.329)$$

where τ_0 parametrizes different solutions and appears as an integration constant when we solve (4.11.328). These solutions to the EOM are called *(anti)instantons* of center τ_0. They are represented in Figure 4.6 for $g = 1/2$ and $\tau_0 = 0$.

We now expand the path integral around these classical solutions, and we consider quantum fluctuations of the path as in (4.7.146). For finite β, the fluctuations $y(\tau)$ also satisfy anti-periodic boundary conditions $y(-u/2) = -y(u/2)$. The operator A appearing in the quadratic action (4.7.148) is

$$\mathsf{A} = -\frac{d^2}{d\tau^2} + V''(q_c(\tau)). \qquad (4.11.330)$$

When $q_c(\tau)$ is one of the solutions in (4.11.329), the operator reads

$$\mathsf{A} = -\frac{d^2}{d\tau^2} + 1 - \frac{3}{2 \cosh^2 \left(\frac{\tau - \tau_0}{2} \right)}. \qquad (4.11.331)$$

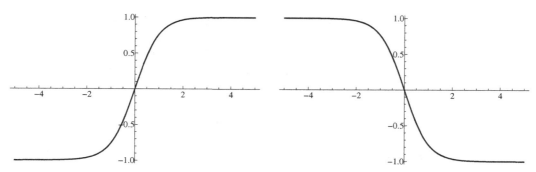

Figure 4.6 (Left) An instanton configuration with center at $\tau_0 = 0$. (Right) An anti-instanton configuration with center at $\tau_0 = 0$. Both are plotted for $g = 1/2$.

It is easy to see, by rescaling $\tau \to \tau/2$, that this operator is proportional to the Pöschl–Teller operator $A_{2,2}$ introduced in (4.8.228):

$$A = \frac{1}{4}A_{2,2}. \qquad (4.11.332)$$

This operator has a zero mode, which is the ground state in the Pöschl–Teller potential. The corresponding eigenfunction is given by:

$$\psi_2^{(0)}(\tau) \propto \mathrm{sech}^2\left(\frac{\tau - \tau_0}{2}\right). \qquad (4.11.333)$$

An alternative way to see the appearance of the zero mode is to consider the EOM (4.7.145) for the classical solution $q_c(\tau)$. By taking an additional derivative w.r.t. τ, we find

$$\frac{\mathrm{d}^2}{\mathrm{d}\tau^2}\dot{q}_c(\tau) - V''(q_c(\tau))\dot{q}_c(\tau) = 0. \qquad (4.11.334)$$

In other words,

$$A\dot{q}_c(\tau) = 0. \qquad (4.11.335)$$

Therefore, the zero mode (4.11.333) is simply the derivative of the classical solution, as one can easily check. The existence of such a zero mode is a problem in the evaluation of the Gaussian integration over the fluctuations, since, as we saw in Section 4.8, this is typically proportional to the inverse of the square root of det(A), and it leads to a divergence.

In order to solve this problem, we need a more conceptual understanding of the appearance of the zero mode. It turns out that the zero mode is explained by time translation invariance. Indeed, this invariance implies that the general classical solution depends on an arbitrary initial time τ_0, as we showed explicitly in (4.11.329). We will then write it as $q_c(\tau; \tau_0)$. This solution solves the EOM for all values of the parameter τ_0,

$$\left.\frac{\delta S}{\delta q(\tau)}\right|_{q(\tau)=q_c(\tau;\tau_0)} = 0. \qquad (4.11.336)$$

This is a general fact: when we solve for a nontrivial saddle point we find in general a *family of solutions*, related by a symmetry transformation. The parameters for such family are called *moduli* or *collective coordinates*. In the case at hand, we have a single modulus, namely the initial time τ_0. If we now take a further derivative of (4.11.336) w.r.t. τ_0, we obtain

$$\int \frac{\delta^2 S}{\delta q(\tau_1)\delta q(\tau_2)}\bigg|_{q(\tau)=q_c(\tau;\tau_0)} \frac{\partial q_c(\tau_2;\tau_0)}{\partial \tau_0} d\tau_2 = 0. \qquad (4.11.337)$$

The second functional derivative of S is the integral kernel of A, as shown in (4.6.125), and we conclude that

$$\xi(\tau;\tau_0) = \frac{\partial q_c(\tau;\tau_0)}{\partial \tau_0}, \qquad (4.11.338)$$

is a zero mode of A. Since $q_c(\tau;\tau_0)$ depends on τ_0 through the difference $\tau - \tau_0$, (4.11.338) equals $-\dot{q}_c(\tau;\tau_0)$.

Let us then revisit the problem of calculating the Gaussian path integral over quantum fluctuations around the saddle point:

$$Z_A = \int \exp\left(-\frac{1}{2}\langle q(\tau) - q_c(\tau;\tau_0)|A|q(\tau) - q_c(\tau;\tau_0)\rangle\right) \mathcal{D}q(\tau), \qquad (4.11.339)$$

in the case in which A has a zero mode (4.11.338) due to a collective coordinate. It is useful to separate the fluctuations

$$y(\tau;\tau_0) = q(\tau) - q_c(\tau;\tau_0), \qquad (4.11.340)$$

into a component proportional to the zero mode, and the components orthogonal to it. To do so, we introduce the product

$$\phi(\tau_0) = -\langle \xi(\tau;\tau_0)|y(\tau;\tau_0)\rangle. \qquad (4.11.341)$$

Then, the condition $\phi(\tau_0) = 0$ selects the fluctuations that are orthogonal to the zero mode. Let us now introduce in the path integral the identity

$$\int d\tau_0\, \phi'(\tau_0)\delta(\phi(\tau_0)) = 1. \qquad (4.11.342)$$

We note that

$$\phi'(\tau_0) = -\left\langle \frac{\partial \xi}{\partial \tau_0}\bigg|y\right\rangle + \langle \xi|\xi\rangle, \qquad (4.11.343)$$

where we used that

$$\frac{\partial y(\tau;\tau_0)}{\partial \tau_0} = -\xi(\tau;\tau_0). \qquad (4.11.344)$$

In addition, we use the Fourier representation of the delta function as

$$\delta(\phi(\tau_0)) = \int \frac{d\alpha}{2\pi} \exp\left(i\alpha\phi(\tau_0)\right), \qquad (4.11.345)$$

and we rewrite Z_A, after multiplying by (4.11.342), as

$$Z_A = \int \frac{d\tau_0 d\alpha}{2\pi} \exp\left(-\frac{1}{2}\langle y|A|y\rangle + i\alpha\,\langle\xi|y\rangle\right)\left(-\left\langle\frac{\partial\xi}{\partial\tau_0}\Big|y\right\rangle + \langle\xi|\xi\rangle\right)\mathcal{D}y(\tau).$$

$$(4.11.346)$$

The linear term in y vanishes, since it is odd under the simultaneous change y, $\alpha \rightarrow -y, -\alpha$. We end up with

$$Z_A = \int \frac{d\tau_0 d\alpha}{2\pi} \exp\left(-\frac{1}{2}\langle y|A|y\rangle + i\alpha\,\langle\xi|y\rangle\right)\langle\xi|\xi\rangle\,\mathcal{D}y(\tau). \qquad (4.11.347)$$

In order to proceed, we introduce a regularization of the operator A. This operator annihilates the functions that are proportional to the zero mode. Let us split the Hilbert space of square integrable functions on $\mathbb{R}$ into the one-dimensional subspace spanned by $\xi(\tau; \tau_0)$, and its orthogonal complement, i.e. $\mathcal{H} = \mathbb{C}|\xi(\tau; \tau_0)\rangle \oplus \mathcal{H}_\perp$. With respect to this splitting, the operator A can be written in matrix form as

$$A = \begin{pmatrix} 0 & 0 \\ 0 & A_\perp \end{pmatrix}. \qquad (4.11.348)$$

It is the zero entry in the diagonal that leads to the vanishing of the determinant. We regularize the operator by introducing a parameter ϵ that we will take to zero at the end of the calculation. The regularized operator A_ϵ is defined as

$$A_\epsilon = \frac{\epsilon}{\langle\xi|\xi\rangle}|\xi\rangle\langle\xi| + A_\perp, \qquad (4.11.349)$$

or, in matrix form,

$$A = \begin{pmatrix} \epsilon & 0 \\ 0 & A_\perp \end{pmatrix}. \qquad (4.11.350)$$

This leads to the regularized partition function

$$Z_{A_\epsilon} = \int \frac{d\tau_0 d\alpha}{2\pi} \exp\left(-\frac{1}{2}\langle y|A_\epsilon|y\rangle + i\alpha\,\langle\xi|y\rangle\right)\langle\xi|\xi\rangle\,\mathcal{D}y(\tau). \qquad (4.11.351)$$

We perform the Gaussian integral over the fluctuations $y(\tau)$ by completing the square:

$$-\frac{1}{2}\langle y|A_\epsilon|y\rangle + i\alpha\,\langle\xi|y\rangle = -\frac{1}{2}\langle y - i\alpha A_\epsilon^{-1}\xi|A_\epsilon|y - i\alpha A_\epsilon^{-1}\xi\rangle - \frac{1}{2}\alpha^2\langle\xi|A_\epsilon^{-1}|\xi\rangle.$$

$$(4.11.352)$$

Since $|\xi\rangle$ is the zero mode, the action of A_ϵ^{-1} on $|\xi\rangle$ is simply multiplication by ϵ^{-1}:

$$\langle\xi|A_\epsilon^{-1}|\xi\rangle = \frac{1}{\epsilon}\,\langle\xi|\xi\rangle. \qquad (4.11.353)$$

The regulated operator A_ϵ does not have a zero mode, so we can integrate over the fluctuations. The result of the integration depends on the boundary conditions. In the case of periodic boundary conditions, we can use (4.8.227) to find

$$Z_{A_\epsilon} = \int \frac{d\tau_0 d\alpha}{2\pi} \frac{1}{\sqrt{\det(A_\epsilon)}} \langle \xi | \xi \rangle \exp\left(-\frac{1}{2\epsilon}\alpha^2 \langle \xi | \xi \rangle\right). \qquad (4.11.354)$$

We can now perform the Gaussian integral over α, to find

$$Z_{A_\epsilon} = \frac{1}{\sqrt{\det(A_\perp)}} \int \frac{d\tau_0}{\sqrt{2\pi}} \langle \xi | \xi \rangle^{1/2}. \qquad (4.11.355)$$

Let us note that $\det(A_\perp)$ is simply the primed determinant (4.8.224). The final result is independent of ϵ, and we finally obtain

$$Z_A = \frac{1}{\sqrt{\det'(A)}} \int \frac{d\tau_0}{\sqrt{2\pi}} \langle \xi | \xi \rangle^{1/2}. \qquad (4.11.356)$$

We conclude that, in the presence of zero modes, we have to replace

$$\frac{1}{\sqrt{\det(A)}} \rightarrow \frac{1}{\sqrt{\det'(A)}} \int \frac{d\tau_0}{\sqrt{2\pi}} \|\xi\|. \qquad (4.11.357)$$

Although the result (4.11.356) was obtained for periodic boundary conditions, it also holds for anti-periodic boundary conditions. To see this, note that when we extract the contribution of the zero mode we have to take the derivative w.r.t. λ of $y_1(u; \lambda) + y_2'(u; \lambda) \mp 2$, for periodic (respectively, anti-periodic) boundary conditions, and both boundary conditions give the same answer, which is (4.11.356).

We are now ready to calculate the path integral $Z_a(\beta)$ in the Gaussian approximation, when β is very large. To calculate the factor $\|\xi\| = \|\dot{q}_c\|$ appearing in (4.11.357), we note that the Euclidean action evaluated at a classical solution is given by

$$S_{E,c} = S_E(q_c(\tau)) = \mathcal{W}(E) - E\beta, \qquad (4.11.358)$$

where

$$\mathcal{W}(E) = \int_{-\beta/2}^{\beta/2} d\tau (\dot{q}_c(\tau))^2 = \|\dot{q}_c\|^2. \qquad (4.11.359)$$

Therefore, for zero energy we have

$$\|\xi\| = \|\dot{q}_c(\tau)\| = S_{E,c}^{1/2}. \qquad (4.11.360)$$

Finally, the integral (4.11.357) over the collective coordinate τ_0 simply gives the length of the interval, β. We then can write,

$$Z_a(\beta) \approx \frac{2}{\sqrt{\det'(A)}} \frac{\beta S_{E,c}^{1/2}}{\sqrt{2\pi}} e^{-S_{E,c}}, \qquad (4.11.361)$$

where the extra factor of 2 is due to the fact that the two solutions $q_\pm(\tau; \tau_0)$ in (4.11.329) give the same contribution. Note that our evaluation in (4.11.361), as explicitly indicated, is an approximate one, since we have only taken into account the Gaussian fluctuations around the classical solution. There are further corrections in $\hbar$, which as we know from (4.11.327), are equivalent to corrections in g^2.

In order to evaluate the expression (4.11.361), we have to calculate the limit of $\det'(A)$ when β is large. In this limit, as we have seen in (4.11.332), A is proportional to a Pöschl–Teller operator. Therefore, we can compute this determinant by using (4.8.240). This result also involves the limit at large β of

$$Z_G(\beta) = \frac{1}{\sqrt{\det(A_0)}}, \qquad (4.11.362)$$

where

$$A_0 = -\frac{d^2}{d\tau^2} + 1. \qquad (4.11.363)$$

This can easily be evaluated from e.g. (4.7.187) with $\omega = 1$, and gives, at large β, $e^{-\beta/2}$.

To extract the energy gap $E_0 - E_1$ we have to take the large β limit of (4.11.361), and, from the relation (4.11.325), we deduce that

$$E_0(g) - E_1(g) = -2\frac{S_{E,c}^{1/2}}{\sqrt{2\pi}}e^{-S_{E,c}}\left[\frac{\det' A}{\det A_0}\right]^{-1/2}. \qquad (4.11.364)$$

Let us now compute the quantities involved in this expression. First of all, the classical action, evaluated at the classical trajectories (4.11.329), is

$$S_{E,c} = \frac{1}{6g^2}. \qquad (4.11.365)$$

For the quotient of determinants we can use the general result for Pöschl–Teller operators,

$$\frac{\det' A}{\det A_0} = 4\frac{\det' A_{2,2}}{\det A_{0,2}} = \frac{1}{12}. \qquad (4.11.366)$$

When these results are plugged into (4.11.364), we recover the result (2.8.339) exactly. We conclude that the nonperturbative effects that in the WKB method were due to the quantum period associated to the tunneling cycle, (2.8.298), are due in the path integral language to nontrivial solutions to the EOM. It is possible to reconstruct the whole sequence of nonperturbative corrections to the energy in the WKB method, (2.8.305), in terms of generalized solutions to the EOM called *multi-instantons*. However, in this problem multi-instanton techniques are computationally less efficient.

In the next chapter, we will study another type of instanton, also called bounce, in the context of resonances.

4.12 Bibliographical Notes

The path integral formulation of quantum mechanics was introduced by Feynman in his Ph.D. thesis, building upon insights of Dirac (1981, 1933). The original references are Feynmans's thesis (reproduced in Brown [2005]) and Feynman (1948). An early, but highly recommended, textbook presentation can be found in Feynman et al. (2010). There are many references on path integrals in quantum mechanics, and virtually every modern book on quantum field theory includes a chapter on path integration. The book by Kleinert (2009) is a comprehensive reference with many applications, and the books by Zinn-Justin (2012) and Baaquie (2014) are useful. Our treatment of the subject has been heavily influenced by Takhtajan (2008). An illuminating discussion of path integrals in statistical mechanics can be found in the book by Feynman (1998), while the first sections of Gross et al. (1982) provide a lucid summary.

The treatment in Section 4.5 is inspired by Schulman (2012). The calculation of the path integral for the partition function in Section 4.7 is due to Takhtajan (2008). A comprehensive survey of solvable path integrals can be found in Khandekar and Lawande (1986).

The book by Ahlfors (1966) has a concise survey of Riemann's zeta function. An excellent presentation of zeta function regularization of determinants in quantum mechanics, and its connection to path integrals, can be found in Takhtajan (2008). The results (4.8.216) and (4.8.223) are proved in Takhtajan (2008). A useful survey of functional determinants can be found in Dunne (2008).

Textbook presentations of instantons in the path integral formulation of quantum mechanics can be found in the books by Zinn-Justin (1996, 2012) and Mariño (2015). The papers by Zinn-Justin (1981, 1983), Zinn-Justin and Jentschura (2004a, 2004b), and Jentschura et al. (2010) present a detailed study of instantons in quantum mechanics, and, in particular, in the double-well potential. They also contain a systematic study of multi-instanton configurations. For the treatment of zero modes in the path integral, we have improved on Mariño (2015) by following the approach of Negele and Orland (1988) and Zinn-Justin (2012). An alternative treatment can be found in Andreassen et al. (2017).

5 Metastable States

5.1 Resonances

In quantum mechanics we often find states that are unstable but relatively long lived. This happens, for example, when we have states that are classically confined inside a potential, but decay through the tunnel effect after a long time. These states are usually called *resonant states* or *resonances*. Mathematically speaking, resonances are, in a sense that we will make precise, the analytic continuation of bound states to the continuum spectrum (in particular, they constitute a discrete set). In this chapter, we will study these states using different techniques. In particular, we will develop techniques to determine their energies, which are complex. We will however restrict ourselves to one-dimensional problems.

To understand resonant states, and their similarities with bound states, let us recall the scattering theory for potentials in one dimension, as developed in Section 1.5. We will assume that the potential goes to zero at infinity, although this assumption might be relaxed, as we will see later. We recall that bound states appear in this theory as zeros of the scattering coefficient $a(k)$ in the positive imaginary axis of the complex plane. Once we analytically continue $a(k)$ to the full complex plane, $a(k)$ might have zeros at other points. *Resonant states* are eigenfunctions of the Schrödinger operator in which the energy is of the form (1.5.151) and the wave vector k satisfies

$$a(k) = 0, \qquad \text{Im}(k) < 0. \tag{5.1.1}$$

We also require that

$$\text{Re}(k) \neq 0, \tag{5.1.2}$$

so that the corresponding energies have a nonzero imaginary part. There might be states satisfying (5.1.1) for which $\text{Re}(k) = 0$, but these have different properties and are called antibound states (we mentioned them briefly in the study of the Pöschl–Teller potential in Example 1.5.3). Resonant states are also sometimes called *Gamow vectors*. Note that the wave vectors satisfying (5.1.1) have a *negative* imaginary part and do not belong to the physical sheet, which is characterized by $\text{Im}(k) > 0$. Let us note that, when k is not real, the first equation in (1.5.160) is replaced by

$$\overline{a(k)} = a\left(-\overline{k}\right). \tag{5.1.3}$$

This implies that, if k corresponds to a resonant state, so does $-\overline{k}$: resonant states appear in pairs. One of the states in the pair has a wave vector belonging to the fourth quadrant of the complex plane, i.e. $\text{Re}(k) > 0$, while for the other state one has $\text{Re}(k) < 0$. The corresponding energies will be denoted by

$$E = \frac{\hbar^2}{2m} \left\{ (\text{Re}(k))^2 - (\text{Im}(k))^2 \mp 2i\,|\text{Re}(k)|\,|\text{Im}(k)| \right\} = E_R \mp i\frac{\Gamma}{2}. \qquad (5.1.4)$$

Resonant states can be also characterized by *purely outgoing* boundary conditions, also known as *Gamow–Siegert* boundary conditions. To understand why is this so, let us recall the asymptotic behavior (see (1.5.163) and (1.5.164)) of the Jost functions $f_i(x, k)$, $i = 1, 2$. Both of them have *outgoing* components, i.e. plane waves $e^{\pm ikx}$ when $x \to \pm\infty$, respectively. They also have an *incoming* component, with coefficient $a(k)$ (in $f_{1,2}(x, k)$, the incoming component occurs at $x \to \mp\infty$, respectively). Therefore, vanishing of $a(k)$ means vanishing of the incoming component.

A third, equivalent characterization of resonances is in terms of wave vectors with a negative imaginary part that lead to *poles* of the resolvent:

$$G(x, y; k)^{-1} = 0, \qquad \text{Im}(k) < 0, \qquad (5.1.5)$$

and this for all $x, y \in \mathbb{R}$. The equivalence of (5.1.1) and (5.1.5) follows from (1.5.169).

What is the physical meaning of resonances? When Γ is small in comparison with E_R, resonant states can be interpreted as "quasi-bound" states, i.e. long-lived metastable states. To understand why, let us consider the resonance with a negative imaginary part for the energy, which is sometimes called the *decaying* Gamow vector. The time evolution of a decaying Gamow vector leads to an exponentially decreasing factor in time,

$$e^{-\Gamma t/2\hbar}, \qquad (5.1.6)$$

which can be interpreted as an exponential decay law for a metastable state. The lifetime of such a state is

$$\tau = \frac{1}{\Gamma}. \qquad (5.1.7)$$

These types of resonant states appear in potentials in which there is a deep well surrounded by high barriers. Let us consider for example the potential

$$V(x) = \frac{1}{2}\frac{x^2}{1 + gx^4}, \qquad (5.1.8)$$

which has the shape shown in Figure 5.1. This potential does not support bound states for any $g > 0$. However, when g is very small, the potential is very close to a quadratic potential. The resonances should then be "quasi-bound" states labeled by a quantum number n, which corresponds to the label for bound states in a harmonic oscillator. These "quasi-bound" states have complex energies of the form

$$E_n = E_{n,R} \pm i\frac{\Gamma_n}{2}, \qquad n = 0, 1, 2, \ldots. \qquad (5.1.9)$$

When $0 < g \ll 1$, the real part of the energy satisfies

$$E_{n,R} \approx \hbar\left(n + \frac{1}{2}\right), \qquad n = 0, 1, 2, \ldots. \qquad (5.1.10)$$

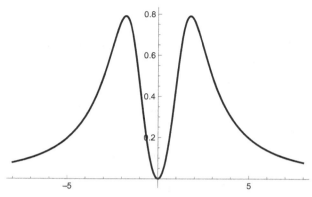

Figure 5.1 The potential (5.1.8) for $g = 1/10$.

The determination of resonances requires an analytic continuation of $a(k)$ to the full complex plane. This can be easily done in exactly solvable potentials, where explicit expressions for $a(k)$ are available. We will now consider some interesting examples that will clarify the role and properties of resonances.

Example 5.1.1 *Resonances in the Pöschl–Teller potential.* The Pöschl–Teller potential (1.5.234) was studied in Example 1.5.3. The zeros of the function $a(k)$ in the complex k-plane were determined in (1.5.250). In the case of the well and the low barrier ($\lambda > 1/2$), there are no resonances, although there are antibound states. In the case of the high barrier, where $\lambda = 1/2 + i\ell$, the zeros $k_1(n)$, $k_2(n)$ read

$$k_{1,2}(n) = \pm\ell - i\left(n + \frac{1}{2}\right), \qquad n = 0, 1, 2, \ldots . \qquad (5.1.11)$$

This is as expected, namely, that resonances come in pairs related by $\overline{k_1(n)} = -k_2(n)$. □

Example 5.1.2 *Resonances in the double delta potential.* The double delta potential was analyzed in some detail in Example 1.5.2. In order to understand resonances in this potential, let us consider the case in which $\beta \gg 1$ with a fixed. Since $\beta = Ba$, this means that the coefficient B in (1.5.227) is very large. In this extreme regime, the double delta potential well is very much like an infinite well of length $2a$, which has a well-known tower of bound states with wave vectors satisfying

$$k_\ell = \frac{\ell\pi}{2a}, \qquad \ell = 1, 2, \ldots, \qquad (5.1.12)$$

Therefore, when β is very large, we expect resonances to be "quasi-bound" states close to the bound states of the infinite well. In particular, the real part of the energy of such resonances will be of the form

$$E_{\ell,R} \approx \frac{\hbar^2 k_\ell^2}{2m}, \qquad \ell = 1, 2, \ldots . \qquad (5.1.13)$$

To verify this, let us look for zeros of the function $a(k)$ in (1.5.231), or equivalently, for zeros of the functions $\phi_0(\mu)$, $\phi_1(\mu)$ defined in (1.5.229), and solve for large β. Let us first consider the zeros of $\phi_0(\mu)$. From (1.5.232), we see that large β requires

$$\mu \approx \frac{\pi}{2} + n\pi, \qquad n \in \mathbb{Z}_{\geq 0}. \tag{5.1.14}$$

These correspond to the odd values of ℓ in (5.1.12). The zeros of $\phi_1(\mu)$ are obtained from (1.5.233). In this case, at large β we must have

$$\mu \approx n\pi, \qquad n \in \mathbb{Z}_{>0}. \tag{5.1.15}$$

These correspond to the even values of ℓ in (5.1.12). Let us write an *ansatz* for the zeros of $\phi_0(\mu)$ at large β of the form

$$\mu = \frac{\pi}{2} + n\pi + \frac{c_1}{\beta} + \frac{c_2}{\beta^2} + \cdots, \tag{5.1.16}$$

and similarly for the zeros of $\phi_1(\mu)$,

$$\mu = n\pi + \frac{d_1}{\beta} + \frac{d_2}{\beta^2} + \cdots. \tag{5.1.17}$$

The coefficients are easily solved. For the zeros of $\phi_0(\mu)$ we find,

$$c_1 = -\frac{\pi}{2}(2n+1), \qquad c_2 = \frac{\pi}{2}(2n+1) - \frac{i\pi^2}{4}(2n+1)^2. \tag{5.1.18}$$

For the zeros of $\phi_1(\mu)$ we find,

$$d_1 = -\pi n = -\frac{\pi}{2}(2n), \qquad d_1 = n\pi - i\pi^2 n^2 = \frac{\pi}{2}(2n) - \frac{\pi^2 i}{4}(2n)^2. \tag{5.1.19}$$

These solutions can be unified into a single formula,

$$\mu = \frac{\pi\ell}{2} - \frac{\pi\ell}{2\beta} + \frac{\pi}{2\beta^2}\left(\ell - \frac{i\pi}{2}\ell^2\right) + \mathcal{O}(\beta^{-3}), \tag{5.1.20}$$

where $\ell = 1, 2, \ldots$ is an arbitrary positive integer. It corresponds to the label of bound states in the infinite well appearing in (5.1.12). Note that the approximation (5.1.20) is very good when β is large. As an example, we plot in Figure 5.2 the imaginary part of μ for $\ell = 1$, as a function of β, as computed numerically from the zeros of $\phi_0(\mu)$ and as given in (5.1.20).

The solutions (5.1.20), obtained at large β, display all the expected properties of resonances. First of all, they are a discrete set of values, labeled by a positive integer ℓ. When β is very large, these resonances are very close to the bound states for a

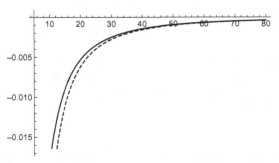

Figure 5.2   (Full line) The imaginary part of μ for $\ell = 1$ as a function of β, obtained numerically from the first zero of $\phi_0(\mu)$. (Dashed line) The first-order approximation in (5.1.20).

particle confined in an infinite well. However, there is a small, *negative* imaginary part for $\mu = ka$, which leads to an imaginary part in the corresponding energies E_ℓ. For large β we have,

$$\mathrm{Im}\, E_\ell \approx -\frac{\pi^3 \ell^3}{4\beta^2 a^2}. \tag{5.1.21}$$

$\square$

5.2 Resonances and Complex Dilatation

In the exactly solvable examples of the previous section, we could find the resonances by looking at complex zeros of $a(k)$ with $\mathrm{Im}\, k < 0$. However, in most cases we do not have explicit formulae of this type, and we need a general method to determine the resonances explicitly (numerically, if needed). It turns out that there is a powerful method, based on complex dilatations, in which one computes the resonances in the same way that one computes energies of bound states in confining potentials.

The basic idea of the complex dilatation method is to consider the group of dilatations on $\mathbb{R}$, with elements U_θ, $\theta \in \mathbb{R}$. This group acts on wavefunctions in $\mathcal{H} = L^2(\mathbb{R})$ by

$$(\mathsf{U}_\theta \psi)(x) = \mathrm{e}^{\theta/2} \psi(\mathrm{e}^\theta x). \tag{5.2.22}$$

The transformation is such that the products of wavefunctions $\langle \psi | \phi \rangle$ remain invariant (as noted above, we restrict ourselves to one-dimensional situations). It is clear that U_θ is a unitary transformation that acts on the Heisenberg operators x, p as

$$\mathsf{U}_\theta \mathsf{x} \mathsf{U}_\theta^{-1} = \mathrm{e}^\theta \mathsf{x}, \qquad \mathsf{U}_\theta \mathsf{p} \mathsf{U}_\theta^{-1} = \mathrm{e}^{-\theta} \mathsf{p}. \tag{5.2.23}$$

If the Hamiltonian H is of the standard form (1.3.83), it will transform as

$$\mathsf{H}(\theta) = \mathsf{U}_\theta \mathsf{H} \mathsf{U}_\theta^{-1} = \mathrm{e}^{-2\theta} \frac{\mathsf{p}^2}{2m} + V(\mathrm{e}^\theta \mathsf{x}). \tag{5.2.24}$$

In particular, for the free theory with $\mathsf{H}_0 = \mathsf{p}^2/2m$ we have

$$\mathsf{H}_0(\theta) = \mathrm{e}^{-2\theta} \mathsf{H}_0. \tag{5.2.25}$$

Let us now allow θ to take values in the complex unit disk D. In this case, the wavefunction $\mathsf{U}_\theta | \psi \rangle$ is not necessarily in $\mathcal{H}$. However, if we restrict ourselves to an appropriate dense subspace $\mathcal{A}$ in $\mathcal{H}$ and to an appropriate range of values of θ, it will still be square integrable. The set of vectors $\mathcal{A}$ can be taken to be the wavefunctions of the form,

$$\psi(x) = P(x)\mathrm{e}^{-\alpha x^2}, \qquad \alpha > 0, \tag{5.2.26}$$

where $P(x)$ is a polynomial. Then, we have that

$$(\mathsf{U}_\theta \psi)(x) = \mathrm{e}^{\theta/2} P(\mathrm{e}^\theta x) \mathrm{e}^{-\alpha \mathrm{e}^{2\theta} x^2}. \tag{5.2.27}$$

If we write

$$\theta = \theta_1 + \mathrm{i}\theta_2, \tag{5.2.28}$$

then, as long as

$$\cos(2\theta_2) > 0, \tag{5.2.29}$$

the function (5.2.27) is still in $L^2(\mathbb{R})$. This means that we should take $|\theta_2| < \pi/4$. In particular, we see from (5.2.25) that the spectrum of the rotated free Hamiltonian is given by the nonnegative real numbers, multiplied by

$$e^{-2i\theta_2}. \tag{5.2.30}$$

We can interpret this in terms of complex values of the wave vector k, which we can write as

$$k = |k|e^{-i\theta_2}. \tag{5.2.31}$$

This means, in particular, that when $\theta_2 > 0$, the complex dilatation is performing an analytic continuation into the lower half plane $\text{Im } k < 0$. Equivalently, the branch cut along the positive real axis in the k plane gets rotated by an angle $-\theta_2$.

An important aspect of the complex dilatation method is that, after the complex rotation, the Jost functions become normalizable wavefunctions precisely when Gamow–Siegert boundary conditions have been imposed. Indeed, after a rotation

$$x \to e^{i\theta}x, \qquad \theta > 0, \tag{5.2.32}$$

we have that

$$\exp(\pm ikx) \to \exp\left(\pm i|k|x\cos(\theta - \varphi) \mp |k|x\sin(\theta - \varphi)\right), \tag{5.2.33}$$

where we have written $k = |k|e^{-i\varphi}$. Therefore, we see from (1.5.163), (1.5.164) that both $f_{1,2}(x, k)$ decay exponentially at infinity if $a(k) = 0$ and $0 < \theta - \varphi < \pi/2$.

We conclude that the complex dilatation $H(\theta)$ of the Hamiltonian H leads to the analytic continuation into the lower half plane of k. Let us now consider a dense subspace of functions $\mathcal{A} \subset \mathcal{H}$ and an appropriate subset $\mathcal{Z}$ of the complex unit disk D such that

$$|\psi_\theta\rangle = U_\theta|\psi\rangle \in \mathcal{H}, \qquad \theta \in \mathcal{Z}. \tag{5.2.34}$$

In this case, the function

$$\langle\psi_\theta|G_\theta(E)|\phi_\theta\rangle, \tag{5.2.35}$$

where

$$G_\theta(E) = \frac{1}{H(\theta) - E}, \tag{5.2.36}$$

provides an analytic continuation of $\langle\psi|G(E)|\phi\rangle$ to the lower half plane

$$-\theta < \text{Im } k < 0. \tag{5.2.37}$$

We can compute the resonances as the poles of (5.2.35). This is equivalent to diagonalizing the complex Hamiltonian $H(\theta)$ on $\mathcal{A}$, a calculation that we can perform numerically, if needed. Let us note that this procedure involves, in principle, a choice of dense subspace $\mathcal{A}$. In most circumstances, however, the resonances obtained in this way are the *bona fide* zeros of $a(k)$ and do not depend on this choice.

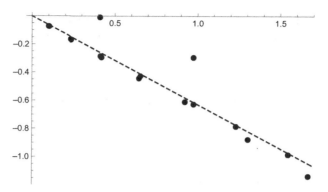

Figure 5.3 The numerical eigenvalues obtained by the complex dilatation method in the potential (5.1.8) with $g = 1/10$ and $\theta = \pi i/10$. Most of the eigenvalues accumulate along the line of slope $-\pi/5$ (shown as a dashed line in the figure). The two resonances (5.2.38) can be seen above the line.

Example 5.2.1 Let us consider the potential (5.1.8), represented in Figure 5.1. This potential clearly supports resonant states, and as $g \to 0$ we should recover the spectrum of bound states of the harmonic oscillator. Let us calculate the spectrum of resonances by using complex dilatation. To diagonalize the Hamiltonian $H(\theta)$ we can use the basis of eigenstates of the harmonic oscillator with $m = \omega = 1$, and we set $\hbar = 1$. The diagonalization of the rotated Hamiltonian for $g = 1/10$ and $\theta = \pi i/10$ leads to the spectrum shown in Figure 5.3, where we used the first 85 levels of the harmonic oscillator. We find an accumulation of eigenvalues along the line of slope $-\pi/5$, corresponding to the rotated branch cut of the continuous spectrum, as well as two resonances, located approximately at

$$E_0 \approx 0.410 - 0.008i, \qquad E_1 \approx 0.97 - 0.3i. \tag{5.2.38}$$

These correspond to the first two levels of the harmonic oscillator, and we see that $E_{0,R} \approx 1/2$, $E_{1,R} \approx 1$, as expected from (5.1.10). Let us note that in order to uncover a resonance where k has an argument $-\varphi$, we need a rotation with an angle $\theta > \varphi$, as is clear from (5.2.33). □

5.3 Resonances in Unbounded Potentials

So far we have discussed resonances in potentials that decrease to zero at infinity. When the potential is unbounded from below a different situation arises: after complex dilatation, there is no continuous spectrum, but just resonances. A typical example of this situation is the inverted quartic oscillator, with the Hamiltonian (2.6.186) and $g < 0$. In order to understand the resonances in this example, it is useful to consider first the case in which $g > 0$. There are two independent solutions to the Schrödinger equation in this case. When $|x|$ is large, they can be approximated by the WKB solutions. These involve the exponents,

$$\int^x p_1(z)\mathrm{d}z = \int^x \sqrt{2\left(gz^4 + z^2/2 - E\right)}\mathrm{d}z \approx \frac{1}{3}\sqrt{2g}|x|^3, \tag{5.3.39}$$

when g is real and positive, $|x| \gg 1$, and we have set $m = \omega = 1$. Therefore, the two possible asymptotic behaviors at infinity are given approximately by

$$\exp\left[\pm\frac{1}{3}\sqrt{2g}x^3\right], \qquad |x| \gg 1. \tag{5.3.40}$$

Let us focus on a solution of fixed parity and consider the behavior as $x \to \infty$. It will be of the form,

$$\psi(x) \approx C_1(E)\exp\left[\frac{1}{3}\sqrt{2g}x^3\right] + C_2(E)\exp\left[-\frac{1}{3}\sqrt{2g}x^3\right]. \tag{5.3.41}$$

Due to the requirement of a fixed parity, the coefficients $C_{1,2}(E)$ are fixed up to an overall constant. If we require this wavefunction to be square integrable, we find the condition

$$C_1(E) = 0, \tag{5.3.42}$$

which fixes the spectrum. Let us now suppose that g is negative, so that $g^{1/2} = \pm i|g|^{1/2}$. Then, the behavior of the analytically continued wavefunction at large x will be given by

$$\psi(x) \approx \widetilde{C}_1(E)\exp\left\{\frac{i}{3}(2|g|)^{1/2}x^3\right\} + \widetilde{C}_2(E)\exp\left\{-\frac{i}{3}(2|g|)^{1/2}x^3\right\}. \tag{5.3.43}$$

The asymptotic behavior corresponds to two rapidly oscillating functions, as expected for an inverted quartic potential. We can regard the first term in the r.h.s. as an outgoing wavefunction, and the second one as an incoming wavefunction. It is then possible to formulate an analogue of the Gamow–Siegert boundary conditions by requiring that the wavefunction is purely outgoing. This boundary condition leads to the quantization condition

$$\widetilde{C}_2(E) = 0. \tag{5.3.44}$$

The complex energies satisfying this condition are the resonances of the problem.

What happens when we implement a complex dilatation? After the transformation $x \to xe^{i\theta}$, the states are of the form

$$\psi(e^{i\theta}x) \approx \widetilde{C}_1(E)\exp\left\{\frac{i}{3}(2|g|)^{1/2}e^{3i\theta}x^3\right\} + \widetilde{C}_2(E)\exp\left\{-\frac{i}{3}(2|g|)^{1/2}e^{3i\theta}x^3\right\}. \tag{5.3.45}$$

If we assume that $\theta > 0$, then the exponentials inherit a real part given by

$$\exp\left\{\mp\frac{1}{3}(2|g|)^{1/2}\sin(3\theta)x^3\right\}, \tag{5.3.46}$$

corresponding to $\widetilde{C}_{1,2}(E)$, respectively. We can now impose square integrability, for nonzero, positive θ. This means that the exponential that increases at infinity should vanish, i.e. its coefficient should be zero. We recover in this way the condition (5.3.44).

Example 5.3.1 The solutions to the quantization condition (5.3.44) can be computed numerically. To do this, we diagonalize the matrix elements of the rotated Hamiltonian

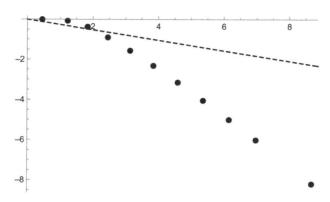

Figure 5.4 Resonant eigenvalues for the quartic oscillator for $g = -1/20$, computed using the complex dilatation method and with $\theta = i\pi/24$. The dashed line with slope $-\pi/12$ corresponds to the position of the branch cut that would appear in a potential that goes to zero at infinity.

$$H_{i\theta} = e^{2i\theta}\left(\frac{\mathsf{p}^2}{2} + \frac{\mathsf{x}^2}{2}\right) + ge^{4i\theta}\mathsf{x}^4, \qquad (5.3.47)$$

calculated in the basis of eigenstates of the harmonic oscillator. Let us consider for example the value $g = -1/20$. For the first resonance, one finds

$$E_0 \approx 0.450336452046007512\ldots - 0.003346640437900065\ldots\mathsf{i}. \qquad (5.3.48)$$

This resonance has been computed by using the first 250 states of the harmonic oscillator, and for two different values of θ, namely $\theta = \pi i/24, 11\pi i/48$. One can check that the resulting value is independent of the choice of θ, as it should be. The resonance with complex energy (5.3.48) can be regarded as the analytic continuation of the ground state of the quartic oscillator with $g > 0$, to negative values of g. In Figure 5.4 we plot the very first resonances for the quartic oscillator for $g = -1/80$ and $\theta = \pi i/24$, as well as the line with slope -2θ (this is the line where we would find a branch cut, in the case of a potential going to zero at infinity). In this case, as we see, there are only resonant poles.   □

A similar situation occurs in the so-called cubic oscillator, with Hamiltonian

$$H = \frac{\mathsf{p}^2}{2} + \frac{\mathsf{x}^2}{2} - g\mathsf{x}^3. \qquad (5.3.49)$$

The resonances in this potential can be also understood in terms of a generalized Gamow–Siegert boundary condition. As in the case of the quartic oscillator, we have to consider the asymptotic behavior of the solutions to the time-independent Schrödinger equation. When $x \to -\infty$ the potential increases as $|x|^3$ and we pick the solution that decays at infinity,

$$\exp\left(-\frac{2}{5}g^{1/2}|x|^{5/2}\right). \qquad (5.3.50)$$

When $x \to \infty$, the solution is given by a linear combination

$$C_+(E)\psi_+(x) + C_-(E)\psi_-(x), \qquad (5.3.51)$$

where

$$\psi_{\pm}(x) \approx \exp\left(\pm\frac{2}{5}g^{1/2}ix^{5/2}\right). \tag{5.3.52}$$

The coefficients $C_{\pm}(E)$ are fixed, up to an overall constant, since we have already imposed a condition on the behavior of the wavefunction at $x \to -\infty$. $\psi_{+}(x)$ can be regarded as an outgoing wavefunction, while $\psi_{-}(x)$ can be regarded as an incoming wavefunction. The analogue of the Gamow–Siegert boundary condition is that

$$C_{-}(E) = 0, \tag{5.3.53}$$

so that there are only outgoing waves at large, positive values of x. The generalized quantization condition (5.3.53) selects a discrete set of resonant energies.

The spectrum of resonances in the cubic oscillator can be determined in various ways. The method of complex dilatation discussed in Section 5.2 can be also applied to the cubic oscillator and provides a precise, numerical determination of the resonances. Note that, as in the case of Jost functions, this method leads to complex rotated eigenfunctions that are square integrable. This can be seen by analyzing the behavior under rotation of the oscillating wavefunction (5.3.51). If we transform $x \to xe^{i\theta}$ in the argument of (5.3.51), with $\theta > 0$, the two oscillating functions develop a real part given by

$$C_{\pm}(E) \exp\left(\mp\frac{2}{5}g^{1/2}\sin(\theta)x^{5/2}\right). \tag{5.3.54}$$

The purely outgoing boundary condition $C_{-}(E) = 0$ therefore leads to a wavefunction that decays exponentially as $x \to +\infty$, and is therefore square-integrable.

A more analytic approach to the resonances of the cubic oscillator can be obtained through the (exact) WKB method. This is similar in many ways to the determination of the energy levels of the double-well potential. The simpler case occurs when there are three real turning points. We show the relevant regions for the WKB analysis in Figure 5.5. This happens when the classical energy is smaller than the height of the barrier in region III. For the Hamiltonian (5.3.49), this happens when

$$0 \le E \le \frac{1}{54g^2}. \tag{5.3.55}$$

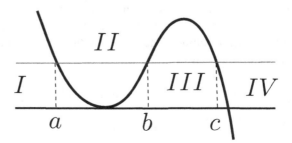

Figure 5.5   The different regions for the WKB analysis of the cubic oscillator.

If we impose Gamow–Siegert boundary conditions as $x \to +\infty$, in region *IV* there is only an outgoing wave. We then have

$$\psi_{IV}(x) = \frac{B}{\sqrt{P(x)}} \exp\left[\frac{i}{\hbar} \int_c^x P(x')dx'\right], \qquad (5.3.56)$$

where we use the notation in Section 2.4. To write down the corresponding wavefunction in region *III*, we can use the connection formula (2.4.122) to obtain

$$\psi_{III}(x) = \frac{B}{\sqrt{P_1(x)}} \left\{ e^{-i\pi/4} \exp\left(\frac{1}{\hbar} \int_x^c P_1(x')dx'\right) \right. \\ \left. + \frac{1}{2} e^{i\pi/4}(1 \pm 1) \exp\left(-\frac{1}{\hbar} \int_x^c P_1(x')dx'\right) \right\}. \qquad (5.3.57)$$

As usual, the choice of sign depends on the choice of lateral Borel resummation. On the other hand, the wavefunction in region *I* is of the form

$$\psi_I(x) = \frac{A}{\sqrt{P_1(x)}} \exp\left(-\frac{1}{\hbar} \int_x^a P_1(x')dx'\right). \qquad (5.3.58)$$

The connection formula gives, for region *II*,

$$\psi_{II}(x) = \frac{2A}{\sqrt{P(x)}} \cos\left(\frac{1}{\hbar} \int_a^x P(x')dx' - \frac{\pi}{4}\right). \qquad (5.3.59)$$

This can be written as

$$\psi_{II}(x) = \frac{2A}{\sqrt{P(x)}} \left\{ \sin\left(\frac{1}{\hbar} \int_a^b P(x')dx'\right) \cos\left(\frac{1}{\hbar} \int_x^b P(x')dx' - \frac{\pi}{4}\right) \right. \\ \left. - \cos\left(\frac{1}{\hbar} \int_a^b P(x')dx'\right) \cos\left(\frac{1}{\hbar} \int_x^b P(x')dx' - \frac{\pi}{4} - \frac{\pi}{2}\right) \right\}. \qquad (5.3.60)$$

Applying the connection formulae from region *II* to region *III*, we find

$$\widetilde{\psi}_{III}(x) = \pm i \frac{A}{\sqrt{P_1(x)}} \exp\left(\mp\frac{i}{\hbar} \int_a^b P(x')dx'\right) \exp\left(-\frac{1}{\hbar} \int_b^x P_1(x')dx'\right) \\ + \frac{2A}{\sqrt{P_1(x)}} \cos\left(\frac{1}{\hbar} \int_a^b P(x')dx'\right) \exp\left(\frac{1}{\hbar} \int_b^x P_1(x')dx'\right). \qquad (5.3.61)$$

Let us write (5.3.57) as

$$\psi_{III}(x) = \frac{B}{\sqrt{P_1(x)}} \left\{ e^{-i\pi/4} \exp\left(\frac{1}{\hbar} \int_b^c P_1(x')dx'\right) \exp\left(-\frac{1}{\hbar} \int_b^x P_1(x')dx'\right) \right. \\ \left. + \frac{1}{2} e^{i\pi/4}(1 \pm 1) \exp\left(-\frac{1}{\hbar} \int_b^c P_1(x')dx'\right) \exp\left(\frac{1}{\hbar} \int_b^x P_1(x')dx'\right) \right\}. \qquad (5.3.62)$$

We can now compare (5.3.62) with (5.3.61). By equating the leading and subleading exponentials, we find two equations that determine A as a function of B and impose a quantization condition,

$$Be^{-i\pi/4} \exp\left(\frac{1}{\hbar} \int_b^c P_1(x')dx'\right) = \pm iA \exp\left(\mp\frac{i}{\hbar} \int_a^b P(x')dx'\right),$$

$$\frac{B}{2} e^{i\pi/4}(1 \pm 1) \exp\left(-\frac{1}{\hbar} \int_b^c P_1(x')dx'\right) = 2A \cos\left(\frac{1}{\hbar} \int_a^b P(x')dx'\right). \qquad (5.3.63)$$

The choice of lower sign leads gives,

$$\cos\left(\frac{1}{\hbar}\int_a^b P(x')dx'\right) = 0 \Rightarrow \int_a^b P(x')dx' = \pi n\hbar\left(1 + \frac{1}{2}\right), \qquad n = 0, 1, 2, \ldots \tag{5.3.64}$$

If we denote

$$v = \frac{1}{\pi\hbar}\int_a^b P(x)dx, \tag{5.3.65}$$

the condition (5.3.64) can be written as,

$$v = n + \frac{1}{2}, \qquad n = 0, 1, \ldots. \tag{5.3.66}$$

This is the all-orders quantization condition for the perturbative quantum period v. Let us now choose the upper sign. We find, after dividing the second equation in (5.3.63) by the first one,

$$1 + \exp\left(\frac{2i}{\hbar}\int_a^b P(x')dx'\right) + \exp\left(-\frac{2}{\hbar}\int_b^c P_1(x')dx'\right) = 0. \tag{5.3.67}$$

In this equation, the last term of the l.h.s. provides a nonperturbative correction to the perturbative quantization condition. This EQC can be written in a form similar to the one for the double-well potential, (2.8.293). Let us call

$$f(v) = \exp\left(-\frac{2}{\hbar}\int_b^c P_1(x)dx\right). \tag{5.3.68}$$

Then, (5.3.67) can be written as

$$-f(v) = 1 + e^{2\pi i v}. \tag{5.3.69}$$

Let us now analyze this condition in some detail, in a similar way to the analysis of Section 2.8. The EQC (5.3.69) says that, if one neglects the contribution of $f(v)$, one obtains the perturbative WKB quantization condition (5.3.64). The leading order term in v can be explicitly computed as a function of E, in terms of elliptic integrals,

$$\text{vol}_0(E, g) = 2\int_a^b \sqrt{2E - x^2 + 2gx^3}dx$$

$$= 4\sqrt{2g}\frac{(b-a)^2\sqrt{c-a}}{15k^4}\left\{(k')^2(k^2 - 2)K(k) + 2\left(k^4 + (k')^2\right)E(k)\right\}, \tag{5.3.70}$$

where

$$k^2 = \frac{b-a}{c-a}, \qquad (k')^2 = 1 - k^2. \tag{5.3.71}$$

The term $f(v)$ gives an imaginary correction to the energy levels that is nonperturbative in $\hbar$. This can be seen by solving the quantization condition (5.3.69) in the same way as for (2.8.301) in the case of the double-well potential. We find that

$$\widehat{v} = v + \Delta v, \tag{5.3.72}$$

where

$$\Delta v = -\frac{i}{2\pi}f(v) + \cdots, \tag{5.3.73}$$

and

$$E_+(\hat{v}) = E(v) + \frac{\partial E}{\partial v} \Delta v + \cdots . \tag{5.3.74}$$

The subindex in $E_+(\hat{v})$ means that this is the calculation of the energy obtained by using the lateral Borel resummation that corresponds to (5.3.69). On the other hand, if one considers the other lateral resummation, corresponding to (5.3.64), we simply obtain

$$E_-(\hat{v}) = E(v). \tag{5.3.75}$$

It follows that the imaginary correction to the energy levels is given by,

$$\text{Im } E(v) = \frac{1}{2i} \left(E_+(\hat{v}) - E_-(\hat{v}) \right) = -\frac{1}{4\pi} \frac{\partial E}{\partial v} f(v) + \cdots . \tag{5.3.76}$$

Physically, this correction is due to the tunneling of the particle through the barrier between $x = a$ and $x = b$. The exponent of $f(v)$ has a power series expansion in $\hbar$, whose coefficients can be computed in terms of elliptic integrals. It is sometimes useful to analyze the imaginary contribution (5.3.76) for small values of the coupling constant g. To do this, one needs to re-express the WKB corrections in terms of perturbative expansions in g, as we did in Section 2.8 in the case of the double-well potential. We will do this in Section 5.4, in the context of the uniform WKB method.

When the classical energy is well above the top of the barrier,

$$E \gg \frac{1}{54g^2}, \tag{5.3.77}$$

the most important contribution to the imaginary part of the resonances does not come from the tunneling contribution, but from the standard WKB quantization condition, analytically continued to complex values. This is only to be expected, since large classical energies correspond to large quantum numbers, and this is where the BS quantization condition should provide a very good approximation to the spectrum. Let us then derive an approximate quantization condition in this regime. For large values of E, the point a has the form

$$a \approx -\zeta, \qquad \zeta = \left(\frac{E}{g} \right)^{1/3}, \tag{5.3.78}$$

while b, c become complex, of the form

$$-e^{\mp \frac{2\pi i}{3}} \zeta, \tag{5.3.79}$$

respectively. The large E limit of the function (5.3.70) can be easily obtained by taking into account that k, the argument of the elliptic integrals, is of the form $k^2 \approx e^{\pi i/3}$, and we find

$$\text{vol}_0(E) \approx \frac{2^{3/2} 3^{3/4}}{5} e^{\pi i/12} K(e^{\pi i/6}) g^{-1/3} E^{5/6}. \tag{5.3.80}$$

Therefore, the complex resonances of the cubic oscillator behave, for large energies, as

$$E_n \approx C e^{-i\pi/5} g^{2/5} \left(n + \frac{1}{2} \right)^{6/5}, \qquad n \gg 1, \tag{5.3.81}$$

where C is a constant that can be read from the above result. This limiting behavior can be also obtained by taking into account that, for large energies, the cubic term in the potential is more important the the quadratic term, and we find

$$\mathrm{vol}_0(E) \approx (2g)^{1/2} \int_{-\zeta}^{-e^{-2\pi i/3}\zeta} \sqrt{\zeta^3 + x^3}\,\mathrm{d}x, \qquad (5.3.82)$$

where ζ is given in (5.3.78).

5.4 Resonances and the Uniform WKB Method

In Section 5.3 we studied resonances in the inverted quartic and cubic oscillators and showed that the exact WKB method can be used to obtain EQCs for the resonances. As in the case of the double-well potential, it is also instructive to rederive these results with the uniform WKB method. This will give us useful results in the limit of a small coupling constant.

We then consider the "unstable" counterpart of the potential (2.3.59), namely

$$V(x) = \frac{1}{2}x^2 - gx^p, \qquad g > 0, \qquad (5.4.83)$$

where $p > 2$ is a nonnegative integer. We will assume that p is odd, although the case in which p is even can be treated in a similar way. As in Example 2.3.1, the Schrödinger equation can be put in the form

$$-\xi^2 \psi''(z) + \left(z^2 - z^p - 2\xi\mathcal{E}\right)\psi = 0, \qquad (5.4.84)$$

where $z, \xi, \mathcal{E}$ are defined by the equations (2.3.60), (2.3.63). Since p is odd, the potential in (5.4.83) has the following behavior,

$$V(x) \to \mp\infty, \qquad x \to \pm\infty. \qquad (5.4.85)$$

Therefore, the natural boundary conditions for $\psi(z)$ are the following: as $z \to -\infty$, the wavefunction decays; as $z \to +\infty$, we will impose Gamow–Siegert boundary conditions, in order to select a discrete tower of resonances.

We now solve (5.4.84) in the uniform WKB approximation. We will assume that the energy is such that there are three turning points, as in Figure 5.5. There are two different regions to consider. Near $x = 0$, there are two turning points, and we should use equation (2.3.53) for $f(\phi)$. The ansatz for the wavefunction is

$$\psi(z) = \frac{1}{\sqrt{u'(z)}} D_{\nu-1/2}\left(-\left(\frac{2}{\xi}\right)^{1/2} u(z)\right), \qquad (5.4.86)$$

which is very similar to (2.3.54). This ansatz involves only one of the parabolic cylinder functions appearing in (2.3.54), and as we will see in a moment this is required by the boundary conditions at $z \to -\infty$. The function $u(z)$ satisfies

$$(u'(z))^2 u^2(z) = z^2 - z^p - 2\xi\left(\mathcal{E} - \nu u'(z)^2\right) + \frac{\xi^2}{2}\{u, z\}, \qquad (5.4.87)$$

to be compared with the corresponding equation in the stable case (2.3.66). Proceeding as in that case, we introduce the function $U(z)$, as in (2.3.67), and make a power series expansion in ξ, as in (2.3.69). We find at leading order

$$U_0'(z) = (z^2 - z^p)^{1/2}, \tag{5.4.88}$$

whose solution is

$$U_0(z) = \int_0^z (s^2 - s^p)^{1/2} ds = \frac{z^2}{2} \, _2F_1\left(-\frac{1}{2}, q; q+1; z^{2/q}\right). \tag{5.4.89}$$

In this equation, q is again given by (2.3.73). As $z \to -\infty$, we have

$$u_0(z) \approx \frac{2}{\sqrt{p+2}} z(-z)^{\frac{p-2}{4}}, \tag{5.4.90}$$

which gives the dominant behavior of the full $u(z)$. We then have that $u(z) \to -\infty$ as $z \to -\infty$, so that the argument of the parabolic cylinder function is large and positive. It follows from (B.43) that (5.4.86) decays as $z \to -\infty$, as required. The functions $u_n(z)$ (or $U_n(z)$) and the coefficients $\mathcal{E}_n(v)$ can be computed recursively, as in Example 2.3.1. For example, the function $U_1(z)$ satisfies equation (2.3.75), and one can integrate it to obtain

$$U_1(z) = v \log(u_0(z)) - \frac{qv}{2} \log \frac{1 - (1 - z^{2/q})^{1/2}}{1 + (1 - z^{2/q})^{1/2}} - qv \log 2, \tag{5.4.91}$$

where the integration constant is fixed by requiring $u_1(z)$ to be smooth at $z = 0$.

The solution obtained above is tailored for the region around $z = 0$ and will not be convenient when we are near the rightmost turning point. There, it is better to use the uniform WKB method with an ansatz of the Airy type, as in (2.4.82). We will write,

$$\psi(z) = \frac{1}{\sqrt{v'(z)}} \mathrm{Ai}^{(\pm)}\left(\xi^{-2/3} v(z)\right), \tag{5.4.92}$$

where the functions $\mathrm{Ai}^{(\pm)}(x)$ are defined in (B.37). This ansatz is essentially identical to (2.4.82), up to a change of sign in $\phi(x)$. After taking into account this change of sign and the change of variables (2.3.60), we find that equation (2.4.84) reads, in this case,

$$v(z)(v'(z))^2 = z^2 - z^p - 2\xi\mathcal{E} + \frac{\xi^2}{2}\{v, z\}. \tag{5.4.93}$$

As in Example 2.3.1, we will solve this equation by setting up a power series expansion in ξ. We first introduce

$$\mathcal{V}(z) = \frac{2}{3} v(z)^{3/2}, \tag{5.4.94}$$

and we use the ansatz:

$$v(z) = \sum_{n \geq 0} v_n(z)\xi^n, \qquad \mathcal{E} = \sum_{n \geq 0} \mathcal{E}_n(v)\left(\frac{\xi^{p/2-1}}{2}\right)^n. \tag{5.4.95}$$

This leads to an expansion of the form

$$\mathcal{V}(z) = \sum_{n \geq 0} \mathcal{V}_n(z)\xi^n, \tag{5.4.96}$$

where

$$\mathcal{V}_0(z) = \frac{2}{3}v_0(z)^{3/2},$$

$$\mathcal{V}_1(z) = v_0(z)^{1/2}v_1(z),$$ (5.4.97)

$$\mathcal{V}_2(z) = v_0(z)^{1/2}v_2(z) + \frac{v_1^2(z)}{4v_0^{1/2}(z)},$$

and so on. The leading order term in $\mathcal{V}(z)$ satisfies the equation,

$$\mathcal{V}_0'(z) = z^2 - z^p,$$ (5.4.98)

which is solved in closed form to give

$$\mathcal{V}_0(z) = \int_z^1 (s^2 - s^p)^{1/2}\mathrm{d}s = \frac{q}{3}(1 - z^{2/q})^{3/2} \; {}_2F_1\left(1 - q, \frac{3}{2}; \frac{5}{2}; 1 - z^{2/q}\right).$$ (5.4.99)

Note that the behavior of $v_0(z)$ as $z \to +\infty$ is of the form

$$v_0(z) \approx -\left(\frac{3}{p+2}\right)^{2/3} z^{\frac{p+2}{3}}.$$ (5.4.100)

Therefore, the asymptotic behavior of $\psi(z)$ along the positive real axis is governed by (B.38). If we impose Gamow–Siegert boundary conditions as $x \to +\infty$, so that there are no outgoing waves, we must pick the function $\mathrm{Ai}^{(+)}$. We can solve recursively for the functions $v_n(z)$ (or $V_n(z)$), as in the previous case. We find, for example,

$$\mathcal{V}_1(z) = \frac{vq}{2}\log\frac{1 - (1 - z^{2/q})^{1/2}}{1 + (1 - z^{2/q})^{1/2}}.$$ (5.4.101)

We have now two different uniform WKB solutions, valid near $z = 0$ and $z = 1$, respectively. In order to find the quantization conditions that determine the resonances, we must match both solutions in the intermediate region. When the coupling constant ξ is real and positive, the matching is performed along a Stokes line of the functions. There are therefore two different possibilities, depending on the choice of lateral Borel resummation.

Let us first consider the asymptotic expansion of the parabolic cylinder function. In the intermediate region, $u(z)$ is positive. Therefore, for real, positive ξ, the argument of the parabolic cylinder function is negative and we have to use (B.44), which in this case reads

$$D_{v-\frac{1}{2}}\left(-\left(\frac{2}{\xi}\right)^{1/2}u\right) \sim \frac{\sqrt{2\pi}}{\Gamma\left(\frac{1}{2} - v\right)}e^{\frac{u^2}{2\xi}}\left(\frac{\xi}{2u^2}\right)^{\frac{1}{4}+\frac{v}{2}} \; {}_2F_0\left(\frac{1}{4} + \frac{v}{2}, \frac{3}{4} - \frac{v}{2}; ; \xi u^{-2}\right)$$

$$\pm \mathrm{i}\,e^{\mp\mathrm{i}\pi v}e^{-\frac{u^2}{2\xi}}\left(\frac{\xi}{2u^2}\right)^{\frac{1}{4}-\frac{v}{2}} \; {}_2F_0\left(\frac{1}{4} - \frac{v}{2}, \frac{3}{4} + \frac{v}{2}; ; -\xi u^{-2}\right).$$ (5.4.102)

Let us now consider the asymptotic expansion of the Airy function. In the intermediate region, $v(z)$ is positive and we are on a Stokes line for $\mathrm{Ai}^{(+)}$. The asymptotic expansion depends on our choice of lateral Borel resummation, and we have to find the correct correlation between this choice and the one made before for the parabolic cylinder function. A useful way of doing this is to give a small imaginary part to ξ, so that the two choices of lateral Borel resummation correspond to the choices of sign of the imaginary part of ξ. A positive (negative) imaginary part corresponds to the negative (positive) sign in (5.4.102). When the argument of ξ is positive, the argument of $\xi^{-2/3}v(z)$ is negative and we should use the asymptotic expansion (B.39). We then have,

$$
\mathrm{Ai}^{(+)}(\xi^{-2/3}v) \sim \frac{1}{\sqrt{\pi}}\xi^{1/6}v^{-1/4}e^{\frac{2v^{3/2}}{3\xi}}\,{}_2F_0\left(\frac{1}{6},\frac{5}{6};;\frac{3\xi}{4}v^{-3/2}\right), \qquad \arg(\xi) > 0.
$$
$$(5.4.103)$$

When the argument of ξ is negative, the argument of $\xi^{-2/3}v(z)$ is positive and we should use the asymptotic expansion (B.41),

$$
\mathrm{Ai}^{(+)}(\xi^{-2/3}v) \sim \frac{1}{\sqrt{\pi}}\xi^{1/6}v^{-1/4}e^{\frac{2v^{3/2}}{3\xi}}\,{}_2F_0\left(\frac{1}{6},\frac{5}{6};;\frac{3\xi}{4}v^{-3/2}\right).
$$
$$
+\frac{i}{\sqrt{\pi}}\xi^{1/6}v^{-1/4}e^{-\frac{2v^{3/2}}{3\xi}}\,{}_2F_0\left(\frac{1}{6},\frac{5}{6};;-\frac{3\xi}{4}v^{-3/2}\right), \qquad \arg(\xi) < 0.
$$
$$(5.4.104)$$

We now have to match these asymptotic expansions, which will lead to quantization conditions for the resonances. Note first that these expansions can be written in a simpler form in terms of the functions $U(z) = u^2(z)/2$, $\mathcal{V}(z) = 2v_0(z)^{3/2}/3$ introduced in (2.3.67) and (5.4.94), respectively. In order to perform the matching, we should note that

$$
\mathcal{V}_0(z) = \int_z^1 (s^2 - s^P)^{1/2}\mathrm{d}s = \int_0^z (s^2 - s^P)^{1/2}\mathrm{d}s + \int_0^1 (s^2 - s^P)^{1/2}\mathrm{d}s
$$
$$(5.4.105)$$
$$
= -U_0(z) + \frac{1}{2}qB(q, 3/2),
$$

where $B(x, y)$ is the beta function. Therefore, the positive exponential in the asymptotics of the Airy function corresponds to the negative exponential in the parabolic cylinder function. The matching depends on the sign of $\arg(\xi)$. Let us first consider the case in which $\arg(\xi) > 0$. In this case, the asymptotic expansion of the Airy function involves only the positive exponential, which becomes a negative exponential near $x = 0$. This means that the term involving the positive exponential in the asymptotic expansion (5.4.102) should vanish. The only way to achieve this is to have

$$
\frac{1}{\Gamma\left(\frac{1}{2} - v\right)} = 0.
$$
$$(5.4.106)$$

This gives the quantization condition (5.3.66) obtained in Section 5.3, in the context of the exact WKB method, for one of the signs of the lateral Borel resummation in the Voros–Silverstone connection formula.

Let us now consider the case in which $\arg(\xi) < 0$. In this case, the asymptotic expansions consist of a leading and a subleading part. The matching condition is implemented by requiring that the quotient of the subleading and the leading part are the same for both expansions. We then obtain the following condition,

$$- \frac{\Gamma\left(\frac{1}{2} - v\right)}{\sqrt{2\pi}\, \mathrm{e}^{\mathrm{i}\pi v}} \mathrm{e}^{-\frac{u^2}{\xi}} \left(\frac{2u^2}{\xi}\right)^v \frac{{}_2F_0\left(\frac{1}{4} - \frac{v}{2}, \frac{3}{4} + \frac{v}{2}; ; -\xi u^{-2}\right)}{{}_2F_0\left(\frac{1}{4} + \frac{v}{2}, \frac{3}{4} - \frac{v}{2}; ; \xi u^{-2}\right)}$$

$$= \mathrm{e}^{\frac{4v^{3/2}}{3\xi}} \frac{{}_2F_0\left(\frac{1}{6}, \frac{5}{6}; ; \frac{3\xi}{4} v^{-3/2}\right)}{{}_2F_0\left(\frac{1}{6}, \frac{5}{6}; ; -\frac{3\xi}{4} v^{-3/2}\right)}. \tag{5.4.107}$$

There is a more convenient way to write this equation, similar to that used in the case of the double-well potential. If we use now the reflection formula for the Gamma function (2.8.331), we find the following equality:

$$2^v \exp\left[-\frac{u^2}{\xi} + v \log\left(\frac{u^2}{\xi}\right) - \frac{4v^{3/2}}{3\xi} + \log \frac{{}_2F_0\left(\frac{1}{4} - \frac{v}{2}, \frac{3}{4} + \frac{v}{2}; ; -\xi u^{-2}\right)}{{}_2F_0\left(\frac{1}{4} + \frac{v}{2}, \frac{3}{4} - \frac{v}{2}; ; \xi u^{-2}\right)} \right.$$

$$\left. + \log \frac{{}_2F_0\left(\frac{1}{6}, \frac{5}{6}; ; -\frac{3\xi}{4} v^{-3/2}\right)}{{}_2F_0\left(\frac{1}{6}, \frac{5}{6}; ; \frac{3\xi}{4} v^{-3/2}\right)} \right] = \frac{\Gamma\left(\frac{1}{2} + v\right)}{\sqrt{2\pi}} \mathrm{e}^{\mathrm{i}\pi} \left(\mathrm{e}^{2\pi \mathrm{i} v} + 1\right). \tag{5.4.108}$$

This suggests introducing the function

$$f(v) = \frac{2^v \sqrt{2\pi}}{\Gamma\left(\frac{1}{2} + v\right)} \exp\left[-\frac{2U}{\xi} + v \log\left(\frac{2U}{\xi}\right) - \frac{2V}{\xi} \right.$$

$$\left. + \log \frac{{}_2F_0\left(\frac{1}{4} - \frac{v}{2}, \frac{3}{4} + \frac{v}{2}; ; -\xi(2U)^{-1}\right)}{{}_2F_0\left(\frac{1}{4} + \frac{v}{2}, \frac{3}{4} - \frac{v}{2}; ; \xi(2U)^{-1}\right)} + \log \frac{{}_2F_0\left(\frac{1}{6}, \frac{5}{6}; ; -\xi(2V)^{-1}\right)}{{}_2F_0\left(\frac{1}{6}, \frac{5}{6}; ; \xi(2V)^{-1}\right)} \right]. \tag{5.4.109}$$

The matching condition then gives a quantization condition of the form

$$-f(v) = \mathrm{e}^{2\pi \mathrm{i} v} + 1. \tag{5.4.110}$$

This is precisely the quantization condition (5.3.69) obtained in Section 5.3, for the other choice of Borel resummation in the Voros–Silverstone connection formulae. We conclude that the uniform WKB method gives the same results as the analysis based on connection formulae. In addition, the uniform WKB method gives an explicit expression for $f(v)$, (5.4.109), as a function of the quantum period v and the coupling constant ξ.

Let us now expand the exponent of (5.4.109) in power series in ξ. The term of order ξ^{-1} can be calculated immediately by using (5.4.105). By using the explicit expressions for U_1, V_1, and u_0 one finds that the ξ-independent term is given by

$$2(U_1 + V_1) - 2v \log u_0 = -2qv \log 2, \tag{5.4.111}$$

and we obtain

$$f(v) = \frac{\sqrt{2\pi} 2^{2qv}}{\Gamma\left(\frac{1}{2} + v\right)} \left(\frac{2}{\xi}\right)^v \exp\left[-\frac{qB(q, 3/2)}{\xi} - \sum_{k \geq 1} c^{(k)}(v) \xi^k \right]. \tag{5.4.112}$$

Note that the structure of this function is very similar to that obtained in the case of the double-well potential, given in equations (2.8.334) and (2.8.335). The coefficients $c^{(k)}(v)$ can be obtained from the functions $u_0, \ldots, u_{k+1}$ and $v_0, \ldots, v_{k+1}$. For example, for $c^{(1)}(v)$ we obtain

$$-c^{(1)}(v) = -\frac{3 + 4v^2}{8u_0^2(z)} + 2v\frac{u_1(z)}{u_0(z)} - u_1^2(z) - 2u_0(z)u_2(z) - \frac{5}{24v_0^{3/2}(z)} - 2\mathcal{V}_2(z).$$

(5.4.113)

Although it is not obvious from the explicit expressions, the $c^{(k)}(v)$ are z-independent, as required by consistency of the whole procedure (and as happened already in the case of the double-well potential).

Example 5.4.1 *Uniform WKB analysis of the cubic oscillator.* Let us work out some of the details in the case $p = 3$, corresponding to the cubic oscillator. The function $U_0(z)$ simplifies to

$$U_0(z) = \frac{2}{15}\left(3z^2\sqrt{1-z} - z\sqrt{1-z} - 2\sqrt{1-z} + 2\right),$$

(5.4.114)

and it is easy to calculate $u_2(z)$ as an expansion around $z = 0$, with the following result,

$$u_2(z) = -\frac{15818v^2 + 3087}{12960} - \frac{5541686v^2 + 1433565}{4976640}z$$

$$- \frac{28892618v^2 + 9730395}{29859840}z^2 + \mathcal{O}(z^3).$$

(5.4.115)

On the other hand, the function $\mathcal{V}_0(z)$ is given by

$$\mathcal{V}_0(z) = \frac{2}{15}(1 - z)^{3/2}(3z + 2).$$

(5.4.116)

In order to compute the coefficients $f^{(k)}(v)$ in (5.4.112), it is useful to expand all functions around a given value of z. In this way one can verify that the result is independent of z, and read the values of the coefficients. A particularly useful value is $z = 0$, since the functions $U_n(z)$ are regular there and their expansions are easy to calculate. The integration constants appearing in the calculation of the functions $\mathcal{V}_n(z)$ are fixed by studying their behavior at $z = 1$. One has, for example,

$$\mathcal{V}_2(z) = \frac{1}{64}\sqrt{1-z}\left(-\frac{2\left(4v^2(7z + 2) + 5z + 6\right)}{z^2} - \frac{64v^2}{z - 1} - \frac{12}{3z + 2}\right).$$

(5.4.117)

This has the expansion around $z = 0$ given by

$$\mathcal{V}_2(z) = -\frac{3 + 4v^2}{16z^2} - \frac{1 + 12v^2}{16z} + \mathcal{O}(z),$$

(5.4.118)

and it is easy to see that this is precisely what is required to cancel the singularities coming from the first two terms in (5.4.113). By pushing the expansion around $z = 0$ of the r.h.s. in (5.4.113) we find that, indeed, $c^{(1)}(v)$ is z-independent and given by

$$c^{(1)}(v) = \frac{77}{128} + \frac{141v^2}{32}.$$

(5.4.119)

This procedure can be extended to compute the coefficients $c^{(k)}(v)$ to the very first orders. More details and results can be found in the references listed in the bibliographical notes (Section 5.7).   □

As a consequence of the above analysis, we can determine the asymptotic form of the imaginary part of the resonances for the cubic oscillator in the limit $g \to 0$ (we recall that we are setting $m = \hbar = 1$). To do this, we use equation (5.3.76) (which is twice the imaginary part), together with (5.4.112), and we take into account that $E(v) = v + \cdots$. For $p = 3$ we have $\xi = 4g^2$, and we obtain

$$\mathrm{Im}\, E_n(g) \sim \mp \frac{2^{3n}}{\sqrt{\pi}n!} g^{-2n-1} \exp\left(-\frac{2}{15g^2}\right). \qquad (5.4.120)$$

As an application of this formula, we can use the general theory developed in Section 2.9 to obtain the large-order behavior of the perturbative expansion for the energy levels of the cubic oscillator. By using standard stationary perturbation theory, the energy levels $E_n(g)$ have a formal expansion in powers of g^2:

$$E_n(g) \sim \sum_{k=0}^{\infty} E_{n,k} g^{2k}. \qquad (5.4.121)$$

This series is not Borel summable, but one can as usual perform lateral Borel resummations. In this way, the energy levels get an imaginary part. Therefore, the Borel resummation of the perturbative series reproduces the spectrum of resonances of the cubic oscillator. For example, for $g = 1/10$, the lateral Borel resummation of the first 75 terms of the perturbative series for $n = 0$ gives

$$E_0 \approx 0.484315997 - i8.0602 \cdot 10^{-6}, \qquad (5.4.122)$$

where we have chosen the negative sign for the imaginary part. This is precisely what is found with the complex dilatation method. The imaginary part of the lateral Borel resummation has the asymptotic form (5.4.120) for small g. Let us focus for simplicity on the resonance with $n = 0$. By comparing (5.4.120) to (2.9.350), we read

$$A = \frac{2}{15}, \qquad b = \frac{1}{2}, \qquad c_0 = \frac{2}{\sqrt{\pi}}. \qquad (5.4.123)$$

Therefore, according to (2.9.357), the coefficients $E_{0,k}$ have the asymptotic behavior

$$E_{0,k} \sim -\frac{1}{\pi^{3/2}} \left(\frac{2}{15}\right)^{-k-1/2} \Gamma\left(k + \frac{1}{2}\right), \qquad k \gg 1. \qquad (5.4.124)$$

This can be explicitly verified by a standard large-order analysis, similar to that undertaken in Figure 2.12 for the double-well potential.

A similar analysis can be done for the unstable potential (5.4.83) when p is even. The result for the imaginary part of the energy is as in (5.3.76), where $f(v)$ is given by (5.4.112) but with an extra factor of 2. For example, in the case of an inverted quartic oscillator, we find

$$\mathrm{Im}\, E_n(g) \sim \mp \frac{2^{2n+1}}{\sqrt{2\pi}n!} g^{-n-1/2} \exp\left(-\frac{1}{3g}\right). \qquad (5.4.125)$$

This formula can also be used to determine the large-order behavior of the perturbative series for the energy levels (inverted) quartic oscillator. In this case, we have a formal power series in powers of g,

$$E_n(g) \sim \sum_{k \geq 0} E_{n,k} g^k. \tag{5.4.126}$$

For the ground state, we find

$$E_{0,k} \sim -\frac{\sqrt{6}}{\pi^{3/2}} 3^k \Gamma\left(k + \frac{1}{2}\right), \qquad k \gg 1. \tag{5.4.127}$$

Clearly, the perturbative series for the conventional quartic oscillator (with potential $x^2/2 + g x^4$, $g > 0$) is obtained from this one by simply changing the sign of g. We obtain in this way the famous Bender–Wu formula

$$E_{0,k} \sim (-1)^{k+1} \frac{\sqrt{6}}{\pi^{3/2}} 3^k \Gamma\left(k + \frac{1}{2}\right), \qquad k \gg 1. \tag{5.4.128}$$

5.5 Analytic Continuation of Eigenvalue Problems

Resonances make it possible to define a spectral problem with an infinite tower of discrete eigenvalues in potentials that do not support bound states. One can extend this idea to more general potentials involving, for example, complex coupling constants, by using the procedure of analytic continuation. This allows one to define the eigenvalues as meromorphic functions of the coupling constant. Resonances and bound states are then particular cases of this general construction, which turns out to have a very rich physical and mathematical structure. The study of analytically continued spectral problems was pioneered by Bender and Wu in 1969.

To understand the procedure of analytic continuation, let us consider the quartic oscillator, following the original analysis of Bender and Wu. The Hamiltonian is given in (2.6.186), where we set $m = \omega = 1$ for simplicity. When $g > 0$, H has a real and discrete spectrum. It follows from (5.3.40) that the eigenfunctions behave at infinity as

$$\psi(x) \approx \exp\left[\mp \frac{1}{3}\sqrt{\frac{g}{2}} x^3\right], \qquad x \to \pm\infty. \tag{5.5.129}$$

In fact, they decrease exponentially in two wedges around the real axis defined by

$$|\arg(\pm x)| < \frac{\pi}{6}, \tag{5.5.130}$$

as shown in Figure 5.6. Let us now suppose that we start rotating g counterclockwise with an angle φ_g:

$$g = |g| e^{i\varphi_g}. \tag{5.5.131}$$

Then, the wedges where the WKB wavefunction decreases at infinity rotate clockwise, and are now defined by

$$\left|\arg(\pm x) + \frac{1}{6}\varphi_g\right| < \frac{\pi}{6}. \tag{5.5.132}$$

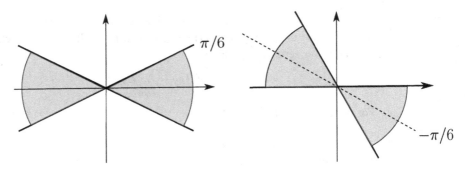

Figure 5.6 (Left) We depict the regions in the complex x plane where the eigenfunctions of the Hamiltonian (2.6.186) with $g > 0$ decay at infinity. As g becomes complex, we can define a spectral problem by requiring that the eigenfunctions decay at infinity in the rotated wedges (5.5.132) of the complex plane. (Right) The relevant wedges for $g < 0$.

The center of these rotated regions is at an angle

$$\varphi_x = -\frac{\varphi_g}{6}. \tag{5.5.133}$$

For example, if $\varphi_g = \pi$, we have to rotate the wedge region an angle $-\pi/6$, as shown in Figure 5.6. We can now consider the following spectral problem, for arbitrary complex g: the eigenfunctions $\psi(x)$ should be solutions of the Schrödinger equation for the quartic potential, and satisfy

$$\lim_{|x|\to\infty} \psi(x) = 0, \tag{5.5.134}$$

inside the wedge (5.5.132). This defines a discrete spectrum of eigenvalues $E_n(g)$, $n = 0, 1, 2, \ldots$, for any complex g.

In practice, the eigenvalues $E_n(g)$ can be computed by complex dilatation, by choosing the rotation angle θ appropriately as we vary the phase of g. The resulting function turns out to display nontrivial properties, as shown in Figure 5.7 for the ground state energy. First of all, there is a periodicity in φ_g of 6π. One also finds that

$$E_n\left(e^{3\pi i}|g|\right) = -E_n\left(|g|\right), \tag{5.5.135}$$

and that

$$\operatorname{Re} E_n\left(e^{3\pi i/2}|g|\right) = \operatorname{Re} E_n\left(e^{7\pi i/2}|g|\right) = 0. \tag{5.5.136}$$

These properties can be explained as follows. Let us consider a slightly more general Hamiltonian of the form

$$H(\alpha, \beta) = \frac{p^2}{2} + \frac{\alpha}{2}x^2 + \beta x^4, \tag{5.5.137}$$

and let us consider the following complex dilatation

$$x \to \lambda x, \qquad p \to \lambda^{-1} p, \tag{5.5.138}$$

which is of the form (5.2.23). Under this transformation, we have

$$H(\alpha, \beta) \to \lambda^{-2} H\left(\lambda^4 \alpha, \lambda^6 \beta\right). \tag{5.5.139}$$

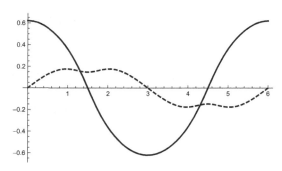

Figure 5.7   (Full line) The real part and (dashed line) the imaginary part of the energy of the ground state of the quartic anharmonic oscillator, as a function of φ_g/π, where φ_g is the argument of the complex coupling constant g. We take $\hbar = 1$, $|g| = 1/4$.

Therefore, if $E(\alpha, \beta)$ is an eigenvalue of (5.5.137), we have that

$$E(\alpha, \beta) = \lambda^{-2} E\left(\lambda^4 \alpha, \lambda^6 \beta\right). \tag{5.5.140}$$

By choosing λ appropriately, we deduce in particular that

$$E(1, \beta) = \beta^{1/3} E(\beta^{-2/3}, 1). \tag{5.5.141}$$

This means that, instead of studying the Hamiltonian (2.6.186) with complex g, we can study instead a Hamiltonian with real g but with a complex coupling α for the quadratic term x^2. This is more convenient in many ways, since one can study the problem in $L^2(\mathbb{R})$ without having to consider rotated contours for the decay of the eigenfunctions. We can now understand the 6π periodicity of $E_n(g)$ by taking $\beta = |g|e^{6\pi i}$ in (5.5.141). Similarly, the property (5.5.135) is obtained by taking $\beta = |g|e^{3\pi i}$, and the properties (5.5.136) are obtained by taking $\beta = |g|e^{3\pi i/2}$ and $\beta = |g|e^{7\pi i/2}$.

The above argument holds as long as one can analytically continue the functions $E_n(g)$ in the complex g plane along paths around the origin. This can be done if $|g|$ is large enough. However, when $|g|$ is sufficiently small, one encounters an infinite set of branch cut singularities of $E_n(g)$, which are called *Bender–Wu branch points*.

A similar analysis can be performed in the case of the cubic oscillator with Hamiltonian

$$H = \frac{p^2}{2} + \frac{kx^2}{2} - gx^3. \tag{5.5.142}$$

We will now assume that g is complex. How do we define a spectral problem in this case? As we saw in Section 5.3, the Schrödinger equation has two linearly independent solutions for any value of g and E. When $x \to +\infty$, the general solution $f(x)$ is a linear combination of the functions $\psi_\pm(x)$ in (5.3.52), of the form (5.3.51). To understand the behavior at $x \to -\infty$, it is useful to consider the function $\widehat{f}(x) = f(-x)$. This function can be written as

$$\widehat{f}(x) = D_+ \phi_+(x) + D_- \phi_-(x), \tag{5.5.143}$$

where

$$\phi_\pm(x) \approx \exp\left(\mp\frac{2}{5}g^{1/2}x^{5/2}\right), \qquad x \to \infty. \tag{5.5.144}$$

Let us analyze the decay properties of these functions when g is complex:

$$g = e^{i\phi_g}|g|. \tag{5.5.145}$$

The function $\psi_+(x)$ in (5.3.52) decays exponentially as $x \to \infty$ in the sector

$$0 < \frac{\phi_g}{2} + \frac{5}{2}\arg(x) < \pi, \tag{5.5.146}$$

around the positive real axis, while $\phi_+(x)$ decays exponentially as $x \to \infty$ in the sector

$$-\frac{\pi}{2} < \frac{\phi_g}{2} + \frac{5}{2}\arg(x) < \frac{\pi}{2}, \tag{5.5.147}$$

around the positive real axis. We will now impose that both $f(x)$ and $\widehat{f}(x)$ decay exponentially when $x \to +\infty$ inside the common sector

$$0 < \frac{\phi_g}{2} + \frac{5}{2}\arg(x) < \frac{\pi}{2}. \tag{5.5.148}$$

This requires that $\widehat{f}(x) = \phi_+(x)$. This fixes the constants $C_\pm(E)$ in (5.3.51) as functions of the energy. In addition, the decay of $f(x)$ when $x \to +\infty$ holds if and only if $C_-(E) = 0$. This leads to an infinite set of discrete (but in general complex) energy eigenvalues, which generalize the resonances of the cubic oscillator to arbitrary complex g.

5.6 Path Integral Formulation of Metastable States

As we have seen, resonances can be regarded as the appropriate generalization of bound states to problems in which strictly speaking there are no actual bound states. In some cases, like the quartic oscillator, resonances appear as we analytically continue the Hamiltonian to values of the coupling for which the Hamiltonian does not support bound states.

It is natural to ask how one can implement the procedure of analytic continuation to understand resonances directly in the path integral formulation of quantum mechanics. In a sense, this is even more natural than in the operator formalism, since the analytic continuation of ordinary integrals to complex values of their parameters is a well-understood process. One could suspect that path integrals can be analytically continued in similar ways, as we will indeed show.

We will then start our exploration of this question by considering ordinary integrals. More precisely, we will consider a toy model for the quartic potential, given by the integral

$$I(g) = \frac{1}{\sqrt{2\pi}} \int_{-\infty}^{+\infty} dz \, e^{-z^2/2 - gz^4/4}. \tag{5.6.149}$$

This integral exists as long as

$$\text{Re}(g) > 0, \tag{5.6.150}$$

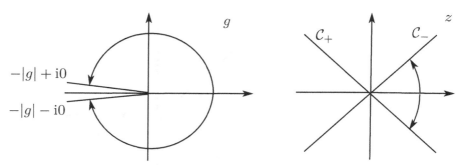

Figure 5.8 The integration contours $\mathcal{C}_\pm$ correspond to the negative values of $g = -|g| \pm i0$.

but we would like to define it for more general, complex values of g by a suitable analytic continuation. In particular we would like to make sense of (5.6.149) for negative values of g. To do this, we will *rotate the contour of integration* for the z variable in such a way that

$$\mathrm{Re}\,(gz^4) > 0, \qquad (5.6.151)$$

and the integral is still convergent. Equivalently, we give a phase to z in such a way that

$$\mathrm{Arg}\,z = -\frac{1}{4}\mathrm{Arg}\,g. \qquad (5.6.152)$$

Obviously, this analytic continuation of the integral is no longer real.

In order to define the integral for negative g, we should rotate g toward the negative real axis. Clearly, this can be done in *two* different ways: clockwise or counterclockwise, as shown in Figure 5.8. The integration contour for z rotates correspondingly. Since the resulting integration contours are complex conjugate to each other, the two integrals defined in this way are also complex conjugate. For $g \to -|g| + i0$, the integration contour is given by

$$\mathcal{C}_+ : \quad \mathrm{Arg}\,z = -\frac{\pi}{4}, \qquad (5.6.153)$$

while for $g \to -|g| - i0$, one has

$$\mathcal{C}_- : \quad \mathrm{Arg}\,z = \frac{\pi}{4}, \qquad (5.6.154)$$

(see Figure 5.8). This means that one can indeed obtain an analytic continuation of the integral $I(g)$ to negative g, but the resulting function will have a *branch cut* along the negative real axis. The discontinuity across the cut is given by

$$I(g + i0) - I(g - i0) = 2i\,\mathrm{Im}\,I(g) = \frac{1}{\sqrt{2\pi}}\int_{\mathcal{C}_+ - \mathcal{C}_-} dz\, e^{-z^2/2 - gz^4/4}. \qquad (5.6.155)$$

The discontinuity (5.6.155) can be computed using saddle-point methods. The saddle points of the integral occur at $z = 0$ or

$$z + gz^3 = 0 \Rightarrow z^2 = -\frac{1}{g}. \qquad (5.6.156)$$

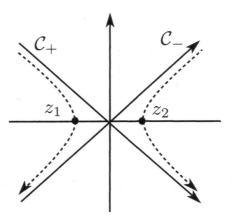

The complex plane for the saddle-point calculation of (5.6.149). Here, $\mathcal{C}^+$ and $\mathcal{C}^-$ are the rotated contours one needs to consider for $g < 0$. Their sum may be evaluated by the contribution of the saddle-point at the origin. Their difference is evaluated by the contribution of the subleading saddle points, here denoted as $\mathcal{S}_1$ and $\mathcal{S}_2$.

Therefore we have two nontrivial saddlepoints $z_{1,2}$

$$z_{1,2} = \mp e^{i(\pi/2 - \phi_g/2)}|g|^{-\frac{1}{2}}, \qquad (5.6.157)$$

where ϕ_g is the phase of g. For $g < 0$, they are on the real axis (see Figure 5.9). The steepest descent trajectories passing through these points are determined by the condition

$$\mathrm{Im}\, f(z) = \mathrm{Im}\, f(z_i), \qquad f(z) = \frac{z^2}{2} + \frac{g}{4} z^4. \qquad (5.6.158)$$

For $g < 0$ these are the hyperbolae

$$x^2 - y^2 = -\frac{1}{g}, \qquad (5.6.159)$$

passing through the saddlepoints $z_{1,2}$ (see Figure 5.9). From this figure it is also clear that the contour $\mathcal{C}_+ - \mathcal{C}_-$ appearing in (5.6.155) can be deformed into the *sum* of the steepest descent trajectories passing through $z_{1,2}$, therefore the imaginary part in (5.6.155) is given by

$$\mathrm{Im}\, I(g) \sim -\frac{1}{\sqrt{2}} \exp\left(\frac{1}{4g}\right), \qquad g \to 0^-. \qquad (5.6.160)$$

The overall factor in (5.6.160) is obtained by doing the Gaussian integrations around the two saddles and adding up the results. Since the integral (5.6.149) is divergent for $g < 0$, the resulting complex function cannot be analytic at $g = 0$. One consequence of this lack of analyticity is that the formal power series expansion of the integral (5.6.149) around $g = 0$, which is worked out in (A.11) and (A.12), has a zero radius of convergence.

The moral of this simple analysis is that, for negative g, the integral $I(g)$ picks an imaginary part that is given by the contribution of the nontrivial saddlepoints. By analogy with this integral, we expect that the path integral of the quartic oscillator will have a similar behavior. Therefore, we expect to find an imaginary part in the canonical partition function for negative coupling $g = -\lambda$, and we also expect this imaginary part to be exponentially suppressed at small λ, just as in (5.6.160). The free energy then reads, when expanded formally in Im Z,

$$F(\beta) = -\frac{1}{\beta} \log Z = -\frac{1}{\beta} \log(\mathrm{Re}\, Z) - \frac{\mathrm{i}}{\beta} \frac{\mathrm{Im} Z}{\mathrm{Re}\, Z} + \cdots, \qquad (5.6.161)$$

Therefore, at leading order in the exponentially suppressed factor, we have

$$\mathrm{Im}\, F(\beta) \approx -\frac{1}{\beta} \frac{\mathrm{Im}\, Z}{\mathrm{Re}\, Z}, \qquad (5.6.162)$$

and

$$\mathrm{Im}\, E(g) = \lim_{\beta \to \infty} \mathrm{Im}\, F(\beta) \approx -\lim_{\beta \to \infty} \frac{1}{\beta} \frac{\mathrm{Im}\, Z}{\mathrm{Re}\, Z}. \qquad (5.6.163)$$

Furthermore, as in (5.6.155), we expect that the quantity

$$\mathrm{disc}\, Z(-\lambda) = Z(-\lambda + \mathrm{i}\epsilon) - Z(-\lambda - \mathrm{i}\epsilon) = 2\mathrm{i}\, \mathrm{Im}\, Z(-\lambda), \qquad (5.6.164)$$

will be given by the sum of the contributions of the *nontrivial* saddle points of the path integral (4.3.37). We will now calculate these contributions.

We will consider quantum-mechanical potentials $V(q)$ that have a relative minimum at $q = 0$. Near this minimum, the potential will be quadratic. We will choose units in which $\hbar = m = 1$. We will also assume that the potential is unstable, as in Figure 5.10. We will study the imaginary part of the thermal free energy, $Z(\beta)$, at large β, in order to extract the imaginary part of the ground state resonance. To do that, we use the Euclidean path integral (4.3.37). The nontrivial saddle points of this path integral are time-dependent, *periodic* solutions of the EOM (4.7.145) for the inverted potential. We also want these solutions to have *finite* Euclidean action (otherwise the semiclassical contribution of such saddles vanishes.) Examples of such nontrivial, periodic saddle points are oscillations around the local minima of $V(q)$, as shown in Figure 5.10. The period of such an oscillation between the turning points q_- and q_+ is given by

$$\beta = 2 \int_{q_-}^{q_+} \frac{\mathrm{d}q}{\sqrt{2(E + V(q))}}. \qquad (5.6.165)$$

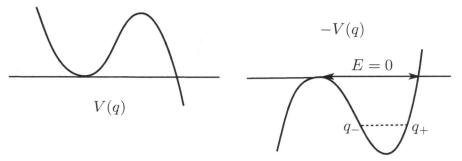

Figure 5.10 A general unstable potential $V(q)$ and the associated inverted potential $-V(q)$. A periodic solution with negative energy moves between the turning points $q_{\pm}$. The trajectory with zero energy, relevant for extracting the imaginary part of the ground state resonance, is also shown.

These trajectories also satisfy the "energy conservation" equation (4.11.328). These solutions are very similar to the instantons that we studied in Section 4.11. We will also refer to the above solutions of the Euclidean EOM as instantons, although in this unstable case some authors prefer to call them *bounces*. Notice that the period (5.6.165) varies between $\beta = \infty$ (corresponding to $E = 0$ in Figure 5.10) and a minimum critical value $\beta_c = 2\pi/\omega$ corresponding to small oscillations with frequency ω around the minimum q_0 of the potential.

Example 5.6.1 Let us consider the inverted quartic oscillator, with potential

$$V(q) = \frac{1}{2}q^2 - \frac{\lambda}{4}q^4, \qquad \lambda > 0, \tag{5.6.166}$$

(see Figure 5.11). We can change variables $q \to \lambda^{-1/2}q$, $E \to E/\lambda$ to set $\lambda = 1$, so that the EOM in the inverted potential reads

$$-\ddot{q}(\tau) + q(\tau) - q^3(\tau) = 0. \tag{5.6.167}$$

The inverted potential has minima at $q = \pm 1$ and zeros at $q = \pm\sqrt{2}$. We will focus on the region $q \geq 0$, since results in the region $q < 0$ follow by the symmetry $q \to -q$ of the problem. There is a solution to the EOM (5.6.167), with energy $-1/4 \leq E \leq 0$, and turning points at

$$q_\pm = \sqrt{1 \pm \sqrt{1 + 4E}}, \tag{5.6.168}$$

which can be written in terms of the Jacobi elliptic function $\mathrm{dn}(u; k)$:

$$q_c(\tau; \tau_0) = q_+ \mathrm{dn}(u; k), \tag{5.6.169}$$

where

$$u = \frac{q_+}{\sqrt{2}}(\tau - \tau_0), \qquad k^2 = 1 - \frac{q_-^2}{q_+^2}. \tag{5.6.170}$$

This solution has a free parameter or modulus τ_0, which corresponds to the initial point of the trajectory. This is similar to the modulus τ_0 appearing in the instanton

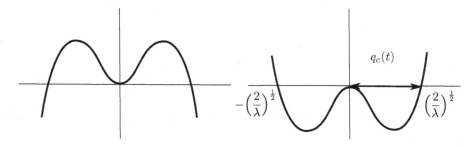

$$V(q) = -\frac{1}{2}q^2 + \frac{\lambda}{4}q^4$$

Figure 5.11 The inverted potential relevant for instanton calculus in the quartic case. The instanton or "bounce" configuration $q_c(\tau)$ leaves the origin at $\tau = -\infty$, reaches the zero $(2/\lambda)^{\frac{1}{2}}$ at $\tau = \tau_0$, and comes back to the origin at $\tau = +\infty$.

solution in (4.11.329). Since τ_0 is defined modulo the period of the motion, we can choose $\tau_0 \in [-\beta/2, \beta/2]$. The period β can be computed from (5.6.165), and is given by

$$\beta = 2\sqrt{2} \left(\frac{2 - k^2}{2}\right)^{1/2} K(k), \tag{5.6.171}$$

where $K(k)$ is the complete elliptic integral of the first kind. In terms of k, the energy reads

$$E = -\frac{1 - k^2}{(2 - k^2)^2}. \tag{5.6.172}$$

The value $k = 0$ corresponds to a particle with minimal energy $E = -1/4$, sitting at the bottom of $V(q)$, and the minimum period is $\beta_c = \sqrt{2}/\pi$ (the frequency of the oscillations around the bottom is $\omega = \sqrt{2}$). The limiting case $k \to 1$ corresponds to a particle with energy $E = 0$ and infinite period $\beta \to \infty$. In this limit the solution (5.6.169) simplifies: the Jacobi function $\mathrm{dn}(u; k)$ becomes $\mathrm{sech}(u)$, and we find

$$q_c(\tau; \tau_0) = \frac{\sqrt{2}}{\cosh(\tau - \tau_0)}. \tag{5.6.173}$$

This trajectory starts at the origin in the infinite past, arrives to the turning point q_+ at $\tau = \tau_0$, and returns to the origin in the infinite future. The trajectory (5.6.173) with $\tau_0 = 0$ is shown in Figure 5.12.   □

Let us now calculate the imaginary part of $Z(\beta)$ by expanding the action around a nontrivial classical solution $q_c(t)$. We write a general path, as in (4.7.146), and at leading order one finds the result (4.7.149). As we know from (4.8.227), the path, integral over fluctuations in (4.7.149) is given by

$$\frac{1}{\sqrt{\det(A)}}, \tag{5.6.174}$$

where A is the operator (4.11.330). However, we also know from the analysis in Section 4.11 that the operator A has a zero mode, and the associated eigenfunction is $\dot{q}_c(t)$. Therefore, the answer (5.6.174) has to be replaced by (4.11.357).

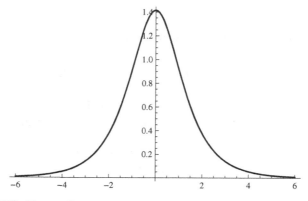

Figure 5.12   The solution (5.6.173) with $\tau_0 = 0$.

In the case of unstable potentials, there is a second subtlety in the calculation of the determinant that is crucial to the physics of the problem. Namely, the operator A has in this case one (and only one) negative eigenvalue. To see this, we note that A has a zero mode $\dot{q}_c(t)$. But $\dot{q}_c(t)$ changes sign at one of the turning points, and there should be an eigenfunction with a lower eigenvalue, which has to be negative. In the limit $\beta \to \infty$, this can be also established by regarding (4.11.330) as a one-dimensional Schrödinger operator. The spectrum of such an operator has the well-known property that the ground state has no nodes, the first excited state has one node, etc. The function $\dot{q}_c(t)$ has one node, so it is the first excited state of A and the ground state must have negative energy. This is the negative mode of A. Due to this negative mode, det' A is negative, and in extracting its square root we will obtain an imaginary result. This is consistent with the fact that this evaluation computes (twice) the imaginary part of the partition function. There is, in addition, a sign ambiguity corresponding to the choice of branch cut of the square root.

Example 5.6.2 Let us again consider the inverted quartic oscillator with potential (5.6.166), in the limit $\beta \to \infty$ and with $\lambda = 1$. The operator A is given in this case by

$$A = -\frac{d^2}{d\tau^2} + 1 - \frac{6}{\cosh^2(\tau - \tau_0)}. \tag{5.6.175}$$

Using translation invariance we can simply set $\tau_0 = 0$ to study the spectrum. It is easy to see that

$$A\psi(\tau) = -3\psi(\tau), \qquad \psi(\tau) = \frac{1}{\cosh^2(\tau)}, \tag{5.6.176}$$

which is the single negative mode of this operator, and $q_0(\tau) \propto \psi(\tau)$.   □

The evaluation of the path integral around $q_c(\tau)$ is very similar to the one leading to (4.11.361). In the limit $\beta \to 0$ we find

$$2i \operatorname{Im} Z \approx e^{-S_{E,c}} \beta Z_G(\beta) \frac{S_{E,c}^{1/2}}{\sqrt{2\pi}} \left[\frac{\det' A}{\det A_0} \right]^{-1/2}. \tag{5.6.177}$$

Since we have assumed that our potential is a perturbed quadratic potential, the real part of $Z(\beta)$ is given, at this order, by

$$\operatorname{Re} Z \approx Z_G(\beta). \tag{5.6.178}$$

Therefore, after we take the limit $\beta \to \infty$, or $E \to 0$, we find the following result:

$$\operatorname{Im} E \approx \pm \frac{S_{E,c}^{1/2}}{2\sqrt{2\pi}} e^{-S_{E,c}} \lim_{\beta \to \infty} \left[-\frac{\det' A}{\det A_0} \right]^{-1/2}. \tag{5.6.179}$$

This formula gives the imaginary part of the ground state energy for general unstable potentials obtained by perturbing a quadratic potential.

Example 5.6.3 *Cubic and quartic potentials.* Let us calculate (5.6.179) in detail for the cubic oscillator, with potential

$$V(x) = \frac{1}{2}x^2 - gx^3. \tag{5.6.180}$$

The turning points for the inverted potential are $q_- = 0$ and

$$q_+ = \frac{1}{2g}. \tag{5.6.181}$$

The instanton solution in the limit $\beta \to \infty$ can be easily found to be

$$q_c(\tau) = \frac{1}{2g \cosh^2\left(\frac{\tau}{2}\right)}, \tag{5.6.182}$$

with Euclidean action

$$S_{E,c} = 2 \int_0^{1/(2g)} (x^2 - 2gx^3)^{\frac{1}{2}}\,dx = \frac{2}{15g^2}. \tag{5.6.183}$$

The operator A reads

$$A = -\frac{d^2}{d\tau^2} + 1 - \frac{3}{\cosh^2\left(\frac{\tau}{2}\right)}. \tag{5.6.184}$$

After rescaling $\tau \to 2\tau$, we find

$$A = \frac{1}{4}A_{3,2}, \qquad A_0 = \frac{1}{4}A_{0,2}. \tag{5.6.185}$$

Therefore,

$$\frac{\det' A}{\det A_0} = 4\frac{\det' A_{3,2}}{\det A_{0,2}} = -\frac{1}{60}, \tag{5.6.186}$$

where we used (4.8.244) and (4.8.240). We then obtain,

$$\mathrm{Im}\, E_0(g) \sim \pm\frac{1}{\sqrt{\pi}\,g^2}e^{-2/(15g^2)}, \tag{5.6.187}$$

in the limit of small g. This agrees precisely with (5.4.120) for $n = 0$.

We can also consider the inverted quartic oscillator with potential (5.6.166). The Euclidean action evaluated on (5.6.173) is

$$S_{E,c} = 2 \int_0^{\sqrt{2/\lambda}} x\sqrt{1 - \frac{\lambda}{2}x^2}\,dx = -\frac{2\left(2 - \lambda x^2\right)^{3/2}}{3\sqrt{2}\lambda}\Bigg|_0^{\sqrt{2/\lambda}} = \frac{4}{3\lambda}. \tag{5.6.188}$$

The operator A appearing in (5.6.175) is the Pöschl–Teller operator $A_{2,1}$, and formula (4.8.239) gives

$$\frac{\det' A}{\det A_0} = -\frac{1}{12}. \tag{5.6.189}$$

Using (5.6.179) we find

$$\mathrm{Im}\, E_0(\lambda) \sim \pm\frac{4}{\sqrt{2\pi\lambda}}e^{-\frac{4}{3\lambda}}, \tag{5.6.190}$$

where we included an additional factor of two due to the fact that there are two instantons related by the symmetry $x \to -x$. The result (5.6.190) agrees with (5.4.125), after setting $g = \lambda/4$. $\qquad\qquad\square$

5.7 Bibliographical Notes

A good introduction to resonances and Gamow states can be found in de la Madrid and Gadella (2002). Resonances in the Pöschl–Teller potential are discussed in, for example, Çevik et al. (2016). A detailed analysis of resonances in the double delta potential can be found in Galindo and Pascual (1990). The potential (5.1.8) is discussed in Konishi and Paffuti (2009). Resonances are subtle, and they have led to an equally subtle mathematical theory sometimes known as the Aguilar–Balslev–Combes–Simon theory of resonances. An excellent introduction to this theory can be found in Hislop and Sigal (2012).

A classical review of the method of complex dilatation can be found in Reinhardt (1982). More recent accounts can be found in Konishi and Paffuti (2009) and Moiseyev (2011). The book Hislop and Sigal (2012) gives a mathematically precise account of the method.

The cubic oscillator and its resonances have been studied in many references. In the paper by Yaris et al. (1978) it was first noted that the method of complex dilatation can be used to calculate the resonances for the cubic oscillator. The results of Yaris et al. (1978) were extended and deepened by Álvarez (1988). The exposition in Section 5.4 is based on the work by Álvarez and Casares (2000a,b).

The analytic continuation of eigenvalue problems in anharmonic oscillators was pioneered by Bender and Wu (1969) and further clarified by Bender and Turbiner (1993). Mathematically rigorous results on this problem for the quartic oscillator can be found in Simon and Dicke (1970). In the case of the cubic oscillator, the spectral problem for the complex coupling constant was analyzed rigorously in Caliceti et al. (1980), and further clarified in Álvarez (1995).

Our exposition of the path integral method for metastable states follows the books by Zinn-Justin (1996) and Mariño (2015). The paper by Collins and Soper (1978) is also useful.

Appendix A Asymptotic Series and Borel Resummation

Let us consider a formal power series of the form,

$$\varphi(z) = \sum_{n=0}^{\infty} a_n z^n. \tag{A.1}$$

We stress that such a formal power series is not a function, but rather a convenient way of collecting the coefficients a_n. We say that this formal power series is asymptotic to the function $f(z)$, in the sense of Poincaré, if, for every N, the remainder after $N + 1$ terms of the series is much smaller than the last retained term as $z \to 0$. More precisely,

$$\lim_{z \to 0} z^{-N} \left(f(z) - \sum_{n=0}^{N} a_n z^n \right) = 0, \tag{A.2}$$

for all $N > 0$. In an asymptotic series, the remainder does not necessarily go to zero as $N \to \infty$ for a fixed z, in contrast to what happens in convergent series. Note that analytic functions might have asymptotic expansions. For example, the Stirling series for the Gamma function

$$\left(\frac{z}{2\pi} \right)^{1/2} \left(\frac{z}{e} \right)^{-z} \Gamma(z) = 1 + \frac{1}{12z} + \frac{1}{288z^2} + \cdots, \tag{A.3}$$

is an asymptotic series for $|z| \to \infty$.

In practice, asymptotic expansions are characterized by the fact that, as we vary N, the partial sums,

$$\varphi_N(z) = \sum_{n=0}^{N} a_n z^n, \tag{A.4}$$

will first approach the true value $f(z)$, and then, for N sufficiently large, will diverge. A natural question is then to find the partial sum that gives the best possible estimate of $f(z)$. To do this, one has to find the N that truncates the asymptotic expansion in an optimal way. This procedure is called *optimal truncation*. Usually, the way to find the optimal value of N is to retain terms up to the smallest term in the series, discarding all terms of higher degree. Let us assume (as it is the case in all interesting examples) that the coefficients a_n in (A.1) grow factorially at large n,

$$a_n \sim A^{-n} n!, \qquad n \gg 1. \tag{A.5}$$

The smallest term in the series, for a fixed $|z|$, is obtained by minimizing N in

$$\left| a_N z^N \right| = cN! \left| \frac{z}{A} \right|^N. \tag{A.6}$$

By using the Stirling approximation, we can rewrite this as

$$c \exp \left\{ N \left(\log N - 1 - \log \left| \frac{A}{z} \right| \right) \right\}. \tag{A.7}$$

This function has a saddle point at large N given by

$$N_* = \left| \frac{A}{z} \right|. \tag{A.8}$$

If $|z|$ is small, the optimal truncation can be performed at large values of N, but as $|z|$ increases, fewer and fewer terms of the series can be used. We can now estimate the error made in the optimal truncation by evaluating the next term in the asymptotics:

$$\epsilon(z) = C_{N_*+1} |z|^{N_*+1} \sim e^{-|A/z|}. \tag{A.9}$$

Therefore, the maximal "resolution" we can expect when we reconstruct a function $f(z)$ from an asymptotic expansion is of order $\epsilon(z)$. This type of ambiguity is sometimes called a *nonperturbative ambiguity*, since it is not seen in perturbation theory. Indeed, the exponential term

$$e^{-A/z}, \tag{A.10}$$

is not analytic at $z = 0$, and therefore it doesn't contribute to the perturbative series. We conclude that, in general, an asymptotic expansion does not determine the function $f(z)$ uniquely, and some additional information is required. Note that the absolute value of A gives the "strength" of the nonperturbative ambiguity.

It is instructive to see optimal truncation at work in a simple example. Let us consider the quartic integral $I(g)$ given in (5.6.149), which is well defined as long as $\mathrm{Re}(g) > 0$. We can find a formal power series expansion of $I(g)$ around $g = 0$ by simply expanding the quartic term inside the integrand and integrating term by term. We obtain,

$$\varphi(g) = \sum_{k=0}^{\infty} a_k g^k, \tag{A.11}$$

where

$$a_k = \frac{(-4)^{-k}}{\sqrt{2\pi}} \int_{-\infty}^{\infty} dz \frac{z^{4k}}{k!} e^{-z^2/2} = (-4)^{-k} \frac{(4k-1)!!}{k!}. \tag{A.12}$$

The formal power series (A.11) has *zero radius of convergence*. Its asymptotic behavior at large k is obtained immediately from Stirling's formula

$$a_k \sim (-4)^k k!. \tag{A.13}$$

It can be easily shown though that $\varphi(g)$ provides an asymptotic expansion of the integral $I(g)$. In view of (A.13) we have that $|A| = 1/4$. In Figure A.1 we plot the difference

$$|I(g) - \varphi_N(g)|, \tag{A.14}$$

as a function of N, for two values of g. The optimal values are seen to be $N_* = 12$ and $N_* = 5$, in agreement with the estimate (A.8).

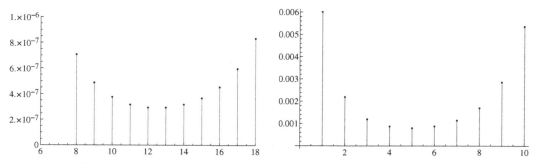

Figure A.1 We illustrate the method of optimal truncation for the quartic integral (5.6.149) by plotting the difference (A.14) between the integral and the partial sum of order N of its asymptotic expansion, as a function of N, for (left) $g = 0.02$ and (right) $g = 0.05$.

Optimal truncation gives a reasonable approximation to the original function for some values of the coupling constant, but it typically becomes a poor approximation for other values. In addition, in optimal truncation only a finite number of terms in the asymptotic expansion are actually used, and the remaining terms cannot be exploited to improve the approximation, as one does with convergent series. In fact, we can do better than optimal truncation and take into account the information contained in *all* the terms of the series. The way to do that is Borel resummation, which we now explain.

Let us consider a series (A.1), where the coefficients a_n behave like (A.5) when n is large (such series are sometimes called *Gevrey-1*). The *Borel transform* of φ, which we will denote by $\widehat{\varphi}(\zeta)$, is defined as the series

$$\widehat{\varphi}(\zeta) = \sum_{n=0}^{\infty} \frac{a_n}{n!} \zeta^n. \tag{A.15}$$

Notice that, due to (A.5), the series $\widehat{\varphi}(\zeta)$ has a finite radius of convergence $\rho = |A|$ and it defines an analytic function in the circle $|\zeta| < |A|$ (see Figure A.2). Typically, there is a singularity at $\zeta = A$, but very often the resulting function can be analytically continued to a wider region of the complex plane. We will now work out some examples of Borel transforms displaying two typical behaviors at the singularity: a simple pole and a logarithmic branch cut.

Example A.1 Let us consider the series

$$\varphi(z) = \sum_{n=0}^{\infty} (-1)^n n! \, z^n, \tag{A.16}$$

which corresponds to a growth of the form (A.5) with $A = -1$. In this case, the Borel transform is

$$\widehat{\varphi}(\zeta) = \sum_{n=0}^{\infty} (-1)^n \zeta^n, \tag{A.17}$$

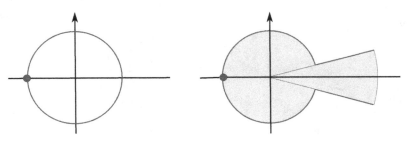

The Borel transform is analytic in a neighborhood of $\zeta = 0$, of radius $\rho = |A|$. Typically we encounter a singularity on the circle $|\zeta| = |A|$, but we can analytically continue the transform to a wider region.

which is a series with radius of convergence $\rho = 1$. However, it is an elementary fact that this series can be analytically continued to a meromorphic function with a single pole at $\zeta = -1$, namely

$$\widehat{\varphi}(\zeta) = \frac{1}{1 + \zeta}. \tag{A.18}$$

In this case, the singularity of the Borel transform is a pole at $\zeta = A = -1$.   □

Example A.2 Consider now the series

$$\varphi(z) = \sum_{k=0}^{\infty} \frac{\Gamma(k + b)}{\Gamma(b)} A^{-k} z^k, \tag{A.19}$$

where b is not an integer. The Borel transform is given by

$$\widehat{\varphi}(\zeta) = \sum_{k=0}^{\infty} \frac{\Gamma(k + b)}{k!\,\Gamma(b)} A^{-k} \zeta^k = (1 - \zeta/A)^{-b}, \tag{A.20}$$

which has a branch cut singularity at $\zeta = A$. Similarly, the series

$$\varphi(z) = \sum_{k=1}^{\infty} \Gamma(k) A^{-k} z^k, \tag{A.21}$$

leads to the Borel transform

$$\widehat{\varphi}(\zeta) = -\log\left(1 - \zeta/A\right), \tag{A.22}$$

which has a logarithmic singularity at $\zeta = A$. This can be regarded as the $b = 0$ case of (A.19).   □

Let us now suppose that the Borel transform $\widehat{\varphi}(\zeta)$ has an analytic continuation to a neighborhood of the positive real axis, in such a way that the Laplace transform

$$s(\varphi)(z) = \int_0^{\infty} e^{-\zeta} \widehat{\varphi}(z\zeta)\, d\zeta, \tag{A.23}$$

exists in some region of the complex z-plane. In this case, we say that the series $\varphi(z)$ is *Borel summable* and $s(\varphi)(z)$ is called the *Borel sum* of $\varphi(z)$. Notice that, by construction, $s(\varphi)(z)$ has an asymptotic expansion around $z = 0$ that coincides with the original series $\widehat{\varphi}(\zeta)$, since

$$s(\varphi)(z) \sim \sum_{n \geq 0} \frac{a_n z^n}{n!} \int_0^\infty d\zeta\, e^{-\zeta} \zeta^n = \sum_{n \geq 0} a_n z^n. \tag{A.24}$$

This procedure makes it possible, in principle, to reconstruct a well-defined function $s(\varphi)(z)$ from the asymptotic series $\varphi(z)$ (at least for some values of z). In many cases of interest in physics, the formal series $\varphi(z)$ is the asymptotic expansion of a well-defined function $f(z)$ (for example, $f(z)$ might be the ground state energy of a quantum system as a function of the coupling z, while $\varphi(z)$ is its asymptotic expansion.) It might then happen that the Borel resummation $s(\varphi)(z)$ agrees with the original function $f(z)$, and in this favorable case, the Borel resummation reconstructs the original nonpertubative answer. There are analyticity conditions on the function $f(z)$ that guarantee that $s(\varphi)(z) = f(z)$ in some region. However, in practice, it is not always easy to use these conditions, since they require detailed information about $f(z)$.

Example A.3 In Example A.1, the Borel transform extends to an analytic function on $\mathbb{C}\backslash\{-1\}$, and the integral (A.23) is

$$s(\varphi)(z) = \int_0^\infty d\zeta\, e^{-\zeta} \frac{1}{1 + z\zeta}, \tag{A.25}$$

which exists for all $z \geq 0$. .   □

In practice, even if the series $\varphi(z)$ is Borel summable, one only knows a few coefficients in its expansion, and this makes it very difficult to analytically continue the Borel transform to a neighborhood of the positive axis. We need a practical method to find accurate approximations to the resulting function. A useful method is to use *Padé approximants*. Given a series

$$\varphi(z) = \sum_{k=0}^\infty a_k z^k, \tag{A.26}$$

its Padé approximant $[l/m]_\varphi$, where l, m are positive integers, is the rational function

$$[l/m]_\varphi(z) = \frac{p_0 + p_1 z + \cdots + p_l z^l}{q_0 + q_1 z + \cdots + q_m z^m}, \tag{A.27}$$

where q_0 is fixed to 1, and one requires that

$$\varphi(z) - [l/m]_\varphi(z) = \mathcal{O}(z^{l+m+1}). \tag{A.28}$$

This fixes the coefficients involved in (A.27).

Given a series $\varphi(z)$, we can use Padé approximants to reconstruct the analytic continuation of its Borel transform. There are various methods to do this, but one simple approach is to use the following Padé approximant,

$$\mathcal{P}_n^\varphi(\zeta) = [[n/2]/[(n+1)/2]]_{\widehat{\varphi}}(\zeta), \tag{A.29}$$

Table A.1. The Borel resummations (A.30), evaluated for the asymptotic series (A.11) to the quartic integral $I(g)$ (5.6.149), for two values of $g = 0.2$ and $g = 0.4$, and for increasing values of n. In the last line we give the numerical result for $I(g)$. All numbers are presented with 10 significant digits. For each value of n, we underline the digits in the result of (A.30) that agree with the numerical result.

n	$s(\varphi)_n(0.2)$	$s(\varphi)_n(0.4)$
10	0.90798_54376	0.8576207823
20	0.9079847776	0.8576086008
30	0.9079847774	0.8576085854
$I(g)$	0.9079847774	0.8576085853

which requires knowledge of the first $n + 1$ coefficients of the original series. The integral

$$s(\varphi)_n(z) = \int_0^\infty d\zeta \, e^{-\zeta} \mathcal{P}_n^\varphi(z\zeta), \qquad (A.30)$$

gives an approximation to the Borel resummation of the series (A.23), which can be systematically improved by increasing n. The combination of Borel resummations with Padé approximants to calculate Borel transforms is often called the *Borel–Padé resummation method*.

Example A.4 *The quartic integral.* A simple example of this procedure is again the quartic integral (5.6.149). The Borel transform of the series (A.11) is given by

$$\widehat{\varphi}(\zeta) = \frac{2K(k)}{\pi(1 + 4\zeta)^{1/4}}, \qquad k^2 = \frac{1}{2} - \frac{1}{2\sqrt{1 + 4\zeta}}. \qquad (A.31)$$

where $K(k)$ is the elliptic integral of the first kind. This function has a branch point at $\zeta = A = -1/4$. The presence of a singularity at this point in the ζ-plane can be also deduced by comparing the asymptotic growth (A.13) with the expression (A.5), and by remembering that the value of A gives the location of the singularity. We can compute (A.30) for increasing values of n and verify that the results give increasingly accurate approximations to the quartic integral (5.6.149): see Table A.1. In this case, the Borel resummation reproduces the original nonperturbative object (5.6.149). On the other hand, for these values of the coupling, optimal truncation is not very good: the best approximation comes from keeping just the first two terms in the series, i.e. $N = 1$, and one finds

$$\varphi_1(0.2) = 0.85, \qquad \varphi_1(0.4) = 0.7. \qquad (A.32)$$

□

Suppose now that $\widehat{\varphi}(\zeta)$ has a singularity at the point $\zeta = \zeta_\omega$ in the complex plane, and let $\theta = \arg(\zeta_\omega)$. Then, the Borel resummation (A.23) is ill-defined when

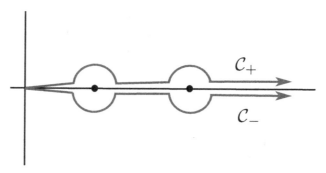

Figure A.3   The paths $C_\pm$ extend in the complex plane slightly above (respectively, below) the positive real axis. In this way they avoid the singularities of the Borel transform.

$\arg(z) = \theta$. A typical example occurs when the singularity of $\widehat{\varphi}(\zeta)$ is on the positive real axis, and then the Borel resummation is not well defined for $z > 0$. One can however define a generalized Borel resummation by deforming the contour in (A.23) appropriately. A useful choice is to consider contours $C_\pm^\theta$ that avoid the singularities by following paths slightly above or below the ray $e^{i\theta}\mathbb{R}_+$. When $\theta = 0$, these paths, which we will simply denote by $C_\pm$, are shown in Figure A.3. We define then the *lateral Borel resummations* by

$$s_\pm(\varphi)(z) = z^{-1} \int_{C_\pm^\theta} d\zeta \, e^{-\zeta/z} \widehat{\varphi}(\zeta). \tag{A.33}$$

One can also use the method of Padé approximants to calculate these lateral resummations with high precision. Due to the singularity along the ray, the lateral resummations are different, and their difference is encoded in the discontinuity

$$\mathrm{disc}(\varphi)(z) = s_+(\varphi)(z) - s_-(\varphi)(z). \tag{A.34}$$

Note that, even if all the coefficients of the original series are real, the lateral Borel resummations along the real axis are in general complex, due to the contour deformation. The discontinuity is in that case purely imaginary.

Example A.5   The quartic integral (5.6.149) leads to a divergent series whose Borel transform has a singularity at $\zeta = -1/4$, i.e. on the negative real axis. The discontinuity $\mathrm{disc}(\varphi)(-g)$, with $g > 0$ is given by (5.6.160).   □

We will now show that the discontinuity (A.34) can be computed in terms of the behavior of the Borel transform near the singularity. We will consider the case in which the singularity occurs at $\zeta = A > 0$ and it is a branch cut, of the form

$$\widehat{\varphi}(A + \xi) = (-\xi)^{-b} \sum_{n \geq 0} \hat{c}_n \xi^n + \cdots, \tag{A.35}$$

where the dots indicate non-singular terms. The Borel transform in (A.20) is a particular example of this, more general, form. We can now plug this expression in

$$s_+(\varphi)(z) - s_-(\varphi)(z) = z^{-1} \oint_\gamma e^{-\zeta/z} \widehat{\varphi}(\zeta) d\zeta, \tag{A.36}$$

where $\gamma = C_+ - C_-$ is a contour that can be deformed around the singularity/branch cut of $\varphi(z)$ at $\zeta = A$. The nonsingular terms do not contribute to the contour integral, and after setting $\zeta = A + \xi$ we have to evaluate

$$\oint_\gamma (-\xi)^{-b} \xi^n e^{-\xi/z} d\xi = \left(e^{\pi i b} - e^{-\pi i b} \right) z^{n-b+1} \int_0^\infty u^{n-b} e^{-u} du$$

$$= 2i \sin(\pi b) z^{n-b+1} \Gamma(n + 1 - b), \tag{A.37}$$

where the discontinuity of $(-\xi)^{-b}$ leads to the difference of phases in the second term of the equation. We conclude that

$$\mathrm{disc}(\varphi)(z) \sim 2i \sin(\pi b)\, e^{-A/z} z^{-b} \sum_{n=0}^\infty c_n z^n, \tag{A.38}$$

as an asymptotic expansion near $z = 0$, where

$$c_n = \Gamma(n + 1 - b)\hat{c}_n. \tag{A.39}$$

A similar calculation can be done when the singularity is a simple pole or a logarithmic branch cut: if

$$\widehat{\varphi}(A + \xi) = -\frac{a}{\xi} - \log(\xi) \sum_{n \geq 0} \hat{c}_n \xi^n + \cdots, \tag{A.40}$$

we find

$$\mathrm{disc}(\varphi)(z) = 2\pi i\, e^{-A/z} \left(\frac{a}{z} + \sum_{n=0}^\infty c_n z^n \right), \tag{A.41}$$

where c_n, $\hat{c}_n$ are related by (A.39) for $b = 0$. Note that, due to this relationship, the series appearing in (A.40) is the Borel transform of the series appearing in the discontinuity. This observation can be extended to the case with arbitrary b by defining the Borel transform of a series with the structure

$$\varphi(z) = \sum_{n \geq 0} a_n z^{n-b}, \tag{A.42}$$

by

$$\widehat{\varphi}(\zeta) = \sum_{n \geq 0} \frac{a_n}{\Gamma(n + 1 - b)} \zeta^{n-b}. \tag{A.43}$$

In the above calculation we have implicitly assumed that $\widehat{\varphi}(\zeta)$ has a single singularity, at $\zeta = A$. However, in general, there will be singularities at $\zeta = A_1, A_2, \ldots$. The discontinuity will then have contributions from all these singularities, which will lead to exponentially small effects of the form $e^{-A_k/z}$. If all the A_k are strictly positive, the largest nonperturbative effect will correspond to the singularity that is closest to the origin.

Appendix B Special Functions

B.1 Airy Functions

Useful resources on the Airy function can be found in Miller (2006); Bender and Orszag (2013); Silverstone et al. (1985b). The Airy functions are obtained as solutions to the differential equation

$$y''(x) = xy(x). \tag{B.1}$$

A very useful approach to understanding these solutions is to write them as Fourier–Laplace integrals:

$$y(x) = \frac{1}{2\pi i} \int_C dz \, e^{xz} q(z). \tag{B.2}$$

Substituting this ansatz into (B.1) we find

$$\frac{1}{2\pi i} \int_C dz \, e^{xz} z^2 q(z) - \frac{1}{2\pi i} \int_C dz \, e^{xz} xq(z) = 0, \tag{B.3}$$

and after integrating by parts and assuming that the boundary terms vanish, we find a solution whenever

$$q'(z) + z^2 q(z) = 0. \tag{B.4}$$

This leads to

$$q(z) = e^{-z^3/3}, \tag{B.5}$$

up to an overall additive constant. For the boundary term to vanish, we need

$$z \to \infty, \qquad \mathrm{Re}\left(-\frac{z^3}{3}\right) < 0. \tag{B.6}$$

Any contour that satisfies these conditions at their asymptotic endpoints leads to a solution to the Airy equation via the Fourier–Laplace integral (B.2). However, contours that can deformed into one another without going into regions in which the integral diverges lead to the same solution. A basis of solutions is then obtained by using the contours γ_j, $j = 1, 2, 3$, shown in Figure B.1. The corresponding functions are given by

$$w_j(x) = \frac{1}{2\pi i} \int_{\gamma_j} e^{xz - z^3/3} \, dz. \tag{B.7}$$

However, these contours are not independent, since

$$\gamma_1 + \gamma_2 + \gamma_3 = 0, \tag{B.8}$$

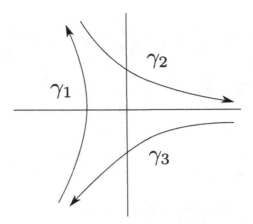

Figure B.1   Three contours that lead to solutions of the Airy equation.

and this translates into

$$w_1(x) + w_2(x) + w_3(x) = 0. \tag{B.9}$$

This leads to two linearly independent solutions to (B.1), as expected from a second-order ODE. The traditional basis of solutions consists of the functions $\mathrm{Ai}(x)$ and $\mathrm{Bi}(x)$, which are defined as

$$\begin{aligned} \mathrm{Ai}(x) &= w_1(x), \\ \mathrm{Bi}(x) &= \mathrm{i}(w_2(x) - w_3(x)). \end{aligned} \tag{B.10}$$

We can deform the contour γ_1 into the imaginary axis, and in this way one obtains the representation for $\mathrm{Ai}(x)$ in (3.7.294).

Let us denote $\omega = \mathrm{e}^{2\pi\mathrm{i}/3}$. The paths γ_j, $j = 1, 2, 3$, are related by rotations of ω, namely

$$\omega\gamma_1 = \gamma_3, \qquad \omega\gamma_3 = \gamma_2. \tag{B.11}$$

It follows that

$$w_1(\omega x) = \omega^{-1}w_3(x), \qquad w_1(\omega^2 x) = \omega^{-2}w_2(x). \tag{B.12}$$

We can then write (B.9) as a relationship for the Ai function,

$$\mathrm{Ai}(x) + \omega\,\mathrm{Ai}(\omega x) + \omega^2\,\mathrm{Ai}(\omega^2 x) = 0, \tag{B.13}$$

which can be also written as

$$\mathrm{Ai}(-x) = \mathrm{e}^{\pi\mathrm{i}/3}\,\mathrm{Ai}\left(\mathrm{e}^{\pi\mathrm{i}/3}x\right) + \mathrm{e}^{-\pi\mathrm{i}/3}\,\mathrm{Ai}\left(\mathrm{e}^{-\pi\mathrm{i}/3}x\right). \tag{B.14}$$

It also follows from the above considerations that

$$\mathrm{Bi}(x) = \mathrm{e}^{\pi\mathrm{i}/6}\,\mathrm{Ai}\left(\mathrm{e}^{2\pi\mathrm{i}/3}x\right) + \mathrm{e}^{-\pi\mathrm{i}/6}\,\mathrm{Ai}\left(\mathrm{e}^{-2\pi\mathrm{i}/3}x\right). \tag{B.15}$$

Another consequence of the above formulae is

$$\mathrm{Bi}(x) = \pm\mathrm{i}\mathrm{Ai}(x) + 2\mathrm{e}^{\mp\mathrm{i}\pi/6}\,\mathrm{Ai}\left(x\mathrm{e}^{\mp2\pi\mathrm{i}/3}\right). \tag{B.16}$$

The asymptotics of $\mathrm{Ai}(x)$ as $|x| \to \infty$, in an angular sector around the positive real axis, can be obtained by using a saddle-point analysis of the integral $w_1(x)$. We set

$$x = r e^{i\kappa} \tag{B.17}$$

and change variables in the integral to

$$z = u r^{1/2}. \tag{B.18}$$

The Airy function can then be written as

$$\mathrm{Ai}(x) = \frac{r^{1/2}}{2\pi i} \int_{\gamma_1} e^{\lambda S_\kappa(u)} du, \tag{B.19}$$

where $\lambda = r^{3/2}$ and

$$S_\kappa(u) = e^{i\kappa} u - \frac{u^3}{3}. \tag{B.20}$$

There are two saddle points:

$$u_0^R = e^{i\kappa/2}, \qquad u_0^L = -e^{i\kappa/2}, \tag{B.21}$$

with

$$S_\kappa(u_0^R) = \frac{2}{3} e^{3i\kappa/2}, \qquad S_\kappa(u_0^L) = -\frac{2}{3} e^{3i\kappa/2}. \tag{B.22}$$

When the angle κ satisfies

$$-\frac{2\pi}{3} < \kappa < \frac{2\pi}{3}, \tag{B.23}$$

the path γ_1 can be deformed into a path of steepest descent through the saddle point at u_0^L. One obtains in this way the leading-order asymptotic behavior for the Airy function:

$$\mathrm{Ai}(x) \sim \frac{1}{2x^{1/4}\sqrt{\pi}} e^{-2x^{3/2}/3}, \qquad |\arg(x)| < \frac{2\pi}{3}. \tag{B.24}$$

In order to obtain the complete asymptotic expansion for $|x| \gg 1$, we solve the original ODE (B.1) with the ansatz

$$y(x) = x^{-1/4} e^{-2x^{3/2}/3} w(x), \tag{B.25}$$

where $w(x)$ is of the form

$$w(x) \sim \sum_{n \geq 0} a_n x^{\alpha n}, \tag{B.26}$$

and $a_0 = 1$, $\alpha < 0$. Then, $w(x)$ solves the equation

$$x^2 w''(x) - \left(2x^{5/2} + \frac{1}{2}x\right) w'(x) + \frac{5}{16} w(x) = 0. \tag{B.27}$$

One finds that

$$\alpha = -\frac{3}{2}, \tag{B.28}$$

and the coefficients a_n satisfy the recursion

$$a_{n+1} = -\frac{3}{4} \frac{(n + 5/6)(n + 1/6)}{n + 1} a_n, \qquad n = 0, 1, \ldots. \tag{B.29}$$

The solution of this recursion relation is

$$a_n = \frac{1}{2\pi} \left(-\frac{3}{4}\right)^n \frac{\Gamma(n + 5/6)\Gamma(n + 1/6)}{n!}. \tag{B.30}$$

We can now write the full asymptotic expansion of $Ai(x)$. Let us consider the following asymptotic power series,

$$\beta(\zeta) = e^\zeta \sum_{k \geq 0} c_k \zeta^{-k}, \qquad c_k = \frac{1}{2\pi} \frac{\Gamma(k + 5/6)\Gamma(k + 1/6)}{2^k k!}. \tag{B.31}$$

Then,

$$Ai(x) \sim \frac{1}{2\sqrt{\pi} x^{1/4}} \beta(-\zeta), \qquad |\arg(x)| < \frac{2\pi}{3}, \tag{B.32}$$

where

$$\zeta = \frac{2}{3} x^{3/2}. \tag{B.33}$$

As we move in the x-plane towards the negative real axis, the asymptotic expansion changes and one has to consider the contribution of both saddle points. As a result, the asymptotic expansion becomes oscillatory. One finds,

$$Ai(-x) \sim \frac{1}{2\sqrt{\pi} x^{1/4}} \left(e^{-\pi i/4} \beta(i\zeta) + e^{\pi i/4} \beta(-i\zeta)\right), \qquad |\arg(x)| < \frac{\pi}{3}. \tag{B.34}$$

In addition, if we perform a Borel resummation of the series in the r.h.s., we obtain an equality (see e.g. Silverstone et al. [1985b]; Delabaere [2006]). There are similar asymptotic expansions for Bi. We have,

$$Bi(x) \sim \frac{1}{2\sqrt{\pi} x^{1/4}} \left(2\beta(\zeta) \pm i\beta(-\zeta)\right), \qquad |\arg(x)| < \frac{2\pi}{3},$$

$$Bi(-x) \sim \frac{1}{2\sqrt{\pi} x^{1/4}} \left(e^{\pi i/4} \beta(i\zeta) + e^{-\pi i/4} \beta(-i\zeta)\right), \qquad |\arg(x)| < \frac{\pi}{3}. \tag{B.35}$$

In the first line, the equality obtained after Borel resummation has to be understood in the following sense. The asymptotic series $\beta(\zeta)$ is not Borel summable in the sector that we have written down, since $\arg(z) = 0$ is a Stokes line. However, one can use the lateral resummations of Borel–Écalle theory. These resummations lead to an imaginary part that is precisely cancelled by the (standard) Borel resummation of $i\beta(-\zeta)$. The $\pm$ sign depends on whether we perform the lateral resummation below or above the positive real axis (i.e. the paths $\mathcal{C}_\pm$). The same result is obtained if we perform an analytic continuation of the Borel transform of $Bi(z)$ from $\arg(z) = \pi$, to the positive real axis. The process of analytic continuation is equivalent to rotating the contour in the Borel integral, from the negative real axis to the positive real axis. Depending on the sign of $\arg(z)$ we get the two different answers, correlated with the two different lateral resummations. We have that $\arg(z) > 0$ (respectively, $\arg(z) < 0$) corresponds to $\mathcal{C}_-$ (respectively, $\mathcal{C}_+$).

A useful consequence of the above asymptotic formulae is the Wronksian identity

$$\text{Ai}(x)\text{Bi}'(x) - \text{Ai}'(x)\text{Bi}(x) = \frac{1}{\pi}. \tag{B.36}$$

The following combinations of Airy functions,

$$\text{Ai}^{(\pm)}(x) = \text{Bi}(x) \pm \text{Ai}(x), \tag{B.37}$$

is sometimes needed. From (B.32) one deduces the following leading-order asymptotic behavior along the negative real axis:

$$\text{Ai}^{(\pm)}(-z) \approx \frac{1}{\sqrt{\pi}} z^{-1/4} \exp\left\{\pm i\left(\zeta + \frac{\pi}{4}\right)\right\}. \tag{B.38}$$

The asymptotic behavior of $\text{Ai}^{(+)}(z)$ in a sector adjacent to the positive real axis, but with a negative argument, is as follows,

$$\text{Ai}^{(+)}(z) \sim \frac{1}{\sqrt{\pi}} z^{-1/4} e^{2z^{3/2}/3} {}_2F_0\left(\frac{1}{6}, \frac{5}{6}; ; \frac{3}{4} z^{-3/2}\right), \qquad -\frac{2\pi}{3} < \arg(z) < 0. \tag{B.39}$$

To write down these asymptotics, we have introduced the following formal series

$$_2F_0(a, b; ; z) = \sum_{k \geq 0} (a)_k (b)_k \frac{z^k}{k!}, \tag{B.40}$$

where $(x)_k$ is the Pochhammer symbol. They are sometimes known as generalized hypergeometric functions. For positive argument we have instead,

$$\text{Ai}^{(+)}(z) \sim \frac{1}{\sqrt{\pi}} z^{-1/4} e^{2z^{3/2}/3} {}_2F_0\left(\frac{1}{6}, \frac{5}{6}; ; \frac{3}{4} z^{-3/2}\right)$$

$$+ \frac{i}{\sqrt{\pi}} z^{-1/4} e^{-2z^{3/2}/3} {}_2F_0\left(\frac{1}{6}, \frac{5}{6}; ; -\frac{3}{4} z^{-3/2}\right), \qquad 0 < \arg(z) < \frac{2\pi}{3}. \tag{B.41}$$

When $\arg(z) = 0$, the two asymptotic expansions (B.39), (B.41) correspond to the two choices of lateral Borel resummation.

B.1.1 Parabolic Cylinder Functions

The *parabolic cylinder functions* $D_\nu(x)$ are solutions to the ODE

$$f''(x) + \left(\nu + \frac{1}{2} - \frac{x^2}{4}\right) f(x) = 0, \tag{B.42}$$

(see e.g. Lebedev [1972] for a detailed treatment). In this book we will only need their asymptotic expansions. The expansion of $D_{\nu-1/2}(x)$ in a sector that includes the positive real axis is given by

$$D_{\nu-\frac{1}{2}}(z) \sim e^{-z^2/4} z^{-\frac{1}{2}+\nu} {}_2F_0\left(\frac{1}{4} - \frac{\nu}{2}, \frac{3}{4} - \frac{\nu}{2}; ; -2z^{-2}\right), \qquad |\arg(z)| < \frac{\pi}{2}. \tag{B.43}$$

The Borel resummation of this asymptotic expansion reconstructs the original function. On the other hand, the asymptotic expansion in a sector including the negative real axis is given by

$$D_{\nu-\frac{1}{2}}(-z) \sim \frac{\sqrt{2\pi}}{\Gamma\left(\frac{1}{2}-\nu\right)} e^{z^2/4} z^{-\frac{1}{2}-\nu} \, {}_2F_0\left(\frac{1}{4}+\frac{\nu}{2}, \frac{3}{4}-\frac{\nu}{2}; ; 2z^{-2}\right)$$
$$\pm \, \mathrm{i} e^{\mp \mathrm{i}\pi\nu} e^{-z^2/4} z^{\nu-1/2} \, {}_2F_0\left(\frac{1}{4}-\frac{\nu}{2}, \frac{3}{4}+\frac{\nu}{2}; ; -2z^{-2}\right), \; 0 < \pm\arg(z) < \frac{\pi}{2}.$$
$$\text{(B.44)}$$

When $\arg(z) = 0$, the sign $\pm$ corresponds to the choice of lateral Borel resummation of the generalized hypergeometric function in the first line of (B.44).

Appendix C Gaussian Integration

In many calculations one needs the Gaussian integral

$$\int_{\mathbb{R}^n} e^{\frac{i}{2}q^T A q + i p \cdot q} \, dq = e^{\frac{\pi i (\nu_+ - \nu_-)}{4}} \frac{(2\pi)^{n/2}}{\sqrt{|\det(A)|}} e^{-\frac{i}{2} p^T A^{-1} p},$$ (C.1)

where $\nu_\pm$ are the number of positive and negative eigenvalues of the matrix A, respectively. This formula can be obtained by starting from the basic Fresnel integral,

$$\int_{\mathbb{R}} e^{i a x^2} dx = \exp\left(\frac{\pi i \, \text{sgn}(a)}{4}\right) \sqrt{\frac{\pi}{|a|}},$$ (C.2)

and diagonalizing A. Since $\nu_+ + \nu_- = n$, we can write the phase in (C.1) as

$$\exp\left[\frac{\pi i n}{4} - \frac{\pi i \nu_-}{2}\right].$$ (C.3)

The r.h.s. of (C.1) is sometimes written as

$$\frac{(2\pi i)^{n/2}}{\sqrt{\det(A)}} e^{-\frac{i}{2} p^T A^{-1} p},$$ (C.4)

where the i in the numerator leads to the first phase in (C.3), while the square root of the determinant leads to the second phase.

We also have the "Euclidean" version,

$$\int_{\mathbb{R}^n} e^{-\frac{1}{2} q^T A q + p \cdot q} \, dq = \frac{(2\pi)^{n/2}}{\sqrt{\det(A)}} e^{\frac{1}{2} p^T A^{-1} p},$$ (C.5)

where the matrix A is positive definite.

References

Ahlfors, L. V. 1966. *Complex analysis*. McGraw Hill.

Álvarez, G. 1988. Coupling-constant behavior of the resonances of the cubic anharmonic oscillator. *Phys. Rev.*, **A37**(11), 4079.

Álvarez, G. 1995. Bender-Wu branch points in the cubic oscillator. *J. Phys.*, **A28**(16), 4589.

Álvarez, G. 2004. Langer–Cherry derivation of the multi-instanton expansion for the symmetric double well. *J. Math. Phys.*, **45**(8), 3095–3108.

Álvarez, G., and Casares, C. 2000a. Exponentially small corrections in the asymptotic expansion of the eigenvalues of the cubic anharmonic oscillator. *J. Phys.*, **A33**(29), 5171.

Álvarez, G., and Casares, C. 2000b. Uniform asymptotic and JWKB expansions for anharmonic oscillators. *J. Phys.*, **A33**(13), 2499.

Álvarez, G., Howls, C. J., and Silverstone, H. J. 2002. Anharmonic oscillator discontinuity formulae up to second-exponentially-small order. *J. Phys.*, **A35**(18), 4003.

Andreassen, A., Farhi, D., Frost, W., and Schwartz, M. D. 2017. Precision decay rate calculations in quantum field theory. *Phys. Rev.*, **D95**(8), 085011.

Arnold, V. I. 1989. *Mathematical methods of classical mechanics*. Springer.

Baaquie, B. E. 2014. *Path integrals and Hamiltonians. Principles and methods*. Cambridge University Press.

Babelon, O., Bernard, D., and Talon, M. 2003. *Introduction to classical integrable systems*. Cambridge Monographs on Mathematical Physics. Cambridge University Press.

Balian, R., Parisi, G., and Voros, A. 1978. Discrepancies from asymptotic series and their relation to complex classical trajectories. *Phys. Rev. Lett.*, **41**, 1141–1144.

Balian, R., Parisi, G., and Voros, A. 1979. Quartic oscillator. Pages 337–360 of: *Feynman Path Integrals*, vol. 106. Springer-Verlag.

Bartlett, M. S., and Moyal, J. E. 1949. The exact transition probabilities of quantum-mechanical oscillators calculated by the phase-space method. *Math. Proc. Camb. Philos. Soc.*, **45**, 545–553.

Başar, G., and Dunne, G. V. 2015. Resurgence and the Nekrasov–Shatashvili limit: Connecting weak and strong coupling in the Mathieu and Lamé systems. *J. High Energy Phys.*, **02**, 160.

Başar, G., Dunne, G. V., and Ünsal, M. 2017. Quantum geometry of resurgent perturbative/non-perturbative relations. *J. High Energy Phys.*, **05**, 087.

Bayen, F., Flato, M., Fronsdal, C., Lichnerowicz, A., and Sternheimer, D. 1978. Deformation theory and quantization. 2. Physical applications. *Annals Phys.*, **111**, 111.

Bender, C. M., and Orszag, S. A. 2013. *Advanced mathematical methods for scientists and engineers I: Asymptotic methods and perturbation theory*. Springer.

Bender, C. M., and Turbiner, A. 1993. Analytic continuation of eigenvalue problems. *Phys. Lett.*, **A173**, 442–446.

Bender, C. M., and Wu, T. T. 1969. Anharmonic oscillator. *Phys. Rev.*, **184**, 1231–1260.

Bender, C. M., and Wu, T. T. 1973. Anharmonic oscillator. 2: A study of perturbation theory in large order. *Phys. Rev.*, **D7**, 1620–1636.

Bender, C. M., Olaussen, K., and Wang, P. S. 1977. Numerological analysis of the WKB approximation in large order. *Phys. Rev.*, **D16**, 1740–1748.

Berezin, F. A., and Shubin, M. 2012. *The Schrödinger equation*. Vol. 66. Springer Science & Business Media.

Berry, M. V. 1977. Semi-classical mechanics in phase space: A study of Wigner's function. *Phil. Trans. Roy. Soc. Lond.*, **A287**, 237–271.

Berry, M. V. 1983. Semiclassical mechanics of regular and irregular motion. *Les Houches lecture series*, **36**, 171–271.

Berry, M. V., and Mount, K. E. 1972. Semiclassical approximations in wave mechanics. *Rep. Prog. Phys.*, **35**(1), 315.

Brezin, E., Parisi, G., and Zinn-Justin, J. 1977. Perturbation theory at large orders for potential with degenerate minima. *Phys. Rev.*, **D16**, 408–412.

Brown, L. M., ed. 2005. *Feynman's thesis: a new approach to quantum theory*. World Scientific.

Caliceti, E., Graffi, S., and Maioli, M. 1980. Perturbation theory of odd anharmonic oscillators. *Commun. Math. Phys.*, **75**(1), 51–66.

Çevik, D., Gadella, M., Kuru, Ş., and Negro, J. 2016. Resonances and antibound states for the Pöschl–Teller potential: Ladder operators and SUSY partners. *Phys. Lett.*, **A380**, 1600–1609.

Cherry, T. M. 1950. Uniform asymptotic formulae for functions with transition points. *Trans. Am. Math. Soc.*, **68**(2), 224–257.

Codesido, S., and Mariño, M. 2018. Holomorphic anomaly and quantum mechanics. *J. Phys.*, **A51**(5), 055402.

Cohen-Tannoudji, C., Dupont-Roc, J., and Grynberg, G. 2012. *Processus d'interaction entre photons et atomes*. EDP Sciences.

Colin de Verdière, Y. 2005. Bohr-Sommerfeld rules to all orders. *Ann. Henri Poincaré*, **6**, 925–936.

Collins, J. C., and Soper, D. E. 1978. Large order expansion in perturbation theory. *Ann. Phys.*, **112**, 209–234.

Curtright, T., Fairlie, D., and Zachos, C. K. 1998. Features of time independent Wigner functions. *Phys. Rev.*, **D58**, 025002.

Curtright, T. L., Fairlie, D. B., and Zachos, C. K. 2014. *A concise treatise on quantum mechanics in phase space*. World Scientific and Imperial College Press.

Curtright, T., Uematsu, T., and Zachos, C. K. 2001. Generating all Wigner functions. *J. Math. Phys.*, **42**, 2396.

Curtright, T., and Zachos, C. K. 2001. Negative probability and uncertainty relations. *Mod. Phys. Lett.*, **A16**, 2381–2385.

de Almeida, A. M. O. 1984. Semiclassical matrix elements. *Revista Brasileira de Física*, **14**(1), 62–85.

de Almeida, A. M. O. 1990. *Hamiltonian systems: Chaos and quantization*. Cambridge University Press.

de Almeida, A. M. O, and Hannay, J. H. 1982. Geometry of two dimensional tori in phase space: Projections, sections and the Wigner function. *Ann. Phys.*, **138**(1), 115–154.

de la Madrid, R., and Gadella, M. 2002. A pedestrian introduction to Gamow vectors. *Am. J. Phys.*, **70**(6), 626–638.

Delabaere, E. 2006. Effective resummation methods for an implicit resurgent function. arXiv:math-ph/0602026

Delabaere, E, Dillinger, H., and Pham, F. 1997. Exact semiclassical expansions for one-dimensional quantum oscillators. *J. Math. Phys.*, **38**(12), 6126–6184.

Delabaere, E., and Pham, F. 1999. Resurgent methods in semi-classical asymptotics. *Ann. Inst. Henri Poincaré*, **71**, 1–94.

Dillinger, H., Delabaere, E., and Pham, F. 1993. Résurgence de Voros et périodes des courbes hyperelliptiques. *Ann. Inst. Fourier*, **43**, 163.

Dirac, P. A. M. 1933. The Lagrangian in quantum mechanics. *Physikalische Zeitschirift der Sowjetunion*, **3**, 312–320.

Dirac, P. A. M. 1981. *The principles of quantum mechanics*. Oxford University Press.

Dorey, P., Dunning, C., and Tateo, R. 2007. The ODE/IM correspondence. *J. Phys.*, **A40**, R205.

Dorey, P., and Tateo, R. 1999. Anharmonic oscillators, the thermodynamic Bethe ansatz, and nonlinear integral equations. *J. Phys.*, **A32**, L419–L425.

Dunham, J. L. 1932. The Wentzel–Brillouin–Kramers method of solving the wave equation. *Phys. Rev.*, **41**, 713–720.

Dunne, G. V. 2008. Functional determinants in quantum field theory. *J. Phys.*, **A41**, 304006.

Dunne, G. V., and Ünsal, M. 2014. Uniform WKB, multi-instantons, and resurgent trans-series. *Phys. Rev.*, **D89**(10), 105009.

Faddeev, L. D., and Takhtajan, L. A. 2015. On the spectral theory of a functional-difference operator in conformal field theory. *Izvestiya: Mathematics*, **79**(2), 388–410

Fairlie, D. B. 1964. The formulation of quantum mechanics in terms of phase space functions. *Proc. Cambridge Phil. Soc.*, **60**, 581–586.

Feynman, R. P. 1948. Space-time approach to nonrelativistic quantum mechanics. *Rev. Mod. Phys.*, **20**, 367–387.

Feynman, R. P. 1998. *Statistical mechanics: A set of lectures*. Westview Press.

Feynman, R. P., Hibbs, A. R., and Styer, D. F. 2010. *Quantum mechanics and path integrals*. Courier Corporation.

Gaiotto, D. 2014. Opers and TBA. ArXiv:1403.6137 [hep-th]

Gaiotto, D., Moore, G. W., and Neitzke, A. 2009. Wall-crossing, Hitchin systems, and the WKB approximation. *Adv. Math.* **234**, 239–403.

Gaiotto, D., Moore, G. W., and Neitzke, A. 2010. Four-dimensional wall-crossing via three-dimensional field theory. *Commun. Math. Phys.*, **299**, 163–224.

Galindo, A., and Pascual, P. 1990. *Quantum mechanics, 1 and 2*. Springer-Verlag.

Galitski, V., Karnakov, B., and Kogan, V. 2013. *Exploring quantum mechanics: A collection of 700+ solved problems for students, lecturers, and researchers.* Oxford University Press.

Gaudin, M., and Pasquier, V. 1992. The periodic Toda chain and a matrix generalization of the Bessel function's recursion relations. *J. Phys.*, **A25**, 5243.

Giannopoulou, K. S., and Makrakis, G. N. 2017. An approximate series solution of the semiclassical Wigner equation. arXiv:1705.06754 [math-ph]

Grammaticos, B., and Voros, A. 1979. Semiclassical approximations for nuclear Hamiltonians. 1. Spin independent potentials. *Ann. Phys.*, **123**, 359.

Grassi, A., Mariño, M., and Zakany, S. 2015. Resumming the string perturbation series. *J. High Energy Phys.*, **1505**, 038.

Gross, D. J., Perry, M. J., and Yaffe, L. G. 1982. Instability of flat space at finite temperature. *Phys. Rev.*, **D25**, 330–355.

Gutzwiller, M. C. 1980. The quantum mechanical Toda lattice. *Ann. Phys.*, **124**, 347.

Gutzwiller, M. C. 1981. The quantum mechanical Toda lattice, II. *Ann. Phys.*, **133**, 304.

Hall, B. C. 2013. *Quantum theory for mathematicians.* Springer-Verlag.

Hillery, M., O'Connell, R. F., Scully, M. O., and Wigner, E. P. 1984. Distribution functions in physics: Fundamentals. *Phys. Rept.*, **106**, 121–167.

Hislop, P. D., and Sigal, I. M. 2012. *Introduction to spectral theory: With applications to Schrödinger operators.* Springer-Verlag.

Hoe, N., D'Etat, B., Grumberg, J., Caby, M., Leboucher, E., and Coulaud, G. 1982. Stark effect of hydrogenic ions. *Phys. Rev.*, **A25**(Feb), 891–906.

Hudson, R. L. 1974. When is the Wigner quasi-probability density non-negative? *Rep. Math. Phys.*, **6**(2), 249–252.

Ito, K., Mariño, M., and Shu, H. 2019. TBA equations and resurgent quantum mechanics. *J. High Energy Phys.*, **01**, 228.

Jentschura, U. D., Surzhykov, A., and Zinn-Justin, J. 2010. Multi-instantons and exact results. III: Unification of even and odd anharmonic oscillators. *Ann. Phys.*, **325**, 1135–1172.

Kashaev, R., and Mariño, M. 2016. Operators from mirror curves and the quantum dilogarithm. *Commun. Math. Phys.*, **346**(3), 967.

Kawai, T., and Takei, Y. 2005. *Algebraic analysis of singular perturbation theory.* American Mathematical Society.

Kenfack, A., and Życzkowski, K. 2004. Negativity of the Wigner function as an indicator of non-classicality. *J. Opt. B: Quantum Semiclass. Opt.*, **6**(10), 396–404.

Khandekar, D. C., and Lawande, S. V. 1986. Feynman path integrals: Some exact results and applications. *Phys. Rep.*, **137**, 115–229.

Kleinert, H. 2009. *Path integrals in quantum mechanics, statistics, polymer physics, and financial markets.* World Scientific.

Konishi, K., and Paffuti, G. 2009. *Quantum mechanics: A new introduction.* Oxford University Press.

Kozlowski, K. K., and Teschner, J. 2010. TBA for the Toda chain. New Trends in Quantum Integrable Systems, Word Scientific, 195–210.

Kurchan, J., Leboeuf, P., and Saraceno, M. 1989. Semiclassical approximations in the coherent-state representation. *Phys. Rev.*, **A40**(Dec), 6800–6813.

Lam, C. S. 1968. Behavior of very high order perturbation diagrams. *Nuovo Cim.*, **A55**, 258–274.

Langer, R. E. 1949. The asymptotic solutions of ordinary linear differential equations of the second order, with special reference to a turning point. *Trans. Am. Math. Soc.*, **67**(2), 461–490.

Lebedev, N. N. 1972. *Special functions and their applications*. Revised ed., Dover Publications

Lee, H. W. 1995. Theory and application of the quantum phase-space distribution functions. *Phys. Rep.*, **259**(3), 147–211.

Mariño, M. 2015. *Instantons and large N: An introduction to non-perturbative methods in quantum field theory*. Cambridge University Press.

Matsuyama, A. 1992. Periodic Toda lattice in quantum mechanics. *Ann. Phys.*, **222**, 300.

Miller, P. D. 2006. *Applied asymptotic analysis*. Vol. 75. American Mathematical Society.

Miller, S. C., Jr., and Good, R. H., Jr. 1953. A WKB-type approximation to the Schrödinger equation. *Phys. Rev.*, **91**(1), 174.

Moiseyev, N. 2011. *Non-Hermitian quantum mechanics*. Cambridge University Press.

Moshinsky, M., and Quesne, C. 1971. Linear canonical transformations and their unitary representations. *J. Math. Phys.*, **12**(8), 1772–1780.

Moyal, A. 2006. *Maverick mathematician. The life and science of J. E. Moyal*. Australian National University Press.

Negele, J. W., and Orland, H. 1988. *Quantum many-particle systems*. Westview.

Nekrasov, N. A., and Shatashvili, S. L. 2009. Quantization of integrable systems and four dimensional gauge theories. In the *16th International Congress on Mathematical Physics, Prague, August 2009*, 265–289, World Scientific 2010.

Nikolaev, N. 2020. Exact solutions for the singularly perturbed Riccati Equation and exact WKB Analysis. arXiv:2008.06492 **[math.CA]**

Reinhardt, W. P. 1982. Complex coordinates in the theory of atomic and molecular structure and dynamics. *Annu. Rev. Phys. Chem.*, **33**(1), 223–255.

Robnik, M., and Romanovski, V. G. 2000. Some properties of the WKB series. *J. Phys.*, **A33**(28), 5093.

Robnik, M., and Salasnich, L. 1997. WKB to all orders and the accuracy of the semiclassical quantization. *J. Phys.*, **A30**(5), 1711.

Schulman, L. S. 2012. *Techniques and applications of path integration*. Courier Corporation.

Serone, M., Spada, G., and Villadoro, G. 2017. The power of perturbation theory. *J. High Energy Phys.*, **05**, 056.

Shen, H., and Silverstone, H. J. 2004. JWKB method as an exact technique. *Int. J. Quantum Chem.*, **99**(4), 336–352.

Shen, H., Silverstone, H. J, and Álvarez, G. 2005. On the bidirectionality of the JWKB connection formula at a linear turning point. *Collect. Czech. Chem. Commun.*, **70**(6), 740–754.

Silverstone, H. J. 1985. JWKB connection-formula problem revisited via Borel summation. *Phys. Rev. Lett.*, **55**(23), 2523.

Silverstone, H. J., Harris, J. G., Čížek, J., and Paldus, J. 1985a. Asymptotics of high-order perturbation theory for the one-dimensional anharmonic oscillator by quasisemiclassical methods. *Phys. Rev.*, **A32**(4), 1965.

Silverstone, H. J., Nakai, S., and Harris, J. G. 1985b. Observations on the summability of confluent hypergeometric functions and on semiclassical quantum mechanics. *Phys. Rev.*, **A32**, 1341–1345.

Simon, B., and Dicke, A. 1970. Coupling constant analyticity for the anharmonic oscillator. *Ann. Phys.*, **58**(1), 76 – 136.

Sklyanin, E. K. 1985. The quantum Toda chain. *Lect. Notes Phys.*, **226**, 196–233.

Sulejmanpasic, T., and Ünsal, M. 2018. Aspects of perturbation theory in quantum mechanics: The BenderWu Mathematica ® package. *Comput. Phys. Commun.*, **228**, 273–289.

Takhtajan, L. A. 2008. *Quantum mechanics for mathematicians*. American Mathematical Society.

Tatarskii, V. I. 1983. The Wigner representation of quantum mechanics. *Physics-Uspekhi*, **26**(4), 311–327.

Temme, N. M. 1990. Asymptotic estimates for Laguerre polynomials. *Z. Angew. Math. Phys.*, **41**(1), 114–126.

Van Vleck, J. H. 1928. The correspondence principle in the statistical interpretation of quantum mechanics. *Proc. Natl. Acad. Sci. U.S.A.*, **14**(2), 178–188.

Veble, G., Robnik, M., and Romanovski, V. 2002. Semiclassical analysis of Wigner functions. *J. Phys.*, **A35**(18), 4151.

Voros, A. 1977. Asymptotic $\hbar$-expansions of stationary quantum states, *Ann. Inst. H. Poincaré*, **A26**, 343–403.

Voros, A. 1980. The zeta function of the quartic oscillator. *Nucl. Phys.*, **B165**, 209–236.

Voros, A. 1981. *Spectre de l'équation de Schrödinger et méthode BKW*. Publications Mathématiques d'Orsay.

Voros, A. 1983. The return of the quartic oscillator. The complex WKB method. *Ann. Inst. Henri Poincaré*, **39**(3), 211–338.

Voros, A. 1989. Wentzel-Kramers-Brillouin method in the Bargmann representation. *Phys. Rev.*, **A40**(Dec), 6814–6825.

Wigner, E. P. 1932. On the quantum correction for thermodynamic equilibrium. *Phys. Rev.*, **40**, 749–760.

Yaris, R., Bendler, J., Lovett, R. A., Bender, C. M., and Fedders, P. A. 1978. Resonance calculations for arbitrary potentials. *Phys. Rev.*, **A18**, 1816.

Zachos, C. K., Fairlie, D. B., and Curtright, T. L. 2005. *Quantum mechanics in phase space: an overview with selected papers*. World Scientific.

Zinn-Justin, J. 1981. Multi-instanton contributions in quantum mechanics. *Nucl. Phys.*, **B192**, 125–140.

Zinn-Justin, J. 1983. Multi-instanton contributions in quantum mechanics. 2. *Nucl. Phys.*, **B218**, 333–348.

Zinn-Justin, J. 1996. *Quantum field theory and critical phenomena*. Clarendon Press.

Zinn-Justin, J. 2012. *Intégrale de chemin en mécanique quantique: Introduction*. EDP sciences.

Zinn-Justin, J., and Jentschura, U. D. 2004a. Multi-instantons and exact results I: Conjectures, WKB expansions, and instanton interactions. *Ann. Phys.*, **313**, 197–267.

Zinn-Justin, J., and Jentschura, U. D. 2004b. Multi-instantons and exact results II: Specific cases, higher-order effects, and numerical calculations. *Ann. Phys.*, **313**, 269–325.

Index